Lecture Notes
in Control and Information Sciences   264

Editors: M. Thoma · M. Morari

T0138108

Springer
London
Berlin
Heidelberg
New York
Barcelona
Hong Kong
Milan
Paris
Singapore
Tokyo

Alfonso Baños, Françoise Lamnabhi-Lagarrigiue
and Francisco J. Montoya (Eds)

Advances in the Control of Nonlinear Systems

With 105 Figures

Springer

Editors

Alfonso Baños, PhD
Dpto. Informatica y Sistemas, Universidad de Murcia, 30071 Murcia, Spain

Françoise Lamnabhi-Lagarrigue, Docteur D'état
Laboratoire des Signaux et Systèmes, CNRS, SUPELEC, 91192 Gif-sur-Yvette, France

Francisco J. Montoya, PhD
Dpto. Informatica y Sistemas, Universidad de Murcia, 30071 Murcia, Spain

ISBN 1-85233-378-2 Springer-Verlag London Berlin Heidelberg

British Library Cataloguing in Publication Data
Advances in the control of nonlinear systems
          - (Lecture notes in control and information
   sciences ; 264)
   1.Nonlinear control theory 2.System analysis
   I.Banos, Alfonso II. Lamnabhi-Lagarrigue, F. (Francoise)
   III.Montoya, Francisco J.
   003.7'5
   ISBN 1852333782

Library of Congress Cataloging-in-Publication Data
Advances in the control of nonlinear systems / Alfonso Baños, Françoise
Lamnabhi-Lagarrigue, and Francisco J. Montoya, (eds.).
       p. cm. -- (Lecture notes in control and information sciences ; 264)
   Includes bibliographical references and index.
   ISBN 1-85233-378-2 (alk. paper)
     1. System analysis. 2. Nonlinear control theory. I. Baños, Alfonso, 1965 II.
   Lamnabhi-Lagarrigue, Françoise, 1953- III. Montoya, Francisco J, 1966- IV. Series.
   TA342.A38 2001
   003'.75--dc21   00-053805

Typesetting: Camera ready by editors
Printed and bound at the Athenæum Press Ltd., Gateshead, Tyne & Wear
69/3830-543210 Printed on acid-free paper SPIN 10776695

Preface

The contributions of this book are based on the lectures of the 2nd NCN Pedagogical School, that was held in Murcia during the last week of September 2000. This School is the second in series of Pedagogical Schools in the framework of the European TMR project "Nonlinear Control Network" (http://www.supelec.fr/lss/NCN). We would like to thank the TMR Program for the financial support which, in particular, helped numerous young researchers from all over Europe in attending the School.

The School was organized around four courses, that are reflected in the four Parts of the present book. The goal of the different courses was to give a pedagogical introduction to four important research areas in the nonlinear control world, and this has been also the spirit in the writing of the different chapters of this book. The book is then organized in four Parts which exhibit a different internal structure, reflecting to a great extent the different styles of its authors. The four Parts are:

1. The Differential Algebraic Approach to Nonlinear Systems
2. Nonlinear Quantitative Feedback Theory
3. Hybrid Systems
4. Physics in Control

Every Part is presented by an informal introduction where the corresponding topic is emphasized as a whole, which somehow simplifies these introductory words. Finally, the editors would like to acknowledge the help of Miguel Moreno and Joaquín Cervera in the organization of the School, and explicitly express their gratitude to all the authors contributing to this volume.

Murcia, Gif–sur–Yvette,
December 2000

Alfonso Baños
Françoise Lamnabhi–Lagarrigue
Francisco J. Montoya

Other books already published in this NCN series:

- Stability and Stabilization of Nonlinear Systems, (D. Aeyels, F. Lamnabhi-Lagarrigue, A.J. van der Schaft, Eds.), LNCIS 246, July 1999, ISBN 1–85233–638–2.
- Nonlinear Control in the Year 2000, (A. Isidori, F. Lamnabhi–Lagarrigue, W. Respondek, Eds.), 2 volumes, LNCIS 258 & 259, November 2000, ISBN 1–85233–363–4 & ISBN 1–85233–364–2.

Contents

Part IV. Physics in Control

Part I

The Differential Algebraic
Approach to Nonlinear Systems

Introduction

Differential algebraic methods have been used in nonlinear control theory since 1985 and have led to a deeper understanding of the underlying concepts, and to definitions of useful new concepts. Although flatness is principally not tied to differential algebra, probably the most important outcome of this approach is the notion of differential flatness. Broadly speaking, differentially flat systems are those that admit a complete, finite and free differential parametrisation. This means that flat systems can be described by a finite set of variables whose trajectories can be assigned independently. The class of differentially flat systems, hence, generalises the class of linear controllable systems. Its importance is due to two facts: On the one hand, many mathematical models of technological processes have been shown to be flat systems; on the other hand, for the flat systems powerful and simple systematic methods are available for the motion planing and the design of feedback laws for stable trajectory tracking. Finally, this notion can be most fruitfully generalised to linear and nonlinear infinite dimensional systems, in particular to boundary controlled distributed parameter systems and linear or nonlinear systems with delays.

These chapters are thought to give an introduction to the differential algebraic approach to nonlinear systems, with an emphasis on differential flatness, and to show the bridges to linear and nonlinear infinite dimensional systems. Herein, the linear systems of finite or infinite dimension are treated in a module theoretic framework, the "linear analogue" of differential algebra. Concepts and methods are illustrated on an important number of technological applications.

Flat Systems, Equivalence and Feedback

Philippe Martin[1], Richard M. Murray[2], and Pierre Rouchon[1]

[1] Centre Automatique et Systèmes, École des Mines de Paris, Paris, France.
[2] Division of Engineering and Applied Science, California Institute of Technology, California, USA.

Abstract. Flat systems, an important subclass of nonlinear control systems introduced *via* differential-algebraic methods, are defined' in a differential geometric framework. We utilize the infinite dimensional geometry developed by Vinogradov and coworkers: a control system is a diffiety, or more precisely, an ordinary diffiety, i.e. a smooth infinite-dimensional manifold equipped with a privileged vector field. After recalling the definition of a Lie-Bäcklund mapping, we say that two systems are equivalent if they are related by a Lie-Bäcklund isomorphism. Flat systems are those systems which are equivalent to a controllable linear one. The interest of such an abstract setting relies mainly on the fact that the above system equivalence is interpreted in terms of endogenous dynamic feedback. The presentation is as elementary as possible and illustrated by the VTOL aircraft.

1 Introduction

Control systems are ubiquitous in modern technology. The use of feedback control can be found in systems ranging from simple thermostats that regulate the temperature of a room, to digital engine controllers that govern the operation of engines in cars, ships, and planes, to flight control systems for high performance aircraft. The rapid advances in sensing, computation, and actuation technologies are continuing to drive this trend and the role of control theory in advanced (and even not so advanced) systems is increasing.

A typical use of control theory in many modern systems is to invert the system dynamics to compute the inputs required to perform a specific task. This inversion may involve finding appropriate inputs to steer a control system from one state to another or may involve finding inputs to follow a desired trajectory for some or all of the state variables of the system. In general, the solution to a given control problem will not be unique, if it exists at all, and so one must trade off the performance of the system for the stability and actuation effort. Often this tradeoff is described as a cost function balancing the desired performance objectives with stability and effort, resulting in an optimal control problem.

This inverse dynamics problem assumes that the dynamics for the system are known and fixed. In practice, uncertainty and noise are always present in systems and must be accounted for in order to achieve acceptable performance of this system. Feedback control formulations allow the system to respond to errors and changing operating conditions in real-time and can substantially affect the operability of the system by stabilizing the system and extending its capabilities. Again, one may formulate the feedback regulation problems as an optimization problem to allow tradeoffs between stability, performance, and actuator effort.

The basic paradigm used in most, if not all, control techniques is to exploit the mathematical structure of the system to obtain solutions to the inverse dynamics and feedback regulation problems. The most common structure to exploit is linear structure, where one approximates the given system by its linearization and then uses properties of linear control systems combined with appropriate cost function to give closed form (or at least numerically computable) solutions. By using different linearizations around different operating points, it is even possible to obtain good results when the system is nonlinear by "scheduling" the gains depending on the operating point.

As the systems that we seek to control become more complex, the use of linear structure alone is often not sufficient to solve the control problems that are arising in applications. This is especially true of the inverse dynamics problems, where the desired task may span multiple operating regions and hence the use of a single linear system is inappropriate.

In order to solve these harder problems, control theorists look for different types of structure to exploit in addition to simple linear structure. In this paper we concentrate on a specific class of systems, called "(differentially) flat systems", for which the structure of the trajectories of the (nonlinear) dynamics can be completely characterized. Flat systems are a generalization of linear systems (in the sense that all linear, controllable systems are flat), but the techniques used for controlling flat systems are much different than many of the existing techniques for linear systems. As we shall see, flatness is particularly well tuned for allowing one to solve the inverse dynamics problems and one builds off of that fundamental solution in using the structure of flatness to solve more general control problems.

Flatness was first defined by Fliess et al. [6,9] using the formalism of differential algebra, see also [13] for a somewhat different approach. In differential algebra, a system is viewed as a differential field generated by a set of variables (states and inputs). The system is said to be flat if one can find a set of variables, called the flat outputs, such that the system is (non-differentially) algebraic over the differential field generated by the set of flat outputs. Roughly speaking, a system is flat if we can find a set of outputs (equal in number to the number of inputs) such that all states and inputs can be determined from these outputs without integration. More precisely, if the system has states $x \in \mathbb{R}^n$, and inputs $u \in \mathbb{R}^m$ then the system is flat if we can find outputs $y \in \mathbb{R}^m$ of the form

$$y = h(x, u, \dot{u}, \ldots, u^{(r)})$$

such that

$$x = \varphi(y, \dot{y}, \ldots, y^{(q)})$$
$$u = \alpha(y, \dot{y}, \ldots, y^{(q)}).$$

More recently, flatness has been defined in a more geometric context, where tools for nonlinear control are more commonly available. One approach is to use exterior differential systems and regard a nonlinear control system as a Pfaffian system on an appropriate space [17]. In this context, flatness can be described in terms of the notion of absolute equivalence defined by E. Cartan [1,2,23].

In this paper we adopt a somewhat different geometric point of view, relying on a Lie-Bäcklund framework as the underlying mathematical structure. This point of view was originally described in [7,10,11] and is related to the work of Pomet et al. [20,19] on "infinitesimal Brunovsky forms" (in the context of feedback linearization). It offers a compact framework in which to describe basic results and is also closely related to the basic techniques that are used to compute the functions that are required to characterize the solutions of flat systems (the so-called flat outputs).

Applications of flatness to problems of engineering interest have grown steadily in recent years. It is important to point out that many classes of systems commonly used in nonlinear control theory are flat. As already noted, all controllable linear systems can be shown to be flat. Indeed, any system that can be transformed into a linear system by changes of coordinates, static feedback transformations (change of coordinates plus nonlinear change of inputs), or dynamic feedback transformations is also flat. Nonlinear control systems in "pure feedback form", which have gained popularity due to the applicability of backstepping [12] to such systems, are also flat. Thus, many of the systems for which strong nonlinear control techniques are available are in fact flat systems, leading one to question how the structure of flatness plays a role in control of such systems.

One common misconception is that flatness amounts to dynamic feedback linearization. It is true that any flat system can be feedback linearized using dynamic feedback (up to some regularity conditions that are generically satisfied). However, flatness is a property of a system and does not imply that one intends to then transform the system, via a dynamic feedback and appropriate changes of coordinates, to a single linear system. Indeed, the power of flatness is precisely that it does not convert nonlinear systems into linear ones. When a system is flat it is an indication that the nonlinear structure of the system is well characterized and one can exploit that structure in designing control algorithms for motion planning, trajectory generation, and stabilization. Dynamic feedback linearization is one such technique, although it is often a poor choice if the dynamics of the system are substantially different in different operating regimes.

Another advantage of studying flatness over dynamic feedback linearization is that flatness is a *geometric* property of a system, independent of coordinate choice. Typically when one speaks of linear systems in a state space context, this does not make sense geometrically since the system is linear only in certain choices of coordinate representations. In particular, it is difficult to discuss the notion of a linear state space system on a manifold since the very definition of linearity requires an underlying linear space. In this way, flatness can be considered the proper geometric notion of linearity, even though the system may be quite nonlinear in almost any natural representation.

Finally, the notion of flatness can be extended to distributed parameters systems with boundary control and is useful even for controlling linear systems, whereas feedback linearization is yet to be defined in that context.

This paper provides a self-contained description of flat systems. Section 2 introduces the fundamental concepts of equivalence and flatness in a simple geometric framework. This is essentially an open-loop point of view.

2 Equivalence and flatness

2.1 Control systems as infinite dimensional vector fields

A system of differential equations

$$\dot{x} = f(x), \quad x \in X \subset \mathbb{R}^n \tag{1}$$

is by definition a pair (X, f), where X is an open set of $\mathbb{R}^n$ and f is a smooth vector field on X. A solution, or *trajectory*, of (1) is a mapping $t \mapsto x(t)$ such that

$$\dot{x}(t) = f(x(t)) \qquad \forall t \geq 0.$$

Notice that if $x \mapsto h(x)$ is a smooth function on X and $t \mapsto x(t)$ is a trajectory of (1), then

$$\frac{d}{dt} h(x(t)) = \frac{\partial h}{\partial x}(x(t)) \cdot \dot{x}(t) = \frac{\partial h}{\partial x}(x(t)) \cdot f(x(t)) \qquad \forall t \geq 0.$$

For that reason the *total derivative*, i.e., the mapping

$$x \mapsto \frac{\partial h}{\partial x}(x) \cdot f(x)$$

is somewhat abusively called the "time-derivative" of h and denoted by $\dot{h}$.

We would like to have a similar description, i.e., a "space" and a vector field on this space, for a control system

$$\dot{x} = f(x, u), \tag{2}$$

where f is smooth on an open subset $X \times U \subset \mathbb{R}^n \times \mathbb{R}^m$. Here f is no longer a vector field on X, but rather an *infinite collection* of vector fields on X parameterized by u: for all $u \in U$, the mapping

$$x \mapsto f_u(x) = f(x, u)$$

is a vector field on X. Such a description is not well-adapted when considering dynamic feedback.

It is nevertheless possible to associate to (2) a vector field with the "same" solutions using the following remarks: given a smooth solution of (2), i.e., a mapping $t \mapsto (x(t), u(t))$ with values in $X \times U$ such that

$$\dot{x}(t) = f(x(t), u(t)) \qquad \forall t \geq 0,$$

we can consider the *infinite* mapping

$$t \mapsto \xi(t) = (x(t), u(t), \dot{u}(t), \dots)$$

taking values in $X \times U \times \mathbb{R}_m^\infty$, where $\mathbb{R}_m^\infty = \mathbb{R}^m \times \mathbb{R}^m \times \dots$ denotes the product of an infinite (countable) number of copies of $\mathbb{R}^m$. A typical point of $\mathbb{R}_m^\infty$ is thus of the form $(u^1, u^2, \dots)$ with $u^i \in \mathbb{R}^m$. This mapping satisfies

$$\dot{\xi}(t) = \big(f(x(t), u(t)), \dot{u}(t), \ddot{u}(t), \dots\big) \qquad \forall t \geq 0,$$

hence it can be thought of as a trajectory of the *infinite* vector field

$$(x, u, u^1, \dots) \mapsto F(x, u, u^1, \dots) = (f(x, u), u^1, u^2, \dots)$$

on $X \times U \times \mathbb{R}_m^\infty$. Conversely, any mapping

$$t \mapsto \xi(t) = (x(t), u(t), u^1(t), \dots)$$

that is a trajectory of this infinite vector field necessarily takes the form $(x(t), u(t), \dot{u}(t), \dots)$ with $\dot{x}(t) = f(x(t), u(t))$, hence corresponds to a solution of (2). Thus F is truly a vector field and no longer a parameterized family of vector fields.

Using this construction, the control system (2) can be seen as the data of the "space" $X \times U \times \mathbb{R}_m^\infty$ together with the "smooth" vector field F on this space. Notice that, as in the uncontrolled case, we can define the "time-derivative" of a smooth function $(x, u, u^1, \dots) \mapsto h(x, u, u^1, \dots, u^k)$ depending on a *finite* number of variables by

$$\dot{h}(x, u, u^1, \dots, u^{k+1}) := Dh \cdot F$$

$$= \frac{\partial h}{\partial x} \cdot f(x, u) + \frac{\partial h}{\partial u} \cdot u^1 + \frac{\partial h}{\partial u^1} \cdot u^2 + \cdots.$$

The above sum is *finite* because h depends on finitely many variables.

Remark 1 *To be rigorous we must say something of the underlying topology and differentiable structure of $\mathbb{R}_m^\infty$ to be able to speak of smooth objects [24]. This topology is the* Fréchet *topology, which makes things look as if we were working on the product of k copies of $\mathbb{R}^m$ for a "large enough" k. For our purpose it is enough to know that a basis of the open sets of this topology consists of infinite products $U_0 \times U_1 \times \ldots$ of open sets of $\mathbb{R}^m$, and that a function is* smooth *if it depends on a finite but arbitrary number of variables and is smooth in the usual sense. In the same way a mapping $\Phi : \mathbb{R}_m^\infty \to \mathbb{R}_n^\infty$ is* smooth *if all of its components are smooth functions.*

$\mathbb{R}_m^\infty$ equipped with the Fréchet topology has very weak properties: useful theorems such as the implicit function theorem, the Frobenius theorem, and the straightening out theorem no longer hold true. This is only because $\mathbb{R}_m^\infty$ is a very big space: indeed the Fréchet topology on the product of k copies of $\mathbb{R}^m$ for any finite k coincides with the usual Euclidian topology.

We can also define manifolds modeled on $\mathbb{R}_m^\infty$ using the standard machinery. The reader not interested in these technicalities can safely ignore the details and won't loose much by replacing "manifold modeled on $\mathbb{R}_m^\infty$" by "open set of $\mathbb{R}_m^\infty$".

We are now in position to give a formal definition of a system:

Definition 1 *A* system *is a pair $(\mathfrak{M}, F)$ where $\mathfrak{M}$ is a smooth manifold, possibly of infinite dimension, and F is a smooth vector field on $\mathfrak{M}$.*

Locally, a control system looks like an open subset of $\mathbb{R}^\alpha$ (α not necessarily finite) with coordinates $(\xi_1, \ldots, \xi_\alpha)$ together with the vector field

$$\xi \mapsto F(\xi) = (F_1(\xi), \ldots, F_\alpha(\xi))$$

where all the components F_i depend only on a finite number of coordinates. A *trajectory* of the system is a mapping $t \mapsto \xi(t)$ such that $\dot{\xi}(t) = F(\xi(t))$.

We saw in the beginning of this section how a "traditional" control system fits into our definition. There is nevertheless an important difference: we lose the notion of *state dimension*. Indeed

$$\dot{x} = f(x, u), \quad (x, u) \in X \times U \subset \mathbb{R}^n \times \mathbb{R}^m \tag{3}$$

and

$$\dot{x} = f(x, u), \quad \dot{u} = v \tag{4}$$

now have the same description $(X \times U \times \mathbb{R}_m^\infty, F)$, with

$$F(x, u, u^1, \ldots) = (f(x, u), u^1, u^2, \ldots),$$

in our formalism: $t \mapsto (x(t), u(t))$ is a trajectory of (3) if and only if $t \mapsto (x(t), u(t), \dot{u}(t))$ is a trajectory of (4). This situation is not surprising since the state dimension is of course not preserved by dynamic feedback. On the other hand we will see there is still a notion of *input dimension*.

Example 1 (The trivial system) *The* trivial system $(\mathbb{R}_m^\infty, F_m)$, *with coordinates* $(y, y^1, y^2, \dots)$ *and vector field*

$$F_m(y, y^1, y^2, \dots) = (y^1, y^2, y^3, \dots)$$

describes any "traditional" system made of m chains of integrators of arbitrary lengths, and in particular the direct transfer $y = u$.

In practice we often identify the "system" $F(x, \bar{u}) := (f(x, u), u^1, u^2, \dots)$ with the "dynamics" $\dot{x} = f(x, u)$ which defines it. Our main motivation for introducing a new formalism is that it will turn out to be a natural framework for the notions of equivalence and flatness we want to define.

Remark 2 *It is easy to see that the manifold $\mathfrak{M}$ is finite-dimensional only when there is no input, i.e., to describe a determined system of differential equations one needs as many equations as variables. In the presence of inputs, the system becomes underdetermined, there are more variables than equations, which accounts for the infinite dimension.*

Remark 3 *Our definition of a system is adapted from the notion of diffiety introduced in [24] to deal with systems of (partial) differential equations. By definition a diffiety is a pair $(\mathfrak{M}, CT\mathfrak{M})$ where $\mathfrak{M}$ is smooth manifold, possibly of infinite dimension, and $CT\mathfrak{M}$ is an involutive finite-dimensional distribution on $\mathfrak{M}$, i.e., the Lie bracket of any two vector fields of $CT\mathfrak{M}$ is itself in $CT\mathfrak{M}$. The dimension of $CT\mathfrak{M}$ is equal to the number of independent variables.*

As we are only working with systems with lumped parameters, hence governed by ordinary differential equations, we consider diffieties with one dimensional distributions. For our purpose we have also chosen to single out a particular vector field rather than work with the distribution it spans.

2.2   Equivalence of systems

In this section we define an equivalence relation formalizing the idea that two systems are "equivalent" if there is an invertible transformation exchanging their trajectories. As we will see later, the relevance of this rather natural equivalence notion lies in the fact that it admits an interpretation in terms of dynamic feedback.

Consider two systems $(\mathfrak{M}, F)$ and $(\mathfrak{N}, G)$ and a smooth mapping $\Psi : \mathfrak{M} \to \mathfrak{N}$ (remember that by definition every component of a smooth mapping depends only on finitely many coordinates). If $t \mapsto \xi(t)$ is a trajectory of $(\mathfrak{M}, F)$, i.e.,

$$\forall \xi, \quad \dot{\xi}(t) = F(\xi(t)),$$

the composed mapping $t \mapsto \zeta(t) = \Psi(\xi(t))$ satisfies the chain rule

$$\dot{\zeta}(t) = \frac{\partial \Psi}{\partial \xi}(\xi(t)) \cdot \dot{\xi}(t) = \frac{\partial \Psi}{\partial \xi}(\xi(t)) \cdot F(\xi(t)).$$

The above expressions involve only finite sums even if the matrices and vectors have infinite sizes: indeed a row of $\frac{\partial \Psi}{\partial \xi}$ contains only a finite number of non zero terms because a component of Ψ depends only on finitely many coordinates. Now, if the vector fields F and G are Ψ-related, i.e.,

$$\forall \xi, \quad G(\Psi(\xi)) = \frac{\partial \Psi}{\partial \xi}(\xi) \cdot F(\xi)$$

then

$$\dot{\zeta}(t) = G(\Psi(\xi(t))) = G(\zeta(t)),$$

which means that $t \mapsto \zeta(t) = \Psi(\xi(t))$ is a trajectory of $(\mathfrak{N}, G)$. If moreover Ψ has a smooth inverse Φ then obviously F, G are also Φ-related, and there is a one-to-one correspondence between the trajectories of the two systems. We call such an invertible Ψ relating F and G an *endogenous transformation*.

Definition 2 *Two systems $(\mathfrak{M}, F)$ and $(\mathfrak{N}, G)$ are equivalent at $(p, q) \in \mathfrak{M} \times \mathfrak{N}$ if there exists an endogenous transformation from a neighborhood of p to a neighborhood of q. $(\mathfrak{M}, F)$ and $(\mathfrak{N}, G)$ are equivalent if they are equivalent at every pair of points (p, q) of a dense open subset of $\mathfrak{M} \times \mathfrak{N}$.*

Notice that when $\mathfrak{M}$ and $\mathfrak{N}$ have the same *finite* dimension, the systems are necessarily equivalent by the straightening out theorem. This is no longer true in infinite dimensions.

Consider the two systems $(X \times U \times \mathbb{R}_m^\infty, F)$ and $(Y \times V \times \mathbb{R}_s^\infty, G)$ describing the dynamics

$$\dot{x} = f(x, u), \quad (x, u) \in X \times U \subset \mathbb{R}^n \times \mathbb{R}^m \tag{5}$$
$$\dot{y} = g(y, v), \quad (y, v) \in Y \times V \subset \mathbb{R}^r \times \mathbb{R}^s. \tag{6}$$

The vector fields F, G are defined by

$$F(x, u, u^1, \dots) = (f(x, u), u^1, u^2, \dots)$$
$$G(y, v, v^1, \dots) = (g(y, v), v^1, v^2, \dots).$$

If the systems are equivalent, the endogenous transformation Ψ takes the form

$$\Psi(x, u, u^1, \dots) = (\psi(x, \overline{u}), \beta(x, \overline{u}), \dot{\beta}(x, \overline{u}), \dots).$$

Here we have used the short-hand notation $\overline{u} = (u, u^1, \dots, u^k)$, where k is some finite but otherwise arbitrary integer. Hence Ψ is completely specified by the mappings ψ and β, i.e, by the expression of y, v in terms of $x, \overline{u}$. Similarly, the inverse Φ of Ψ takes the form

$$\Phi(y, v, v^1, \dots) = (\varphi(y, \overline{v}), \alpha(y, \overline{v}), \dot{\alpha}(y, \overline{v}), \dots). \; \bullet$$

As Ψ and Φ are inverse mappings we have

$$\psi\big(\varphi(y,\overline{v}),\overline{\alpha}(y,\overline{v})\big) = y \qquad \varphi\big(\psi(x,\overline{u}),\overline{\beta}(x,\overline{u})\big) = x$$

$$\text{and}$$

$$\beta\big(\varphi(y,\overline{v}),\overline{\alpha}(y,\overline{v})\big) = v \qquad \alpha\big(\psi(x,\overline{u}),\overline{\beta}(x,\overline{u})\big) = u.$$

Moreover F and G Ψ-related implies

$$f\big(\varphi(y,\overline{v}),\alpha(y,\overline{v})\big) = D\varphi(y,\overline{v}) \cdot \overline{g}(y,\overline{v})$$

where $\overline{g}$ stands for $(g, v^1, \ldots, v^k)$, i.e., a truncation of G for some large enough k. Conversely,

$$g\big(\psi(x,\overline{u}),\beta(y,\overline{u})\big) = D\psi(x,\overline{u}) \cdot \overline{f}(y,\overline{u}).$$

In other words, whenever $t \mapsto (x(t), u(t))$ is a trajectory of (5)

$$t \mapsto (y(t), v(t)) = \big(\varphi(x(t),\overline{u}(t)),\ \alpha(x(t),\overline{u}(t))\big)$$

is a trajectory of (6), and vice versa.

Example 2 (The PVTOL) *The system generated by*

$$\ddot{x} = -u_1 \sin\theta + \varepsilon u_2 \cos\theta$$
$$\ddot{z} = u_1 \cos\theta + \varepsilon u_2 \sin\theta - 1$$
$$\ddot{\theta} = u_2.$$

is globally equivalent to the systems generated by

$$\ddot{y}_1 = -\xi\sin\theta, \qquad \ddot{y}_2 = \xi\cos\theta - 1,$$

where ξ and θ are the control inputs. Indeed, setting

$$X := (x, z, \dot{x}, \dot{z}, \theta, \dot{\theta}) \qquad Y := (y_1, y_2, \dot{y}_1, \dot{y}_2)$$

$$\text{and}$$

$$U := (u_1, u_2) \qquad V := (\xi, \theta)$$

and using the notations in the discussion after definition 2, we define the mappings $Y = \psi(X,\overline{U})$ and $V = \beta(X,\overline{U})$ by

$$\psi(X,\overline{U}) := \begin{pmatrix} x - \varepsilon\sin\theta \\ z + \varepsilon\cos\theta \\ \dot{x} - \varepsilon\dot{\theta}\cos\theta \\ \dot{z} - \varepsilon\dot{\theta}\sin\theta \end{pmatrix} \qquad \text{and} \qquad \beta(X,\overline{U}) := \begin{pmatrix} u_1 - \varepsilon\dot{\theta}^2 \\ \theta \end{pmatrix}$$

to generate the mapping Ψ. The inverse mapping Φ is generated by the mappings $X = \varphi(Y,\overline{V})$ and $U = \alpha(Y,\overline{V})$ defined by

$$\varphi(Y,\overline{V}) := \begin{pmatrix} y_1 + \varepsilon\sin\theta \\ y_2 - \varepsilon\cos\theta \\ \dot{y}_1 + \varepsilon\dot{\theta}\cos\theta \\ \dot{y}_2 - \varepsilon\dot{\theta}\sin\theta \\ \theta \\ \dot{\theta} \end{pmatrix} \qquad \text{and} \qquad \alpha(Y,\overline{V}) := \begin{pmatrix} \xi + \varepsilon\dot{\theta}^2 \\ \ddot{\theta} \end{pmatrix}$$

An important property of endogenous transformations is that they preserve the input dimension:

Theorem 1 *If two systems $(X \times U \times \mathbb{R}_m^\infty, F)$ and $(Y \times V \times \mathbb{R}_s^\infty, G)$ are equivalent, then they have the same number of inputs, i.e., $m = s$.*

Proof. Consider the truncation Φ_μ of Φ on $X \times U \times (\mathbb{R}^m)^\mu$,

$$\Phi_\mu : X \times U \times (\mathbb{R}^{m+k})^\mu \to Y \times V \times (\mathbb{R}^s)^\mu$$
$$(x, u, u^1, \dots, u^{k+\mu}) \mapsto (\varphi, \alpha, \dot{\alpha}, \dots, \alpha^{(\mu)}),$$

i.e., the first $\mu + 2$ blocks of components of Ψ; k is just a fixed "large enough" integer. Because Ψ is invertible, Ψ_μ is a submersion for all μ. Hence the dimension of the domain is greater than or equal to the dimension of the range,

$$n + m(k + \mu + 1) \geq s(\mu + 1) \qquad \forall \mu > 0,$$

which implies $m \geq s$. Using the same idea with Ψ leads to $s \geq m$.

Remark 4 *Our definition of equivalence is adapted from the notion of equivalence between diffieties. Given two diffieties $(\mathfrak{M}, CT\mathfrak{M})$ and $(\mathfrak{N}, CT\mathfrak{N})$, we say that a smooth mapping Ψ from (an open subset of) $\mathfrak{M}$ to $\mathfrak{N}$ is Lie-Bäcklund if its tangent mapping $T\Psi$ satisfies $T\Phi(CT\mathfrak{M}) \subset CT\mathfrak{N}$. If moreover Ψ has a smooth inverse Φ such that $T\Psi(CT\mathfrak{N}) \subset CT\mathfrak{M}$, we say it is a Lie-Bäcklund isomorphism. When such an isomorphism exists, the diffieties are said to be equivalent. An endogenous transformation is just a special Lie-Bäcklund isomorphism, which preserves the time parameterization of the integral curves. It is possible to define the more general concept of orbital equivalence [7,5] by considering general Lie-Bäcklund isomorphisms, which preserve only the geometric locus of the integral curves.*

2.3   Differential Flatness

We single out a very important class of systems, namely systems equivalent to a trivial system $(\mathbb{R}_s^\infty, F_s)$ (see example 1):

Definition 3 *The system $(\mathfrak{M}, F)$ is flat at $p \in \mathfrak{M}$ (resp. flat) if it equivalent at p (resp. equivalent) to a trivial system.*

We specialize the discussion after definition 2 to a flat system $(X \times U \times \mathbb{R}_m^\infty, F)$ describing the dynamics

$$\dot{x} = f(x, u), \quad (x, u) \in X \times U \subset \mathbb{R}^n \times \mathbb{R}^m.$$

By definition the system is equivalent to the trivial system $(\mathbb{R}_s^\infty, F_s)$ where the endogenous transformation Ψ takes the form

$$\Psi(x, u, u^1, \dots) = (h(x, \overline{u}), \dot{h}(x, \overline{u}), \ddot{h}(x, \overline{u}), \dots), \tag{7}$$

In other words Ψ is the infinite prolongation of the mapping h. The inverse Φ of Ψ takes the form

$$\Psi(\overline{y}) = (\psi(\overline{y}), \beta(\overline{y}), \dot{\beta}(\overline{y}), \dots).$$

As Φ and Ψ are inverse mappings we have in particular

$$\varphi\big(\overline{h}(x, \overline{u})\big) = x \quad \text{and} \quad \alpha\big(\overline{h}(x, \overline{u})\big) = u.$$

Moreover F and G Φ-related implies that whenever $t \mapsto y(t)$ is a trajectory of $y = v$ –i.e., nothing but an *arbitrary* mapping–

$$t \mapsto \big(x(t), u(t)\big) = \big(\psi(\overline{y}(t)), \ \beta(\overline{y}(t))\big)$$

is a trajectory of $\dot{x} = f(x, u)$, and vice versa.

We single out the importance of the mapping h of the previous example:

Definition 4 *Let $(\mathfrak{M}, F)$ be a flat system and Ψ the endogenous transformation putting it into a trivial system. The first block of components of Ψ, i.e., the mapping h in (7), is called a flat (or linearizing) output.*

With this definition, an obvious consequence of theorem 1 is:

Corollary 1 *Consider a flat system. The dimension of a flat output is equal to the input dimension, i.e., $s = m$.*

Example 3 (The PVTOL) *The system studied in example 2 is flat, with*

$$y = h(X, \overline{U}) := (x - \varepsilon \sin \theta, z + \varepsilon \cos \theta)$$

as a flat output. Indeed, the mappings $X = \varphi(\overline{y})$ and $U = \alpha(\overline{y})$ which generate the inverse mapping Φ can be obtained from the implicit equations

$$(y_1 - x)^2 + (y_2 - z)^2 = \varepsilon^2$$
$$(y_1 - x)(\ddot{y}_2 + 1) - (y_2 - z)\ddot{y}_1 = 0$$
$$(\ddot{y}_2 + 1)\sin \theta + \ddot{y}_1 \cos \theta = 0.$$

We first solve for x, z, θ,

$$x = y_1 + \varepsilon \frac{\ddot{y}_1}{\sqrt{\ddot{y}_1^2 + (\ddot{y}_2 + 1)^2}}$$

$$z = y_2 + \varepsilon \frac{(\ddot{y}_2 + 1)}{\sqrt{\ddot{y}_1^2 + (\ddot{y}_2 + 1)^2}}$$

$$\theta = \arg(\ddot{y}_1, \ddot{y}_2 + 1),$$

and then differentiate to get $\dot{x}, \dot{z}, \dot{\theta}, u$ in function of the derivatives of y. Notice the only singularity is $\ddot{y}_1^2 + (\ddot{y}_2 + 1)^2 = 0$.

2.4 Application to motion planning

We now illustrate how flatness can be used for solving control problems. Consider a nonlinear control system of the form

$$\dot{x} = f(x, u) \qquad x \in \mathbb{R}^n, u \in \mathbb{R}^m$$

with flat output

$$y = h(x, u, \dot{u}, \dots, u^{(r)}).$$

By virtue of the system being flat, we can write all trajectories $(x(t), u(t))$ satisfying the differential equation in terms of the flat output and its derivatives:

$$x = \varphi(y, \dot{y}, \dots, y^{(q)})$$
$$u = \alpha(y, \dot{y}, \dots, y^{(q)}).$$

We begin by considering the problem of steering from an initial state to a final state. We parameterize the components of the flat output y_i, $i = 1, \dots, m$ by

$$y_i(t) := \sum_j A_{ij} \lambda_j(t), \tag{8}$$

where the $\lambda_j(t)$, $j = 1, \dots, N$ are basis functions. This reduces the problem from finding a function in an infinite dimensional space to finding a finite set of parameters.

Suppose we have available to us an initial state x_0 at time τ_0 and a final state x_f at time τ_f. Steering from an initial point in state space to a desired point in state space is trivial for flat systems. We have to calculate the values of the flat output and its derivatives from the desired points in state space and then solve for the coefficients A_{ij} in the following system of equations:

$$y_i(\tau_0) = \sum_j A_{ij} \lambda_j(\tau_0) \qquad y_i(\tau_f) = \sum_j A_{ij} \lambda_j(\tau_f)$$
$$\vdots \qquad\qquad\qquad \vdots \tag{9}$$
$$y_i^{(q)}(\tau_0) = \sum_j A_{ij} \lambda_j^{(q)}(\tau_0) \qquad y_i^{(q)}(\tau_f) = \sum_j A_{ij} \lambda_j^{(q)}(\tau_f).$$

To streamline notation we write the following expressions for the case of a *one*-dimensional flat output only. The multi-dimensional case follows by repeatedly applying the one-dimensional case, since the algorithm is decoupled in the component of the flat output. Let $\Lambda(t)$ be the $q + 1$ by N matrix $\Lambda_{ij}(t) = \lambda_j^{(i)}(t)$ and let

$$\bar{y}_0 = (y_1(\tau_0), \dots, y_1^{(q)}(\tau_0))$$
$$\bar{y}_f = (y_1(\tau_f), \dots, y_1^{(q)}(\tau_f)) \tag{10}$$
$$\bar{y} = (\bar{y}_0, \bar{y}_f).$$

Then the constraint in equation (9) can be written as

$$\bar{y} = \begin{pmatrix} \Lambda(\tau_0) \\ \Lambda(\tau_f) \end{pmatrix} A =: \Lambda A. \tag{11}$$

That is, we require the coefficients A to be in an affine sub-space defined by equation (11). The only condition on the basis functions is that Λ is full rank, in order for equation (11) to have a solution.

The implications of flatness is that the trajectory generation problem can be reduced to simple algebra, in theory, and computationally attractive algorithms in practice. In the case of the towed cable system [16], a reasonable state space representation of the system consists of approximately 128 states. Traditional approaches to trajectory generation, such as optimal control, cannot be easily applied in this case. However, it follows from the fact that the system is flat that the feasible trajectories of the system are completely characterized by the motion of the point at the bottom of the cable. By converting the input constraints on the system to constraints on the curvature and higher derivatives of the motion of the bottom of the cable, it is possible to compute efficient techniques for trajectory generation.

2.5 Motion planning with singularities

In the previous section we assumed the endogenous transformation

$$\Psi(x, u, u_1, \ldots) := \left(h(x, \bar{u}), \dot{h}(x, \bar{u}), \ddot{h}(x, \bar{u}), \ldots \right)$$

generated by the flat output $y = h(x, \bar{u})$ everywhere nonsingular, so that we could invert it and express x and u in function of y and its derivatives,

$$(y, \dot{y}, \ldots, y^{(q)}) \mapsto (x, u) = \phi(y, \dot{y}, \ldots, y^{(q)}).$$

But it may well be that a singularity is in fact an interesting point of operation. As ϕ is not defined at such a point, the previous computations do not apply. A way to overcome the problem is to "blow up" the singularity by considering trajectories $t \mapsto y(t)$ such that

$$t \mapsto \phi\big(y(t), \dot{y}(t), \ldots, y^{(q)}(t)\big)$$

can be prolonged into a smooth mapping at points where ϕ is not defined. To do so requires a detailed study of the singularity. A general statement is beyond the scope of this paper and we simply illustrate the idea with an example.

Example 4 *Consider the flat dynamics*

$$\dot{x}_1 = u_1, \quad \dot{x}_2 = u_2 u_1, \quad \dot{x}_3 = x_2 u_1,$$

with flat output $y := (x_1, x_3)$. *When* $u_1 = 0$, *i.e.*, $\dot{y}_1 = 0$ *the endogenous transformation generated by the flat output is singular and the inverse mapping*

$$(y, \dot{y}, \ddot{y}) \stackrel{\phi}{\longmapsto} (x_1, x_2, x_3, u_1, u_2) = \left(y_1, \frac{\dot{y}_2}{\dot{y}_1}, y_2, \dot{y}_1, \frac{\ddot{y}_2 \dot{y}_1 - \ddot{y}_1 \dot{y}_2}{\dot{y}_1^3} \right),$$

is undefined. But if we consider trajectories $t \mapsto y(t) := (\sigma(t), p(\sigma(t)))$, *with* σ *and* p *smooth functions, we find that*

$$\frac{\dot{y}_2(t)}{\dot{y}_1(t)} = \frac{\frac{dp}{d\sigma}(\sigma(t)) \cdot \dot{\sigma}(t)}{\dot{\sigma}(t)} \quad and \quad \frac{\ddot{y}_2 \dot{y}_1 - \ddot{y}_1 \dot{y}_2}{\dot{y}_1^3} = \frac{\frac{d^2 p}{d\sigma^2}(\sigma(t)) \cdot \dot{\sigma}^3(t)}{\dot{\sigma}^3(t)},$$

hence we can prolong $t \mapsto \phi(y(t), \dot{y}(t), \ddot{y}(t))$ *everywhere by*

$$t \mapsto \left(\sigma(t), \frac{dp}{d\sigma}(\sigma(t)), p(\sigma(t)), \dot{\sigma}(t), \frac{d^2 p}{d\sigma^2}(\sigma(t)) \right).$$

The motion planning can now be done as in the previous section: indeed, the functions σ *and* p *and their derivatives are constrained at the initial (resp. final) time by the initial (resp. final) point but otherwise arbitrary.*

For a more substantial application see [21,22,9], where the same idea was applied to nonholonomic mechanical systems by taking advantage of the "natural" geometry of the problem.

3   Feedback design with equivalence

3.1   From equivalence to feedback

The equivalence relation we have defined is very natural since it is essentially a $1-1$ correspondence between trajectories of systems. We had mainly an open-loop point of view. We now turn to a closed-loop point of view by interpreting equivalence in terms of feedback. For that, consider the two dynamics

$$\dot{x} = f(x, u), \quad (x, u) \in X \times U \subset \mathbb{R}^n \times \mathbb{R}^m$$
$$\dot{y} = g(y, v), \quad (y, v) \in Y \times V \subset \mathbb{R}^r \times \mathbb{R}^s.$$

They are described in our formalism by the systems $(X \times U \times \mathbb{R}_m^\infty, F)$ and $(Y \times V \times \mathbb{R}_s^\infty, G)$, with F and G defined by

$$F(x, u, u^1, \dots) := (f(x, u), u^1, u^2, \dots)$$
$$G(y, v, v^1, \dots) := (g(y, v), v^1, v^2, \dots).$$

Assume now the two systems are equivalent, i.e., they have the same trajectories. Does it imply that it is possible to go from $\dot{x} = f(x,u)$ to $\dot{y} = g(y,v)$ by a (possibly) dynamic feedback

$$\dot{z} = a(x,z,v), \quad z \in Z \subset \mathbb{R}^q$$
$$u = \kappa(x,z,v),$$

and *vice versa*? The question might look stupid at first glance since such a feedback can only increase the state dimension. Yet, we can give it some sense if we agree to work "up to pure integrators" (remember this does not change the system in our formalism, see the remark after definition 1).

Theorem 2 *Assume $\dot{x} = f(x,u)$ and $\dot{y} = g(y,v)$ are equivalent. Then $\dot{x} = f(x,u)$ can be transformed by (dynamic) feedback and coordinate change into*

$$\dot{y} = g(y,v), \quad \dot{v} = v^1, \quad \dot{v}^1 = v^2, \quad \dots \quad , \quad \dot{v}^\mu = w$$

for some large enough integer μ. Conversely, $\dot{y} = g(y,v)$ can be transformed by (dynamic) feedback and coordinate change into

$$\dot{x} = f(x,u), \quad \dot{u} = u^1, \quad \dot{u}^1 = u^2, \quad \dots \quad , \quad \dot{u}^\nu = w$$

for some large enough integer ν.

Proof (Proof [13].*).* Denote by F and G the infinite vector fields representing the two dynamics. Equivalence means there is an invertible mapping

$$\Phi(y,\overline{v}) = (\varphi(y,\overline{v}), \alpha(y,\overline{v}), \dot{\alpha}(y,\overline{v}), \dots)$$

such that

$$F(\Phi(y,\overline{v})) = D\Phi(y,\overline{v}).G(y,\overline{v}). \tag{12}$$

Let $\tilde{y} := (y,v,v^1,\dots,v^\mu)$ and $w := v^{\mu+1}$. For μ large enough, φ (resp. α) depends only on $\tilde{y}$ (resp. on $\tilde{y}$ and w). With these notations, Φ reads

$$\Phi(\tilde{y},\overline{w}) = (\varphi(\tilde{y}), \alpha(\tilde{y},w), \dot{\alpha}(y,\overline{w}), \dots),$$

and equation (12) implies in particular

$$f(\varphi(\tilde{y}), \alpha(\tilde{y},w)) = D\varphi(\tilde{y}).\tilde{g}(\tilde{y},w), \tag{13}$$

where $\tilde{g} := (g,v^1,\dots,v^k)$. Because Φ is invertible, φ is full rank hence can be completed by some map π to a coordinate change

$$\tilde{y} \mapsto \phi(\tilde{y}) = (\varphi(\tilde{y}), \pi(\tilde{y})).$$

Consider now the dynamic feedback

$$u = \alpha(\phi^{-1}(x,z),w))$$
$$\dot{z} = D\pi(\phi^{-1}(x,z)).\tilde{g}(\phi^{-1}(x,z),w)),$$

which transforms $\dot{x} = f(x, u)$ into

$$\begin{pmatrix} \dot{x} \\ \dot{z} \end{pmatrix} = \tilde{f}(x, z, w) := \begin{pmatrix} f(x, \alpha(\phi^{-1}(x, z), w)) \\ D\pi(\phi^{-1}(x, z)).\tilde{g}(\phi^{-1}(x, z), w)) \end{pmatrix}.$$

Using (13), we have

$$\tilde{f}(\phi(\tilde{y}), w) = \begin{pmatrix} f(\varphi(\tilde{y}), \alpha(\tilde{y}, w)) \\ D\pi(\tilde{y}).\tilde{g}(\tilde{y}, w) \end{pmatrix} = \begin{pmatrix} D\varphi(\tilde{y}) \\ D\pi(\tilde{y}) \end{pmatrix} \cdot \tilde{g}(\tilde{y}, w) = D\phi(\tilde{y}).\tilde{g}(\tilde{y}, w).$$

Therefore $\tilde{f}$ and $\tilde{g}$ are ϕ-related, which ends the proof. Exchanging the roles of f and g proves the converse statement.

As a flat system is equivalent to a trivial one, we get as an immediate consequence of the theorem:

Corollary 2 *A flat dynamics can be linearized by (dynamic) feedback and coordinate change.*

Remark 5 *As can be seen in the proof of the theorem there are many feedbacks realizing the equivalence, as many as suitable mappings π. Notice all these feedback explode at points where φ is singular (i.e., where its rank collapses).*

Further details about the construction of a linearizing feedback from an output and the links with extension algorithms can be found in [14].

Example 5 (The PVTOL) *We know from example 3 that the dynamics*

$$\ddot{x} = -u_1 \sin\theta + \varepsilon u_2 \cos\theta$$
$$\ddot{z} = u_1 \cos\theta + \varepsilon u_2 \sin\theta - 1$$
$$\ddot{\theta} = u_2$$

admits the flat output

$$y = (x - \varepsilon\sin\theta, z + \varepsilon\cos\theta).$$

It is transformed into the linear dynamics

$$y_1^{(4)} = v_1, \qquad y_2^{(4)} = v_2$$

by the feedback

$$\ddot{\xi} = -v_1\sin\theta + v_2\cos\theta + \xi\dot{\theta}^2$$
$$u_1 = \xi + \varepsilon\dot{\theta}^2$$
$$u_2 = \frac{-1}{\xi}(v_1\cos\theta + v_2\sin\theta + 2\dot{\xi}\dot{\theta})$$

and the coordinate change

$$(x, z, \theta, \dot{x}, \dot{z}, \dot{\theta}, \xi, \dot{\xi}) \mapsto (y, \dot{y}, \ddot{y}, y^{(3)}).$$

The only singularity of this transformation is $\xi = 0$, i.e., $\ddot{y}_1^2 + (\ddot{y}_2 + 1)^2 = 0$. Notice the PVTOL is not linearizable by static feedback.

3.2   Endogenous feedback

Theorem 2 asserts the existence of a feedback such that

$$\dot{x} = f(x, \kappa(x, z, w))$$
$$\dot{z} = a(x, z, w). \tag{14}$$

reads, up to a coordinate change,

$$\dot{y} = g(y, v), \quad \dot{v} = v^1, \quad \ldots \quad, \quad \dot{v}^\mu = w. \tag{15}$$

But (15) is trivially equivalent to $\dot{y} = g(y, v)$ (see the remark after definition 1), which is itself equivalent to $\dot{x} = f(x, u)$. Hence, (14) is equivalent to $\dot{x} = f(x, u)$. This leads to

Definition 5 *Consider the dynamics $\dot{x} = f(x, u)$. We say the feedback*

$$u = \kappa(x, z, w)$$
$$\dot{z} = a(x, z, w)$$

is endogenous *if the open-loop dynamics $\dot{x} = f(x, u)$ is equivalent to the closed-loop dynamics*

$$\dot{x} = f(x, \kappa(x, z, w))$$
$$\dot{z} = a(x, z, w).$$

The word "endogenous" reflects the fact that the feedback variables z and w are in loose sense "generated" by the original variables $x, \overline{u}$ (see [13,15] for further details and a characterization of such feedbacks)

Remark 6 *It is also possible to consider at no extra cost "generalized" feedbacks depending not only on w but also on derivatives of w.*

We thus have a more precise characterization of equivalence and flatness:

Theorem 3 *Two dynamics $\dot{x} = f(x, u)$ and $\dot{y} = g(y, v)$ are equivalent if and only if $\dot{x} = f(x, u)$ can be transformed by endogenous feedback and coordinate change into*

$$\dot{y} = g(y, v), \quad \dot{v} = v^1, \quad \ldots \quad, \quad \dot{v}^\mu = w. \tag{16}$$

for some large enough integer ν, and vice versa.

Corollary 3 *A dynamics is flat if and only if it is linearizable by endogenous feedback and coordinate change.*

Another trivial but important consequence of theorem 2 is that an endogenous feedback can be "unraveled" by another endogenous feedback:

Corollary 4 *Consider a dynamics*

$$\dot{x} = f(x, \kappa(x, z, w))$$
$$\dot{z} = a(x, z, w)$$

where

$$u = \kappa(x, z, w)$$
$$\dot{z} = a(x, z, w)$$

is an endogenous feedback. Then it can be transformed by endogenous feedback and coordinate change into

$$\dot{x} = f(x, u), \quad \dot{u} = u^1, \quad \ldots \quad , \quad \dot{u}^\mu = w. \tag{17}$$

for some large enough integer μ.

This clearly shows which properties are preserved by equivalence: properties that are preserved by adding pure integrators and coordinate changes, in particular controllability.

An endogenous feedback is thus truly "reversible", up to pure integrators. It is worth pointing out that a feedback which is *invertible* in the sense of the standard –but maybe unfortunate– terminology [18] is not necessarily endogenous. For instance the invertible feedback $\dot{z} = v$, $u = v$ acting on the scalar dynamics $\dot{x} = u$ is not endogenous. Indeed, the closed-loop dynamics $\dot{x} = v$, $\dot{z} = v$ is no longer controllable, and there is no way to change that by another feedback!

3.3   Tracking: feedback linearization

One of the central problems of control theory is *trajectory tracking*: given a dynamics $\dot{x} = f(x, u)$, we want to design a controller able to track any reference trajectory $t \mapsto \big(x_r(t), u_r(t)\big)$. If this dynamics admits a flat output $y = h(x, \overline{u})$, we can use corollary 2 to transform it by (endogenous) feedback and coordinate change into the linear dynamics $y^{(\mu+1)} = w$. Assigning then

$$v := y_r^{(\mu+1)}(t) - K\Delta\tilde{y}$$

with a suitable gain matrix K, we get the stable closed-loop error dynamics

$$\Delta y^{(\mu+1)} = -K\Delta\tilde{y},$$

where $y_r(t) := (x_r(t), \overline{u}_r(t))$ and $\tilde{y} := (y, \dot{y}, \ldots, y^\mu)$ and $\Delta\xi$ stands for $\xi - \xi_{r(t)}$. This control law meets the design objective. Indeed, there is by the definition of flatness an invertible mapping

$$\Phi(\overline{y}) = (\varphi(\overline{y}), \alpha(\overline{y}), \dot{\alpha}(\overline{y}), \ldots)$$

relating the infinite dimension vector fields $F(x, \overline{u}) := (f(x, u), u, u^1, \dots)$ and $G(\overline{y}) := (y, y^1, \dots)$. From the proof of theorem 2, this means in particular

$$x = \varphi(\tilde{y}_r(t) + \Delta \tilde{y})$$
$$= \varphi(\tilde{y}_r(t)) + R_\varphi(y_r(t), \Delta \tilde{y}).\Delta \tilde{y}$$
$$= x_r(t) + R_\varphi(y_r(t), \Delta \tilde{y}).\Delta \tilde{y}$$

and

$$u = \alpha(\tilde{y}_r(t) + \Delta \tilde{y}, -K \Delta \tilde{y})$$
$$= \alpha(\tilde{y}_r(t)) + R_\alpha(y_r^{(\mu+1)}(t), \Delta \tilde{y}) \cdot \begin{pmatrix} \Delta \tilde{y} \\ -K \Delta \tilde{y} \end{pmatrix}$$
$$= u_r(t) + R_\alpha(\tilde{y}_r(t), y_r^{(\mu+1)}(t), \Delta \tilde{y}, \Delta w) \cdot \begin{pmatrix} \Delta \tilde{y} \\ -K \Delta \tilde{y} \end{pmatrix},$$

where we have used the fundamental theorem of calculus to define

$$R_\varphi(Y, \Delta Y) := \int_0^1 D\varphi(Y + t\Delta Y)dt$$

$$R_\alpha(Y, w, \Delta Y, \Delta w) := \int_0^1 D\alpha(Y + t\Delta Y, w + t\Delta w)dt.$$

Since $\Delta y \to 0$ as $t \to \infty$, this means $x \to x_r(t)$ and $u \to u_r(t)$. Of course the tracking gets poorer and poorer as the ball of center $\tilde{y}_r(t)$ and radius Δy approaches a singularity of φ. At the same time the control effort gets larger and larger, since the feedback explodes at such a point (see the remark after theorem 2). Notice the tracking quality and control effort depend only on the mapping Φ, hence on the flat output, and not on the feedback itself.

We end this section with some comments on the use of feedback linearization. A linearizing feedback should always be fed by a *trajectory generator*, even if the original problem is not stated in terms of tracking. For instance, if it is desired to *stabilize* an equilibrium point, applying directly feedback linearization without first planning a reference trajectory yields very large control effort when starting from a distant initial point. The role of the trajectory generator is to define an *open-loop* "reasonable" trajectory –i.e., satisfying some state and/or control constraints– that the linearizing feedback will then track.

3.4   Tracking: singularities and time scaling

Tracking by feedback linearization is possible only far from singularities of the endogenous transformation generated by the flat output. If the reference trajectory passes through or near a singularity, then feedback linearization cannot be directly applied, as is the case for motion planning, see section 2.5. Nevertheless, it can be used after a *time scaling*, at least in the presence of "simple" singularities. The interest is that it allows exponential tracking, though in a new "singular" time.

Example 6 *Take a reference trajectory $t \mapsto y_r(t) = (\sigma(t), p(\sigma(t))$ for example 4. Consider the dynamic time-varying compensator $u_1 = \xi\dot\sigma(t)$ and $\dot\xi = v_1\dot\sigma(t)$. The closed loop system reads*

$$x_1' = \xi, \quad x_2' = u_2\xi, \quad x_3' = x_2\xi \quad \xi' = v_1.$$

where ' stands for $d/d\sigma$, the extended state is (x_1, x_2, x_3, ξ), the new control is (v_1, v_2). An equivalent second order formulation is

$$x_1'' = v_1, \quad x_3'' = u_2\xi^2 + x_2 v_1.$$

When ξ is far from zero, the static feedback $u_2 = (v_2 - x_2 v_1)/\xi^2$ linearizes the dynamics,

$$x_1'' = v_1, \quad x_3'' = v_2$$

in σ scale. When the system remains close to the reference, $\xi \approx 1$, even if for some t, $\dot\sigma(t) = 0$. Take

$$\begin{aligned}
v_1 &= 0 - sign(\sigma)a_1(\xi - 1) - a_2(x_1 - \sigma) \\
v_2 &= \tfrac{d^2 p}{d\sigma^2} - sign(\sigma)a_1\left(x_2\xi - \tfrac{dp}{d\sigma}\right)) - a_2(x_3 - p)
\end{aligned} \tag{18}$$

with $a_1 > 0$ and $a_2 > 0$, then the error dynamics becomes exponentially stable in σ-scale (the term $sign(\sigma)$ is for dealing with $\dot\sigma < 0$).

Similar computations for trailer systems can be found in [8,5].

Notice that linearizing controller can be achieved via quasi-static feedback as proposed in [4].

3.5   Tracking: flatness and backstepping

Some drawbacks of feedback linearization We illustrate on two simple (and caricatural) examples that feedback linearization may not lead to the best tracking controller in terms of control effort.

Example 7 *Assume we want to track any trajectory $t \mapsto (x_r(t), u_r(t))$ of*

$$\dot x = -x - x^3 + u, \quad x \in \mathbb{R}.$$

The linearizing feedback

$$\begin{aligned}
u &= x + x^3 - k\Delta x + \dot x_r(t) \\
&= u_r(t) + 3x_r(t)\Delta x^2 + \left(1 + 3x_r^2(t) - k\right)\Delta x + \Delta x^3
\end{aligned}$$

meets this objective by imposing the closed-loop dynamics $\Delta\dot x = -k\Delta x$.

But a closer inspection shows the open-loop error dynamics

$$\Delta \dot{x} = -\left(1 + 3x_r^2(t)\right)\Delta x - \Delta x^3 + 3x_r(t)\Delta x^2 + \Delta u$$
$$= -\Delta x\left(1 + 3x_r^2(t) - 3x_r(t)\Delta x + \Delta x^2\right) + \Delta u$$

is naturally stable when the open-loop control $u := u_r(t)$ is applied (indeed $1 + 3x_r^2(t) - 3x_r(t)\Delta x + \Delta x^2$ is always strictly positive). In other words, the linearizing feedback does not take advantage of the natural damping effects.

Example 8 *Consider the dynamics*

$$\dot{x}_1 = u_1, \qquad \dot{x}_2 = u_2(1 - u_1),$$

for which it is required to track an arbitrary trajectory $t \mapsto (x_r(t), u_r(t))$ (notice $u_r(t)$ may not be so easy to define because of the singularity $u_1 = 1$). The linearizing feedback

$$u_1 = -k\Delta x_1 + \dot{x}_{1r}(t)$$
$$u_2 = \frac{-k\Delta x_2 + \dot{x}_{2r}(t)}{1 + k\Delta x_1 - \dot{x}_{1r}(t)}$$

meets this objective by imposing the closed-loop dynamics $\Delta \dot{x} = -k\Delta x$. Unfortunately u_2 grows unbounded as u_1 approaches one. This means we must in practice restrict to reference trajectories such that $|1 - u_{1r}(t)|$ is always "large" –in particular it is impossible to cross the singularity– and to a "small" gain k.

A smarter control law can do away with these limitations. Indeed, considering the error dynamics

$$\Delta \dot{x}_1 = \Delta u_1$$
$$\Delta \dot{x}_2 = (1 - u_{1r}(t) - \Delta u_1)\Delta u_2 - u_{2r}(t)\Delta u_1,$$

and differentiating the positive function $V(\Delta x) := \frac{1}{2}(\Delta x_1^2 + \Delta x_2^2)$ we get

$$\dot{V} = \Delta u_1(\Delta x_1 - u_{2r}(t)\Delta x_2) + (1 - u_{1r}(t) - \Delta u_1)\Delta u_1 \Delta u_2.$$

The control law

$$\Delta u_1 = -k(\Delta x_1 - u_{2r}(t)\Delta x_2)$$
$$\Delta u_2 = -(1 - u_{1r}(t) - \Delta u_1)\Delta x_2$$

does the job since

$$\dot{V} = -\left(\Delta x_1 - u_{2r}(t)\Delta x_2\right)^2 - \left((1 - u_{1r}(t) - \Delta u_1)\Delta x_2\right)^2 \le 0.$$

Moreover, when $u_{1r}(t) \ne 0$, $\dot{V}$ is zero if and only if $\|\Delta x\|$ is zero. It is thus possible to cross the singularity –which has been made an unstable equilibrium of the closed-loop error dynamics– and to choose the gain k as large as desired. Notice the singularity is overcome by a "truly" multi-input design.

It should not be inferred from the previous examples that feedback linearization necessarily leads to inefficient tracking controllers. Indeed, when the trajectory generator is well-designed, the system is always close to the reference trajectory. Singularities are avoided by restricting to reference trajectories which stay away from them. This makes sense in practice when singularities do not correspond to interesting regions of operations. In this case, designing a tracking controller "smarter" than a linearizing feedback often turns out to be rather complicated, if possible at all.

Backstepping The previous examples are rather trivial because the control input has the same dimension as the state. More complicated systems can be handled by *backstepping*. Backstepping is a versatile design tool which can be helpful in a variety of situations: stabilization, adaptive or output feedback, etc ([12] for a complete survey). It relies on the simple yet powerful following idea: consider the system

$$\dot{x} = f(x, \xi), \quad f(x_0, \xi_0) = 0$$
$$\dot{\xi} = u,$$

where $(x, \xi) \in \mathbb{R}^n \times \mathbb{R}$ is the state and $u \in \mathbb{R}$ the control input, and assume we can asymptotically stabilize the equilibrium x_0 of the subsystem $\dot{x} = f(x, \xi)$, i.e., we know a control law $\xi = \alpha(x)$, $\alpha(x_0) = \xi_0$ and a positive function $V(x)$ such that

$$\dot{V} = DV(x).f(x, \alpha(x)) \leq 0.$$

A key observation is that the "virtual" control input ξ can then "backstepped" to stabilize the equilibrium (x_0, ξ_0) of the complete system. Indeed, introducing the positive function

$$W(x, \xi) := V(x) + \frac{1}{2}(\xi - \alpha(x))^2$$

and the error variable $z := \xi - \alpha(x)$, we have

$$\dot{W} = DV(x).f(x, \alpha(x) + z) + z\big(u - \dot{\alpha}(x, \xi)\big)$$
$$= DV(x).\big(f(x, \alpha(x)) + R(x, z).z\big) + z\big(u - D\alpha(x).f(x, \xi)\big)$$
$$= \dot{V} + z\big(u - D\alpha(x).f(x, \xi) + DV(x).R(x, z)\big),$$

where we have used the fundamental theorem of calculus to define

$$R(x, h) := \int_0^1 \frac{\partial f}{\partial \xi}(x, x + th)dt$$

(notice $R(x, h)$ is trivially computed when f is linear in ξ). As $\dot{V}$ is negative by assumption, we can make $\dot{W}$ negative, hence stabilize the system, by choosing for instance

$$u := -z + D\alpha(x).f(x, \xi) - DV(x).R(x, z).$$

Blending equivalence with backstepping Consider a dynamics $\dot{y} = g(y, v)$ for which we would like to solve the tracking problem. Assume it is equivalent to another dynamics $\dot{x} = f(x, u)$ for which we can solve this problem, i.e., we know a tracking control law together with a Lyapunov function. How can we use this property to control $\dot{y} = g(y, v)$? Another formulation of the question is: assume we know a controller for $\dot{x} = f(x, u)$. How can we derive a controller for

$$\dot{x} = f(x, \kappa(x, z, v))$$
$$\dot{z} = a(x, z, v),$$

where $u = \kappa(x, z, v), \dot{z} = a(x, z, v)$ is an endogenous feedback? Notice backstepping answers the question for the elementary case where the feedback in question is a pure integrator.

By theorem 2, we can transform $\dot{x} = f(x, u)$ by (dynamic) feedback and coordinate change into

$$\dot{y} = g(y, v), \quad \dot{v} = v^1, \quad \ldots \quad , \quad \dot{v}^\mu = w. \tag{19}$$

for some large enough integer μ. We can then trivially backstep the control from v to w and change coordinates. Using the same reasoning as in section 3.3, it is easy to prove this leads to a control law solving the tracking problem for $\dot{x} = f(x, u)$. In fact, this is essentially the method we followed in section 3.3 on the special case of a flat $\dot{x} = f(x, u)$. We illustrated in section 3.5 potential drawbacks of this approach.

However, it is often possible to design better –though in general more complicated– tracking controllers by suitably using backstepping. This point of view is extensively developed in [12], though essentially in the single-input case, where general equivalence boils down to equivalence by coordinate change. In the multi-input case new phenomena occur as illustrated by the following examples.

Example 9 (The PVTOL) *We know from example 2 that*

$$\ddot{x} = -u_1 \sin\theta + \varepsilon u_2 \cos\theta$$
$$\ddot{z} = u_1 \cos\theta + \varepsilon u_2 \sin\theta - 1 \tag{20}$$
$$\ddot{\theta} = u_2$$

is globally equivalent to

$$\ddot{y}_1 = -\xi \sin\theta, \qquad \ddot{y}_2 = \xi \cos\theta - 1,$$

where $\xi = u_1 + \varepsilon\dot{\theta}^2$. This latter form is rather appealing for designing a tracking controller and leads to the error dynamics

$$\Delta\ddot{y}_1 = -\xi \sin\theta + \xi_r(t) \sin\theta_r(t)$$
$$\Delta\ddot{y}_2 = \xi \cos\theta - \xi_r(t) \cos\theta_r(t)$$

Clearly, if θ were a control input, we could track trajectories by assigning

$$-\xi \sin \theta = \alpha_1(\Delta y_1, \Delta \dot{y}_1) + \ddot{y}_{1r}(t)$$
$$\xi \cos \theta = \alpha_2(\Delta y_2, \Delta \dot{y}_2) + \ddot{y}_{2r}(t)$$

for suitable functions α_1, α_2 and find a Lyapunov function $V(\Delta y, \Delta \dot{y})$ for the system. In other words, we would assign

$$\xi = \Xi(\Delta y, \Delta \dot{y}, \ddot{y}_r(t)) := \sqrt{(\alpha_1 + \ddot{y}_{1r})^2 + (\alpha_2 + \ddot{y}_{2r})^2}$$
$$\theta = \Theta(\Delta y, \Delta \dot{y}, \ddot{y}_r(t)) := \arg(\alpha_1 + \ddot{y}_{1r}, \alpha_2 + \ddot{y}_{2r}). \tag{21}$$

The angle θ is a priori not defined when $\xi = 0$, i.e., at the singularity of the flat output y. We will not discuss the possibility of overcoming this singularity and simply assume we stay away from it. Aside from that, there remains a big problem: how should the "virtual" control law (21) be understood? Indeed, it seems to be a differential equation: because y depends on θ, hence Ξ and Θ are in fact functions of the variables

$$x, \dot{x}, z, \dot{z}, \theta, \dot{\theta}, y_r(t), \dot{y}_r(t), \ddot{y}_r(t).$$

Notice ξ is related to the actual control u_1 by a relation that also depends on $\dot{\theta}$.

Let us forget this apparent difficulty for the time being and backstep (21) the usual way. Introducing the error variable $\kappa_1 := \theta - \Theta(\Delta y, \Delta \dot{y}, \ddot{y}_r(t))$ and using the fundamental theorem of calculus, the error dynamics becomes

$$\Delta \ddot{y}_1 = \alpha_1(\Delta y_1, \Delta \dot{y}_1) - \kappa_1 \, R_{\sin}(\Theta(\Delta y, \Delta \dot{y}, \ddot{y}_r(t)), \kappa_1) \, \Xi(\Delta y, \Delta \dot{y}, \ddot{y}_r(t))$$
$$\Delta \ddot{y}_2 = \alpha_2(\Delta y_1, \Delta \dot{y}_1) + \kappa_1 \, R_{\cos}(\Theta(\Delta y, \Delta \dot{y}, \ddot{y}_r(t)), \kappa_1) \, \Xi(\Delta y, \Delta \dot{y}, \ddot{y}_r(t))$$
$$\dot{\kappa}_1 = \dot{\theta} - \dot{\Theta}(\kappa_1, \Delta y, \Delta \dot{y}, \ddot{y}_r(t), y_r^{(3)}(t))$$

Notice the functions

$$R_{\sin}(x, h) = \sin x \frac{\cos h - 1}{h} + \cos x \frac{\sin h}{h}$$
$$R_{\cos}(x, h) = \cos x \frac{\cos h - 1}{h} - \sin x \frac{\sin h}{h}$$

are bounded and analytic. Differentiate now the positive function

$$V_1(\Delta y, \Delta \dot{y}, \kappa_1) := V(\Delta y, \Delta \dot{y}) + \frac{1}{2}\kappa_1^2$$

to get

$$\dot{V}_1 = \frac{\partial V}{\partial \Delta y_1} \Delta \dot{y}_1 + \frac{\partial V}{\partial \Delta \dot{y}_1}(\alpha_1 - \kappa_1 R_{\sin} \Xi) +$$
$$\frac{\partial V}{\partial \Delta y_2} \Delta \dot{y}_2 + \frac{\partial V}{\partial \Delta \dot{y}_2}(\alpha_2 + \kappa_1 R_{\cos} \Xi) + \kappa_1 \, (\dot{\theta} - \dot{\Theta})$$
$$= \dot{V} + \kappa_1 \left(\dot{\theta} - \dot{\Theta} + \kappa_1 \left(R_{\cos} \frac{\partial V}{\partial \Delta y_1} - R_{\sin} \frac{\partial V}{\partial \Delta y_2} \right) \Xi \right),$$

where we have omitted arguments of all the functions for the sake of clarity. If $\dot{\theta}$ were a control input, we could for instance assign

$$\dot{\theta} := -\kappa_1 + \dot{\Theta} - \kappa_1 \left(R_{\cos} \frac{\partial V}{\partial \Delta y_1} - R_{\sin} \frac{\partial V}{\partial \Delta y_2} \right) \Xi$$

$$:= \Theta_1 \left(\kappa_1, \Delta y, \Delta \dot{y}, \ddot{y}_r(t), y_r^{(3)}(t) \right),$$

to get $\dot{V}_1 = \dot{V} - \kappa_1^2 \leq 0$. We thus backstep this "virtual" control law: we introduce the error variable

$$\kappa_2 := \dot{\theta} - \Theta_1 \left(\kappa_1, \Delta y, \Delta \dot{y}, \ddot{y}_r(t), y_r^{(3)}(t) \right)$$

together with the positive function

$$V_2(\Delta y, \Delta \dot{y}, \kappa_1, \kappa_2) := V_1(\Delta y, \Delta \dot{y}, \kappa_1) + \frac{1}{2} \kappa_2^2.$$

Differentiating

$$\dot{V}_2 = \dot{V} + \kappa_1(-\kappa_1 + \kappa_2) + \kappa_2(v_2 - \dot{\Theta}_1)$$

$$= \dot{V}_1 + \kappa_2(u_2 - \dot{\Theta}_1 + \kappa_2),$$

and we can easily make $\dot{V}_1$ negative by assigning

$$u_2 := \Theta_2 \left(\kappa_1, \kappa_2, \Delta y, \Delta \dot{y}, \ddot{y}_r(t), y_r^{(3)}(t), y_r^{(4)}(t) \right) \tag{22}$$

for some suitable function Θ_2.

A key observation is that Θ_2 and V_2 are in fact functions of the variables

$$x, \dot{x}, z, \dot{z}, \theta, \dot{\theta}, y_r(t), \dots, y_r^{(4)}(t),$$

which means (22) makes sense. We have thus built a static control law

$$u_1 = \Xi \left(x, \dot{x}, z, \dot{z}, \theta, \dot{\theta}, y_r(t), \dot{y}_r(t), \ddot{y}_r(t) \right) + \varepsilon \dot{\theta}^2$$

$$u_2 = \Theta_2 \left(x, \dot{x}, z, \dot{z}, \theta, \dot{\theta}, y_r(t), \dots, y_r^{(4)}(t) \right)$$

that does the tracking for (20). Notice it depends on $y_r(t)$ up to the fourth derivative.

Example 10 *The dynamics*

$$\dot{x}_1 = u_1, \quad \dot{x}_2 = x_3(1 - u_1), \quad \dot{x}_3 = u_2,$$

admits (x_1, x_2) as a flat output. The corresponding endogenous transformation is singular, hence any linearizing feedback blows up, when $u_1 = 1$. However, it is easy to backstep the controller of example 8 to build a globally tracking static controller

Remark 7 *Notice that none the of two previous examples can be linearized by static feedback. Dynamic feedback is necessary for that. Nevertheless we were able to derive static tracking control laws for them. An explanation of why this is possible is that a flat system can in theory be linearized by a quasistatic feedback [3] –provided the flat output does not depend on derivatives of the input–.*

Backstepping and time-scaling Backstepping can be combined with linearization and time-scaling, as illustrated in the following example.

Example 11 *Consider example 4 and its tracking control defined in example 6. Assume, for example, that $\dot{\sigma} \geq 0$. With the dynamic controller*

$$\dot{\xi} = v_1 \dot{\sigma}, \quad u_1 = \xi \dot{\sigma}, \quad u_2 = (v_2 - x_2 v_1)/\xi^2$$

where v_1 and v_2 are given by equation (18), we have, for the error $e = y - y_r$, a Lyapunov function $V(e, de/d\sigma)$ satisfying

$$dV/d\sigma \leq -aV \tag{23}$$

with some constant $a > 0$. Remember that $de/d\sigma$ corresponds to $(\xi - 1, x_2 \xi - dp/d\sigma)$. Assume now that the real control is not (u_1, u_2) but $(\dot{u}_1 := w_1, u_2)$. With the extended Lyapunov function

$$W = V(e, de/d\sigma) + \frac{1}{2}(u_1 - \xi \dot{\sigma})^2$$

we have

$$\dot{W} = \dot{V} + (w_1 - \dot{\xi}\dot{\sigma} - \xi\ddot{\sigma})((u_1 - \xi\dot{\sigma}).$$

Some manipulations show that

$$\dot{V} = (u_1 - \dot{\sigma}\xi)\left(\frac{\partial V}{\partial e_1} + \frac{\partial V}{\partial e_2}x_2 + \frac{\partial V}{\partial e_2'}u_2\xi\right) + \dot{\sigma}\frac{dV}{d\sigma}$$

(remember $\dot{\xi} = v_1\dot{\sigma}$ and (v_1, v_2) are given by (18)). The feedback $(b > 0)$

$$w_1 = -\left(\frac{\partial V}{\partial e_1} + \frac{\partial V}{\partial e_2}x_2 + \frac{\partial V}{\partial e_2'}u_2\xi\right) + \dot{\xi}\dot{\sigma} + \xi\ddot{\sigma} - b(u_1 - \xi\dot{\sigma})$$

achieves asymptotic tracking since $\dot{W} \leq -a\dot{\sigma}V - b(u_1 - \xi\dot{\sigma})^2$.

Conclusion It is possible to generalize the previous examples to prove that a control law can be backstepped "through" any endogenous feedback. In particular a flat dynamics can be seen as a (generalized) endogenous feedback acting on the flat output; hence we can backstep a control law for the flat output through the whole dynamics. In other words the flat output serves as a first "virtual" control in the backstepping process. It is another illustration of the fact that a flat output "summarizes" the dynamical behavior.

Notice also that in a tracking problem the knowledge of a flat output is extremely useful not only for the tracking itself (i.e., the closed-loop problem) but also for the trajectory generation (i.e., the open-loop problem)

References

1. E. Cartan. Sur l'équivalence absolue de certains systèmes d'équations différentielles et sur certaines familles de courves. *Bull. Soc. Math. France*, 42:12–48, 1914. Also in Œuvres Complètes, part II, vol. 2, pp.1133–1168, CNRS, Paris, 1984.

2. E. Cartan. Sur l'intégration de certains systèmes indéterminés d'équations différentielles. *J. für reine und angew. Math.*, 145:86–91, 1915. Also in Œuvres Complètes, part II, vol. 2, pp.1164–1174, CNRS, Paris, 1984.

3. E. Delaleau and J. Rudolph. Decoupling and linearization by quasi-static feedback of generalized states. In *Proc. of the 3rd European Control Conf.*, pages 1069–1074, Rome, 1995.

4. E. Delaleau and J. Rudolph. Control of flat systems by quasi-static feedback of generalized states. *Int. Journal of Control*, 71:745–765, 1998.

5. M. Fliess, J. levine, P. Martin, F. Ollivier, and P. Rouchon. Controlling nonlinear systems by flatness. In C.I. Byrnes, B.N. Datta, D.S. Gilliam, and C.F. Martin, editors, *Systems and control in the Twenty-First Century*, Progress in Systems and Control Theory. Birkhäuser, 1997.

6. M. Fliess, J. Lévine, Ph. Martin, and P. Rouchon. Sur les systèmes non linéaires différentiellement plats. *C.R. Acad. Sci. Paris*, I–315:619–624, 1992.

7. M. Fliess, J. Lévine, Ph. Martin, and P. Rouchon. Linéarisation par bouclage dynamique et transformations de Lie-Bäcklund. *C.R. Acad. Sci. Paris*, I-317:981–986, 1993.

8. M. Fliess, J. Lévine, Ph. Martin, and P. Rouchon. Design of trajectory stabilizing feedback for driftless flat systems. In *Proc. of the 3rd European Control Conf.*, pages 1882–1887, Rome, 1995.

9. M. Fliess, J. Lévine, Ph. Martin, and P. Rouchon. Flatness and defect of nonlinear systems: introductory theory and examples. *Int. J. Control*, 61(6):1327–1361, 1995.

10. M. Fliess, J. Lévine, Ph. Martin, and P. Rouchon. Nonlinear control and diffieties, with an application to physics. *Contemporary Mathematics*, 219:81–92, 1998.

11. M. Fliess, J. Lévine, Ph. Martin, and P. Rouchon. A Lie-Bäcklund approach to equivalence and flatness of nonlinear systems. *IEEE AC*, 44:922–937, 1999.

12. M. Krstić, I. Kanellakopoulos, and P. Kokotović. *Nonlinear and Adaptive Control Design*. John Wiley & Sons, Inc., 1995.

13. Ph. Martin. *Contribution à l'étude des systèmes différentiellement plats*. PhD thesis, École des Mines de Paris, 1992.

14. Ph. Martin. An intrinsic condition for regular decoupling. *Systems & Control Letters*, 20:383–391, 1993.

15. Ph. Martin. Endogenous feedbacks and equivalence. In *Systems and Networks: Marthematical Theory and Applications (MTNS'93)*, volume II, pages 343–346. Akademie Verlag, Berlin, 1994.

16. R. M. Murray. Trajectory generation for a towed cable flight control system. In *Proc. IFAC World Congress*, pages 395–400, San Francisco, 1996.

17. M. van Nieuwstadt, M. Rathinam, and R.M. Murray. Differential flatness and absolute equivalence. In *Proc. of the 33rd IEEE Conf. on Decision and Control*, pages 326–332, Lake Buena Vista, 1994.

18. H. Nijmeijer and A.J. van der Schaft. *Nonlinear Dynamical Control Systems*. Springer-Verlag, 1990.
19. J.B. Pomet. A differential geometric setting for dynamic equivalence and dynamic linearization. In *Workshop on Geometry in Nonlinear Control, Banach Center Publications, Warsaw*, 1993.
20. J.B. Pomet, C. Moog, and E. Aranda. A non-exact Brunovsky form and dynamic feedback linearization. In *Proc. of the 31st IEEE Conf. on Decision and Control*, pages 2012–2017, 1992.
21. P. Rouchon, M. Fliess, J. Lévine, and Ph. Martin. Flatness and motion planning: the car with n-trailers. In *Proc. ECC'93, Groningen*, pages 1518–1522, 1993.
22. P. Rouchon, M. Fliess, J. Lévine, and Ph. Martin. Flatness, motion planning and trailer systems. In *Proc. of the 32nd IEEE Conf. on Decision and Control*, pages 2700–2705, San Antonio, 1993.
23. W.M. Sluis. *Absolute Equivalence and its Application to Control Theory*. PhD thesis, University of Waterloo, Ontario, 1992.
24. V.V. Zharinov. *Geometrical Aspects of Partial Differential Equations*. World Scientific, Singapore, 1992.

Flat Systems: open problems, infinite dimensional extension, symmetries and catalog

Philippe Martin[1], Richard M. Murray[2], and Pierre Rouchon[1]

[1] Centre Automatique et Systèmes, École des Mines de Paris, Paris, France.
[2] Division of Engineering and Applied Science, California Institute of Technology, California, USA.

Abstract. This chapter is devoted to open problems and new perspectives on flat systems, including developments on symmetries and distributed parameters systems based on examples of physical interest. It contains a representative catalog of flat systems arising in various fields of engineering.

1 Checking flatness: an overview

1.1 The general problem

Devising a general computable test for checking whether $\dot{x} = f(x, u), x \in \mathbb{R}^n, u \in \mathbb{R}^m$ is flat remains up to now an open problem. This means there are no systematic methods for constructing flat outputs. This does not make flatness a useless concept: for instance Lyapunov functions and uniform first integrals of dynamical systems are extremely helpful notions both from a theoretical and practical point of view though they cannot be systematically computed.

The main difficulty in checking flatness is that a candidate flat output $y = h(x, u, \ldots, u^{(r)})$ may a priori depend on derivatives of u of arbitrary order r. Whether this order r admits an upper bound (in terms of n and m) is at the moment completely unknown. Hence we do not know whether a finite bound exists at all. In the sequel, we say a system is r-flat if it admits a flat output depending on derivatives of u of order at most r.

To illustrate this upper bound might be at least linear in the state dimension, consider the system

$$x_1^{(\alpha_1)} = u_1, \quad x_2^{(\alpha_2)} = u_2, \quad \dot{x}_3 = u_1 u_2$$

with $\alpha_1 > 0$ and $\alpha_2 > 0$. It admits the flat output

$$y_1 = x_3 + \sum_{i=1}^{\alpha_1} (-1)^i x_1^{(\alpha_1 - i)} u_2^{(i-1)}, \quad y_2 = x_2,$$

hence is r-flat with $r := \min(\alpha_1, \alpha_2) - 1$. We suspect (without proof) there is no flat output depending on derivatives of u of order less than $r - 1$.

If such a bound $\kappa(n,m)$ were known, the problem would amount to checking p-flatness for a *given* $p \leq \kappa(n,m)$ and could be solved in theory. Indeed, it consists [36] in finding m functions $h_1, \ldots, h_m$ depending on $(x, u, \ldots, u^{(p)})$ such that

$$\dim \operatorname{span} \left\{ dx_1, \ldots, dx_n, du_1, \ldots, du_m, dh_1^{(\mu)}, \ldots, dh_m^{(\mu)} \right\}_{0 \leq \mu \leq \nu} = m(\nu + 1),$$

where $\nu := n + pm$. This means checking the integrability of the partial differential system with a transversality condition

$$dx_i \wedge dh \wedge \ldots \wedge dh^{(\nu)} = 0, \qquad\qquad i = 1, \ldots, n$$

$$du_j \wedge dh \wedge \ldots \wedge dh^{(\nu)} = 0, \qquad\qquad j = 1, \ldots, m$$

$$dh \wedge \ldots \wedge dh^{(\nu)} \neq 0,$$

where $dh^{(\mu)}$ stands for $dh_1^{(\mu)} \wedge \ldots \wedge dh_m^{(\mu)}$. It is in theory possible to conclude by using a computable criterion [3,59], though this seems to lead to practically intractable calculations. Nevertheless it can be hoped that, due to the special structure of the above equations, major simplifications might appear.

1.2   Known results

Systems linearizable by static feedback. A system which is linearizable by static feedback and coordinate change is clearly flat. Hence the geometric necessary and sufficient conditions in [26,25] provide sufficient conditions for flatness. Notice a flat system is in general not linearizable by static feedback, with the major exception of the single-input case.

Single-input systems. When there is only one control input flatness reduces to static feedback linearizability [6] and is thus completely characterized by the test in [26,25].

Affine systems of codimension 1. A system of the form

$$\dot{x} = f_0(x) + \sum_{j=1}^{n-1} u_j g_j(x), \qquad x \in \mathbb{R}^n,$$

i.e., with one input less than states and linear w.r.t. the inputs is 0-flat as soon as it is controllable [6] (more precisely strongly accessible for almost every x).

The picture is much more complicated when the system is not linear w.r.t. the control, see [37] for a geometric sufficient condition.

Affine systems with 2 inputs and 4 states. Necessary and sufficient conditions for 1-flatness of the system can be found in [58]. They give a good idea of the complexity of checking r-flatness even for r small.

Driftless systems. For driftless systems of the form $\dot{x} = \sum_{i=1}^{m} f_i(x) u_i$ additional results are available.

Theorem 1 (Driftless systems with two inputs [39]) *The system*

$$\dot{x} = f_1(x) u_1 + f_2(x) u_2$$

is flat if and only if the generic rank of E_k is equal to $k+2$ for $k = 0, \ldots, n-2n$ where $E_0 := span\{f_1, f_2\}$, $E_{k+1} := span\{E_k, [E_k, E_k]\}$, $k \geq 0$.

A flat two-input driftless system is always 0-flat. As a consequence of a result in [49], a flat two-input driftless system satisfying some additional regularity conditions can be put by *static* feedback and coordinate change into the *chained system* [50]

$$\dot{x}_1 = u_1, \quad \dot{x}_2 = u_2, \quad \dot{x}_3 = x_2 u_1, \quad \ldots, \quad \dot{x}_n = x_{n-1} u_1.$$

Theorem 2 (Driftless systems, n states, and $n-2$ inputs [40,41])

$$\dot{x} = \sum_{i=1}^{n} u_i f_i(x), \quad x \in \mathbb{R}^n$$

is flat as soon as it is controllable (i.e., strongly accessible for almost every x). More precisely it is 0-flat when n is odd, and 1-flat when n is even.

All the results mentioned above rely on the use of exterior differential systems. Additional results on driftless systems, with applications to nonholonomic systems, can be found in [76,75,72].

Mechanical systems. For mechanical systems with one control input less than configuration variables, [63] provides a geometric characterization, in terms of the metric derived form the kinetic energy and the control codistribution, of flat outputs depending only on the configuration variables.

A necessary condition. Because it is not known whether flatness can be checked with a finite test, see section 1.1, it is very difficult to prove that a system is *not* flat. The following result provides a simple necessary condition.

Theorem 3 (The ruled-manifold criterion [66,12]) *Assume $\dot{x} = f(x, u)$ is flat. The projection on the p-space of the submanifold $p = f(x, u)$, where x is considered as a parameter, is a ruled submanifold for all x.*

The criterion just means that eliminating u from $\dot{x} = f(x, u)$ yields a set of equations $F(x, \dot{x}) = 0$ with the following property: for all (x, p) such that $F(x, p) = 0$, there exists $a \in \mathbb{R}^n$, $a \neq 0$ such that

$$\forall \lambda \in \mathbb{R}, \quad F(x, p + \lambda a) = 0.$$

$F(x, p) = 0$ is thus a ruled manifold containing straight lines of direction a.

The proof directly derives from the method used by Hilbert [23] to prove the second order Monge equation $\frac{d^2z}{dx^2} = \left(\frac{dy}{dx}\right)^2$ is not solvable without integrals.

A restricted version of this result was proposed in [73] for systems linearizable by a special class of dynamic feedbacks.

As crude as it may look, this criterion is up to now the only way –except for two-input driftless systems– to prove a multi-input system is not flat.

Example 1 *The system*

$$\dot{x}_1 = u_1, \quad \dot{x}_2 = u_2, \quad \dot{x}_3 = (u_1)^2 + (u_2)^3$$

is not flat, since the submanifold $p_3 = p_1^2 + p_2^3$ is not ruled: there is no $a \in \mathbb{R}^3$, $a \neq 0$, such that

$$\forall \lambda \in \mathbb{R}, p_3 + \lambda a_3 = (p_1 + \lambda a_1)^2 + (p_2 + \lambda a_2)^3.$$

Indeed, the cubic term in λ implies $a_2 = 0$, the quadratic term $a_1 = 0$ hence $a_3 = 0$.

Example 2 *The system $\dot{x}_3 = \dot{x}_1^2 + \dot{x}_2^2$ does not define a ruled submanifold of $\mathbb{R}^3$: it is not flat in $\mathbb{R}$. But it defines a ruled submanifold in $\mathbb{C}^3$: in fact it is flat in $\mathbb{C}$, with the flat output*

$$y = \left(x_3 - (\dot{x}_1 - \dot{x}_2\sqrt{-1})(x_1 + x_2\sqrt{-1}), \; x_1 + x_2\sqrt{-1}\right).$$

Example 3 (The ball and beam [21]) *We now prove by the ruled manifold criterion that*

$$\ddot{r} = -Bg\sin\theta + Br\dot{\theta}^2$$
$$(mr^2 + J + J_b)\ddot{\theta} = \tau - 2mr\dot{r}\dot{\theta} - mgr\cos\theta,$$

where $(r, \dot{r}, \theta, \dot{\theta})$ is the state and τ the input, is not flat (as it is a single-input system, we could also prove it is not static feedback linearizable, see section 1.2). Eliminating the input τ yields

$$\dot{r} = v_r, \quad \dot{v}_r = -Bg\sin\theta + Br\dot{\theta}^2, \quad \dot{\theta} = v_\theta$$

which defines a ruled manifold in the $(\dot{r}, \dot{v}_r, \dot{\theta}, \dot{v}_\theta)$-space for any r, v_r, θ, v_θ, and we cannot conclude directly. Yet, the system is obviously equivalent to

$$\dot{r} = v_r, \quad \dot{v}_r = -Bg\sin\theta + Br\dot{\theta}^2,$$

which clearly does not define a ruled submanifold for any (r, v_r, θ). Hence the system is not flat.

2 Infinite dimension "flat" systems

The idea underlying equivalence and flatness –a one-to-one correspondence between trajectories of systems– is not restricted to control systems described by *ordinary* differential equations. It can be adapted to delay differential systems and to partial differential equations with boundary control. Of course, there are many more technicalities and the picture is far from clear. Nevertheless, this new point of view seems promising for the design of control laws. In this section, we sketch some recent developments in this direction.

2.1 Delay systems

Consider for instance the simple differential delay system

$$\dot{x}_1(t) = x_2(t), \quad \dot{x}_2(t) = x_1(t) - x_2(t) + u(t-1).$$

Setting $y(t) := x_1(t)$, we can clearly explicitly parameterize its trajectories by

$$x_1(t) = y(t), \quad x_2(t) = \dot{y}(t), \quad u(t) = \ddot{y}(t+1) + \dot{y}(t+1) - y(t+1).$$

In other words, $y(t) := x_1(t)$ plays the role of a"flat" output. This idea is investigated in detail in [43], where the class of δ-*free* systems is defined (δ is the delay operator). More precisely, [43,45] considers linear differential delay systems

$$M(d/dt, \delta)w = 0$$

where M is a $(n-m) \times n$ matrix with entries polynomials in d/dt and δ and $w = (w_1, \dots, w_n)$ are the system variables. Such a system is said to be δ-free if it can be related to the "free" system $y = (y_1, \dots, y_m)$ consisting of arbitrary functions of time by

$$w = P(d/dt, \delta, \delta^{-1})y$$
$$y = Q(d/dt, \delta, \delta^{-1})w,$$

where P (resp. Q) is a $n \times m$ (resp. $m \times n$) matrix the entries of which are polynomial in d/dt, δ and δ^{-1}.

Many linear delay systems are δ-free. For example, $\dot{x}(t) = Ax(t) + Bu(t-1)$, (A, B) controllable, is δ-free, with the Brunovski output of $\dot{x} = Ax + Bv$ as a "δ-free" output.

The following systems, commonly used in process control,

$$z_i(s) = \sum_{j=1}^{m} \left\{ \frac{K_i^j \exp(-s\delta_i^j)}{1 + \tau_i^j s} \right\} u_j(s), \quad i = 1, \dots p$$

(s Laplace variable, gains K_i^j, delays δ_i^j and time constants τ_i^j between u_j and z_i) are δ-free [56]. Other interesting examples of δ-free systems arise from partial differential equations:

Example 4 (Torsion beam system) *The torsion motion of a beam (figure 1) can be modeled in the linear elastic domain by*

$$\partial_t^2 \theta(x,t) = \partial_x^2 \theta(x,t), \quad x \in [0,1]$$
$$\partial_x \theta(0,t) = u(t)$$
$$\partial_x \theta(1,t) = \partial_t^2 \theta(1,t),$$

where $\theta(x,t)$ is the torsion of the beam and $u(t)$ the control input. From d'Alembert's formula, $\theta(x,t) = \phi(x+t) + \psi(x-t)$, we easily deduce

$$2\theta(t,x) = \dot{y}(t+x-1) - \dot{y}(t-x+1) + y(t+x-1) + y(t-x+1)$$
$$2u(t) = \ddot{y}(t+1) + \ddot{y}(t-1) - \dot{y}(t+1) + \dot{y}(t-1),$$

where we have set $y(t) := \theta(1,t)$. This proves the system is δ-free with $\theta(1,t)$ as a "δ-flat" output. See [46,15,18] for details and an application to motion planning.

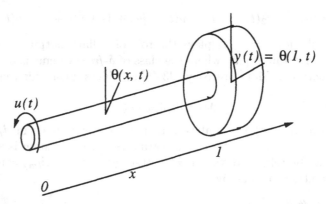

Fig. 1. torsion of a flexible beam

Many examples of delay systems derived from the 1D-wave equation can be treated via such techniques (see [8] for tank filled with liquid, [14] for the telegraph equation and [57] for two physical examples with delay depending on control).

2.2   Distributed parameters systems

For partial differential equations with boundary control and mixed systems of partial and ordinary differential equations, it seems possible to describe the one-to-one correspondence via series expansion, though a sound theoretical framework is yet to be found. We illustrate this original approach to control design on the following two "flat" systems.

Example 5 (Heat equation) *Consider as in [30] the linear heat equation*

$$\partial_t \theta(x,t) = \partial_x^2 \theta(x,t), \quad x \in [0,1] \tag{1}$$
$$\partial_x \theta(0,t) = 0 \tag{2}$$
$$\theta(1,t) = u(t), \tag{3}$$

where $\theta(x,t)$ is the temperature and $u(t)$ is the control input. We claim that

$$y(t) := \theta(0,t)$$

is a "flat" output. Indeed, the equation in the Laplace variable s reads

$$s\hat{\theta}(x,s) = \hat{\theta}''(x,s) \quad \text{with} \quad \hat{\theta}'(0,s) = 0, \quad \hat{\theta}(1,s) = \hat{u}(s)$$

(' stands for ∂_x and ^ for the Laplace transform), and the solution is clearly $\hat{\theta}(x,s) = \cosh(x\sqrt{s})\hat{u}(s)/\cosh(\sqrt{s})$. As $\hat{\theta}(0,s) = \hat{u}(s)/\cosh(\sqrt{s})$, this implies

$$\hat{u}(s) = \cosh(\sqrt{s})\,\hat{y}(s) \quad \text{and} \quad \hat{\theta}(x,s) = \cosh(x\sqrt{s})\,\hat{y}(s).$$

Since $\cosh\sqrt{s} = \sum_{i=0}^{+\infty} s^i/(2i)!$, we eventually get

$$\theta(x,t) = \sum_{i=1}^{+\infty} x^{2i} \frac{y^{(i)}(t)}{(2i)!} \tag{4}$$

$$u(t) = \sum_{i=1}^{+\infty} \frac{y^{(i)}(t)}{(2i)!}. \tag{5}$$

In other words, whenever $t \mapsto y(t)$ is an arbitrary function (i.e., a trajectory of the trivial system $y = v$), $t \mapsto (\theta(x,t), u(t))$ defined by (4)-(5) is a (formal) trajectory of (1)-(3), and vice versa. This is exactly the idea underlying the definition of flatness. Notice these calculations have been known for a long time, see [77, pp. 588 and 594].

To make the statement precise, we now turn to convergence issues. On the one hand, $t \mapsto y(t)$ must be a smooth function such that

$$\exists\, K, M > 0, \quad \forall i \geq 0, \forall t \in [t_0, t_1], \quad |y^{(i)}(t)| \leq M(Ki)^{2i}$$

to ensure the convergence of the series (4)-(5).

On the other hand $t \mapsto y(t)$ cannot in general be analytic. Indeed, if the system is to be steered from an initial temperature profile $\theta(x,t_0) = \alpha_0(x)$ at time t_0 to a final profile $\theta(x,t_1) = \alpha_1(x)$ at time t_1, equation (1) implies

$$\forall t \in [0,1], \forall i \geq 0, \quad y^{(i)}(t) = \partial_t^i \theta(0,t) = \partial_x^{2i} \theta(0,t),$$

and in particular

$$\forall i \geq 0, \quad y^{(i)}(t_0) = \partial_x^{2i}\alpha_0(0) \quad \text{and} \quad y^{(i)}(t_1) = \partial_x^{2i}\alpha_1(1).$$

If for instance $\alpha_0(x) = c$ for all $x \in [0,1]$ (i.e., uniform temperature profile), then $y(t_0) = c$ and $y^{(i)}(t_0) = 0$ for all $i \geq 1$, which implies $y(t) = c$ for all t when the function is analytic. It is thus impossible to reach any final profile but $\alpha_1(x) = c$ for all $x \in [0,1]$.

Smooth functions $t \in [t_0, t_1] \mapsto y(t)$ that satisfy

$$\exists\, K, M > 0, \quad \forall i \geq 0, \qquad |y^{(i)}(t)| \leq M(Ki)^{\sigma i}$$

are known as Gevrey-Roumieu functions of order σ [62] (they are also closely related to class S functions [20]). The Taylor expansion of such functions is convergent for $\sigma \leq 1$ and divergent for $\sigma > 1$ (the larger σ is, the "more divergent" the Taylor expansion is). Analytic functions are thus Gevrey-Roumieu of order ≤ 1.

In other words we need a Gevrey-Roumieu function on $[t_0, t_1]$ of order > 1 but ≤ 2, with initial and final Taylor expansions imposed by the initial and final temperature profiles. With such a function, we can then compute open-loop control steering the system from one profile to the other by the formula (4).

For instance, we steered the system from uniform temperature 0 at $t = 0$ to uniform temperature 1 at $t = 1$ by using the function

$$\mathbb{R} \ni t \mapsto y(t) := \begin{cases} 0 & \text{if } t < 0 \\ 1 & \text{if } t > 1 \\ \dfrac{\int_0^t \exp\big(-1/(\tau(1-\tau))^\gamma\big)d\tau}{\int_0^1 \exp\big(-1/(\tau(1-\tau))^\gamma\big)d\tau} & \text{if } t \in [0,1], \end{cases}$$

with $\gamma = 1$ (this function is Gevrey-Roumieu functions of order $1 + 1/\gamma$). The evolution of the temperature profile $\theta(x,t)$ is displayed on figure 2 (the Matlab simulation is available upon request at rouchon@cas.ensmp.fr).

Similar but more involved calculations with convergent series corresponding to Mikunsiński operators are used in [16,17] to control a chemical reactor and a flexible rod modeled by an Euler-Bernoulli equation. For nonlinear systems, convergence issues are more involved and are currently under investigation. Yet, it is possible to work –at least formally– along the same line.

Example 6 (Flexion beam system) *Consider with [29] the mixed system*

$$\rho\partial_t^2 u(x,t) = \rho\omega^2(t)u(x,t) - EI\partial_x^4 u(x,t), \quad x \in [0,1]$$
$$\dot\omega(t) = \frac{\Gamma_3(t) - 2\omega(t)<u, \partial_t u>(t)}{I_d + <u, u>(t)}$$

with boundary conditions

$$u(0,t) = \partial_x u(0,t) = 0, \qquad \partial_x^2 u(1,t) = \Gamma_1(t), \qquad \partial_x^3 u(1,t) = \Gamma_2(t),$$

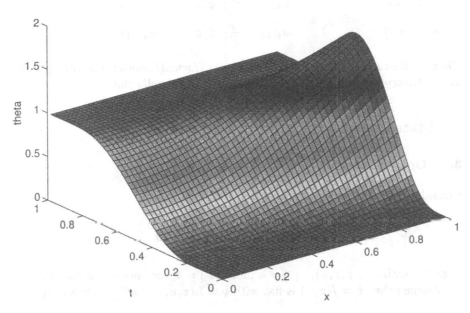

Fig. 2. evolution of the temperature profile for $t \in [0,1]$.

where ρ, EI, I_d are constant parameters, $u(x,t)$ is the deformation of the beam, $\omega(t)$ is the angular velocity of the body and $<f,g>(t) := \int_0^1 \rho f(x,t)g(x,t)dx$. The three control inputs are $\Gamma_1(t)$, $\Gamma_2(t)$, $\Gamma_3(t)$. We claim that

$$y(t) := \left(\partial_x^2 u(0,t),\ \partial_x^3 u(0,t), \omega(t)\right)$$

is a "flat" output. Indeed, $\omega(t)$, $\Gamma_1(t)$, $\Gamma_2(t)$ and $\Gamma_3(t)$ can clearly be expressed in terms of $y(t)$ and $u(x,t)$, which transforms the system into the equivalent Cauchy-Kovalevskaya form

$$EI\partial_x^4 u(x,t) = \rho y_3^2(t)u(x,t) - \rho\partial_t^2 u(x,t) \qquad and \qquad \begin{cases} u(0,t) = 0 \\ \partial_x u(0,t) = 0 \\ \partial_x^2 u(0,t) = y_1(t) \\ \partial_x^3 u(0,t) = y_2(t). \end{cases}$$

Set then formally $u(x,t) = \sum_{i=0}^{+\infty} a_i(t)\frac{x^i}{i!}$, plug this series into the above system and identify term by term. This yields

$$a_0 = 0, \qquad a_1 = 0, \qquad a_2 = y_1, \qquad a_3 = y_2,$$

and the iterative relation $\forall i \geq 0$, $EIa_{i+4} = \rho y_3^2 a_i - \rho \ddot{a}_i$. *Hence for all* $i \geq 1$,

$$a_{4i} = 0 \qquad\qquad a_{4i+2} = \frac{\rho}{EI}(y_3^2 a_{4i-2} - \ddot{a}_{4i-2})$$

$$a_{4i+1} = 0 \qquad\qquad a_{4i+3} = \frac{\rho}{EI}(y_3^2 a_{4i-1} - \ddot{a}_{4i-1}).$$

There is thus a 1–1 correspondence between (formal) solutions of the system and arbitrary mappings $t \mapsto y(t)$: *the system is formally flat.*

3   State constraints and optimal control

3.1   Optimal control

Consider the standard optimal control problem

$$\min_u J(u) = \int_0^T L(x(t), u(t)) dt$$

together with $\dot{x} = f(x, u)$, $x(0) = a$ and $x(T) = b$, for known a, b and T.
Assume that $\dot{x} = f(x, u)$ is flat with $y = h(x, u, \dots, u^{(r)})$ as flat output,

$$x = \varphi(y, \dots, y^{(q)}), \quad u = \alpha(y, \dots, y^{(q)}).$$

A numerical resolution of $\min_u J(u)$ a priori requires a discretization of the state space, i.e., a finite dimensional approximation. A better way is to discretize the flat output space. Set $y_i(t) = \sum_1^N A_{ij}\lambda_j(t)$. The initial and final conditions on x provide then initial and final constraints on y and its derivatives up to order q. These constraints define an affine sub-space V of the vector space spanned by the the A_{ij}'s. We are thus left with the nonlinear programming problem

$$\min_{A \in V} J(A) = \int_0^T L(\varphi(y, \dots, y^{(q)}), \alpha(y, \dots, y^{(q)})) dt,$$

where the y_i's must be replaced by $\sum_1^N A_{ij}\lambda_j(t)$.

This methodology is used in [53] for trajectory generation and optimal control. It should also be very useful for predictive control. The main expected benefit is a dramatic improvement in computing time and numerical stability. Indeed the exact quadrature of the dynamics –corresponding here to exact discretization via well chosen input signals through the mapping α– avoids the usual numerical sensitivity troubles during integration of $\dot{x} = f(x, u)$ and the problem of satisfying $x(T) = b$. A systematic method exploiting flatness for predictive control is proposed in [13]. Se also [55] for an industrial application of such methodology on a chemical reactor.

3.2   State constraints and predictive control

In the previous section, we did not consider state constraints. We now turn to the problem of planning a trajectory steering the state from a to b while satisfying the constraint $k(x, u, \ldots, u^{(p)}) \leq 0$. In the flat output "coordinates" this yields the following problem: find $T > 0$ and a smooth function $[0, T] \ni t \mapsto y(t)$ such that $(y, \ldots, y^{(q)})$ has prescribed value at $t = 0$ and T and such that $\forall t \in [0, T]$, $K(y, \ldots, y^{(\nu)})(t) \leq 0$ for some ν. When $q = \nu = 0$ this problem, known as the *piano mover problem*, is already very difficult.

Assume for simplicity sake that the initial and final states are equilibrium points. Assume also there is a quasistatic motion strictly satisfying the constraints: there exists a *path* (not a trajectory) $[0, 1] \ni \sigma \mapsto Y(\sigma)$ such that $Y(0)$ and $Y(1)$ correspond to the initial and final point and for any $\sigma \in [0, 1]$, $K(Y(\sigma), 0, \ldots, 0) < 0$. Then, there exists $T > 0$ and $[0, T] \ni t \mapsto y(t)$ solution of the original problem. It suffices to take $Y(\eta(t/T))$ where T is large enough, and where η is a smooth increasing function $[0, 1] \ni s \mapsto \eta(s) \in [0, 1]$, with $\eta(0) = 0$, $\eta(1) - 1$ and $\frac{d^i \eta}{ds^i}(0, 1) = 0$ for $i = 1, \ldots, \max(q, \nu)$.

In [65] this method is applied to a two-input chemical reactor. In [61] the minimum-time problem under state constraints is investigated for several mechanical systems. [70] considers, in the context of non holonomic systems, the path planning problem with obstacles. Due to the nonholonomic constraints, the above quasistatic method fails: one cannot set the y-derivative to zero since they do not correspond to time derivatives but to arc-length derivatives. However, several numerical experiments clearly show that sorting the constraints with respect to the order of y-derivatives plays a crucial role in the computing performance.

4   Symmetries

4.1   Symmetry preserving flat output

Consider the dynamics $\dot{x} = f(x, u)$, $(x, u) \in X \times U \subset \mathbb{R}^n \times \mathbb{R}^m$. It generates a system $(F, \mathfrak{M})$, where $\mathfrak{M} := X \times U \times \mathbb{R}_m^\infty$ and $F(x, u, u^1, \ldots) := (f(x, u), u^1, u^2, \ldots)$. At the heart of our notion of equivalence are endogenous transformations, which map solutions of a system to solutions of another system. We single out here the important class of transformations mapping solutions of a system onto solutions of the *same* system:

Definition 1 *An endogenous transformation* $\Phi_g : \mathfrak{M} \longmapsto \mathfrak{M}$ *is a* symmetry *of the system* $(F, \mathfrak{M})$ *if*

$$\forall \xi := (x, u, u^1, \ldots) \in \mathfrak{M}, \quad F(\Phi_g(\xi)) = D\Phi_g(\xi) \cdot F(\xi).$$

More generally, we can consider a *symmetry group*, i.e., a collection $\left(\Phi_g\right)_{g \in G}$ of symmetries such that $\forall g_1, g_2 \in G, \Phi_{g_1} \circ \Phi_{g_2} = \Phi_{g_1 * g_2}$, where $(G, *)$ is a group.

Assume now the system is flat. The choice of a flat output is by no means unique, since any endogenous transformation on a flat output gives rise to another flat output.

Example 7 (The kinematic car) *The system generated by*

$$\dot{x} = u_1 \cos\theta, \quad \dot{y} = u_1 \sin\theta, \quad \dot{\theta} = u_2,$$

admits the 3-parameter symmetry group of planar (orientation-preserving) isometries: for all translation $(a, b)'$ and rotation α, the endogenous mapping generated by

$$X = x \cos\alpha - y \sin\alpha + a$$
$$Y = x \sin\alpha + y \cos\alpha + b$$
$$\Theta = \theta + \alpha$$
$$U^1 = u^1$$
$$U^2 = u^2$$

is a symmetry, since the state equations remain unchanged,

$$\dot{X} = U_1 \cos\Theta, \quad \dot{Y} = U_1 \sin\Theta, \quad \dot{\Theta} = U_2.$$

This system is flat $z := (x, y)$ as a flat output. Of course, there are infinitely many other flat outputs, for instance $\tilde{z} := (x, y + \dot{x})$. Yet, z is obviously a more "natural" choice than $\tilde{z}$, because it "respects" the symmetries of the system. Indeed, each symmetry of the system induces a transformation on the flat output z

$$\begin{pmatrix} z_1 \\ z_2 \end{pmatrix} \longmapsto \begin{pmatrix} Z_1 \\ Z_2 \end{pmatrix} = \begin{pmatrix} X \\ Y \end{pmatrix} = \begin{pmatrix} z_1 \cos\alpha - z_2 \sin\alpha + a \\ z_1 \sin\alpha + z_2 \cos\alpha + b \end{pmatrix}$$

which does not involve derivatives *of z, i.e., a point* transformation. *This point transformation generates an endogenous transformation $(z, \dot{z}, \dots) \mapsto (Z, \dot{Z}, \dots)$. Following [19], we say such an endogenous transformation which is the total prolongation of a point transformation is* holonomic.
On the contrary, the induced transformation on $\tilde{z}$

$$\begin{pmatrix} \tilde{z}_1 \\ \tilde{z}_2 \end{pmatrix} \longmapsto \begin{pmatrix} \tilde{Z}_1 \\ \tilde{Z}_2 \end{pmatrix} = \begin{pmatrix} X \\ Y + \dot{X} \end{pmatrix} = \begin{pmatrix} \tilde{z}_1 \cos\alpha + (\dot{\tilde{z}}_1 - \tilde{z}_2)\sin\alpha + a \\ \tilde{z}_1 \sin\alpha + \tilde{z}_2 \cos\alpha + (\ddot{\tilde{z}}_1 - \dot{\tilde{z}}_2)\sin\alpha + b \end{pmatrix}$$

is not *a point transformation (it involves derivatives of $\tilde{z}$) and does not give to a holonomic transformation.*

Consider the system $(F, \mathfrak{M})$ admitting a symmetry Φ_g (or a symmetry group $(\Phi_g)_{g \in G}$). Assume moreover the system is flat with h as a flat output and denotes by $\Psi := (h, \dot{h}, \ddot{h}, \dots)$ the endogenous transformation generated by h. We then have:

Definition 2 (Symmetry-preserving flat output) *The flat output h preserves the symmetry Φ_g if the composite transformation $\Psi \circ \Phi_g \circ \Psi^{-1}$ is holonomic.*

This leads naturally to a fundamental question: assume a flat system admits the symmetry group $\left(\Phi_g\right)_{g \in G}$. Is there a flat output which preserves $\left(\Phi_g\right)_{g \in G}$?

This question can in turn be seen as a special case of the following problem: view a dynamics $\dot{x} - f(x, u) = 0$ as an *underdetermined differential system* and assume it admits a symmetry group; can it then be reduced to a "smaller" differential system? Whereas this problem has been studied for a long time and received a positive answer in the *determined* case, the underdetermined case seems to have been barely untouched [54]. Some connected question relative to invariant tracking are sketched in [69].

4.2   Flat outputs as potentials and gauge degree of freedom

Symmetries and the quest for potentials are at the heart of physics. To end the paper, we would like to show that flatness fits into this broader scheme.

Maxwell's equations in an empty medium imply that the magnetic field H is divergent free, $\nabla \cdot H = 0$. In Euclidian coordinates (x_1, x_2, x_3), it gives the underdetermined partial differential equation

$$\frac{\partial H_1}{\partial x_1} + \frac{\partial H_2}{\partial x_2} + \frac{\partial H_3}{\partial x_3} = 0$$

A key observation is that the solutions to this equation derive from a vector potential $H = \nabla \times A$: the constraint $\nabla \cdot H = 0$ is automatically satisfied whatever the potential A. This potential parameterizes all the solutions of the underdetermined system $\nabla \cdot H = 0$, see [60] for a general theory. A is a priori not uniquely defined, but up to an arbitrary gradient field, the gauge degree of freedom. The symmetries of the problem indicate how to use this degree of freedom to fix a "natural" potential.

The picture is similar for flat systems. A flat output is a "potential" for the underdetermined differential equation $\dot{x} - f(x, u) = 0$. Endogenous transformations on the flat output correspond to gauge degrees of freedom. The "natural" flat output is determined by symmetries of the system. Hence controllers designed from this flat output can also preserve the physics.

A slightly less esoteric way to convince the reader that flatness is an interesting notion is to take a look at the following small catalog of flat systems.

5   A catalog of flat systems

We give here a (partial) list of flat systems encountered in applications.

5.1 Holonomic mechanical systems

Example 8 (Fully actuated holonomic systems) *The dynamics of a holonomic system with as many independent inputs as configuration variables is*

$$\frac{\mathrm{d}}{\mathrm{d}t}\left(\frac{\partial L}{\partial \dot{q}}\right) - \frac{\partial L}{\partial q} = M(q)u + D(q,\dot{q}),$$

with $M(q)$ invertible. It admits q as a flat output –even when $\frac{\partial^2 L}{\partial \dot{q}^2}$ is singular–: indeed, u can be expressed in function of $q, \dot{q}$ by the computed torque formula

$$u = M(q)^{-1}\left(\frac{\mathrm{d}}{\mathrm{d}t}\left(\frac{\partial L}{\partial \dot{q}}\right) - \frac{\partial L}{\partial q} - D(q,\dot{q})\right).$$

If q is constrained by $c(q) = 0$ the system remains flat, and the flat output corresponds to the configuration point in $c(q) = 0$.

Example 9 (Planar rigid body with forces) *Consider a planar rigid body moving in a vertical plane under the influence of gravity and controlled by two forces having lines of action that are fixed with respect to the body and intersect at a single point (see figure 3) (see [78]).*

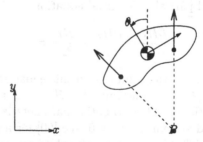

Fig. 3. A rigid body controlled by two body fixed forces.

Let (x, y) represent the horizontal and vertical coordinates of center of mass G of the body with respect to a stationary frame, and let θ be the counterclockwise orientation of a body fixed line through the center of mass. Take m as the mass of the body and J as the moment of inertia. Let $g \approx 9.8 \ m/sec^2$ represent the acceleration due to gravity.

Without loss of generality, we will assume that the lines of action for F_1 and F_2 intersect the y axis of the rigid body and that F_1 and F_2 are perpendicular. The equations of motion for the system can be written as

$$m\ddot{x} = F_1 \cos\theta - F_2 \sin\theta$$
$$m\ddot{y} = F_1 \sin\theta + F_2 \cos\theta - mg$$
$$J\ddot{\theta} = rF_1.$$

The flat output of this system corresponds to Huyghens center of oscillation [12]

$$\left(x - \frac{J}{mr}\sin\theta, \quad y + \frac{J}{mr}\cos\theta\right).$$

This example has some practical importance. The PVTOL system, the gantry crane and the robot $2k\pi$ (see below) are of this form, as is the simplified planar ducted fan [52]. Variations of this example can be formed by changing the number and type of the inputs [48].

Example 10 (PVTOL aircraft) A simplified Vertical Take Off and Landing aircraft moving in a vertical Plane [22] can be described by

$$\ddot{x} = -u_1 \sin\theta + \varepsilon u_2 \cos\theta$$
$$\ddot{z} = u_1 \cos\theta + \varepsilon u_2 \sin\theta - 1$$
$$\ddot{\theta} = u_2.$$

A flat output is $y = (x - \varepsilon \sin\theta, z + \varepsilon \cos\theta)$, see [38] more more details and a discussion in relation with unstable zero dynamics.

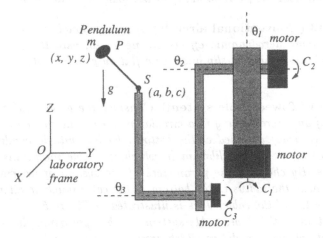

Fig. 4. The robot $2k\pi$ carrying its pendulum.

Example 11 (The robot $2k\pi$ of Ecole des Mines) In [31] a robot arm carrying a pendulum is considered, see figure 4. The control objective is to flip the pendulum from its natural downward rest position to the upward position and maintains it there. The first three degrees of freedom (the angles $\theta_1, \theta_2, \theta_3$) are actuated by electric motors, while the two degrees of freedom of the pendulum are not actuated.

The position $P = (x, y, z)$ of the pendulum oscillation center is a flat output. Indeed, it is related to the position $S = (a, b, c)$ of the suspension *point by*

$$(x - a)(\ddot{z} + g) = \ddot{x}(z - c)$$
$$(y - b)(\ddot{z} + g) = \ddot{y}(z - c)$$
$$(x - a)^2 + (y - b)^2 + (z - c)^2 = l^2,$$

where l is the distance between S and P. On the other hand the geometry of the robot defines a relation $(a, b, c) = \mathcal{T}(\theta_1, \theta_2, \theta_3)$ between the position of S and the robot configuration. This relation is locally invertible for almost all configurations but is not globally invertible.

Example 12 (Gantry crane [12,35,33]) *A direct application of Newton's laws provides the implicit equations of motion*

$$m\ddot{x} = -T \sin\theta \qquad\qquad x = R\sin\theta + D$$
$$m\ddot{z} = -T \cos\theta + mg \qquad\qquad z = R\cos\theta,$$

where x, z, θ are the configuration variables and T is the tension in the cable. The control inputs are the trolley position D and the cable length R. This system is flat, with the position (x, z) of the load as a flat output.

Example 13 (Conventional aircraft) *A conventional aircraft is flat, provided some small aerodynamic effects are neglected, with the coordinates of the center of mass and side-slip angle as a flat output. See [36] for a detailed study.*

Example 14 (Towed cable system) *Consider the dynamics of a system consisting of an aircraft flying in a circular pattern while towing a cable with a tow body (drogue) attached at the bottom. Under suitable conditions, the cable reaches a relative equilibrium in which the cable maintains its shape as it rotates. By choosing the parameters of the system appropriately, it is possible to make the radius at the bottom of the cable much smaller than the radius at the top of the cable. This is illustrated in Figure 5.*

The motion of the towed cable system can be approximately represented using a finite element model in which segments of the cable are replaced by rigid links connected by spherical joints. The forces acting on the segment (tension, aerodynamic drag and gravity) are lumped and applied at the end of each rigid link. In addition to the forces on the cable, we must also consider the forces on the drogue and the towplane. The drogue is modeled as a sphere and essentially acts as a mass attached to the last link of the cable, so that the forces acting on it are included in the cable dynamics. The external forces on the drogue again consist of gravity and aerodynamic drag. The towplane is attached to the top of the cable and is subject to drag, gravity, and the force of the attached cable. For simplicity, we simply model the towplane as a pure force applied at the top of the cable. Our goal is to generate trajectories for

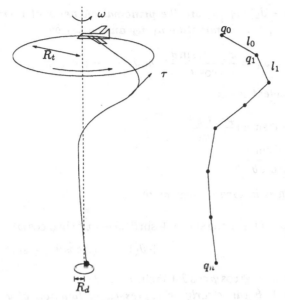

Fig. 5. Towed cable system and finite link approximate model.

this system that allow operation away from relative equilibria as well as transition between one equilibrium point and another. Due to the high dimension of the model for the system (128 states is typical), traditional approaches to solving this problem, such as optimal control theory, cannot be easily applied. However, it can be shown that this system is differentially flat using the position of the bottom of the cable as the differentially flat output. Thus all feasible trajectories for the system are characterized by the trajectory of the bottom of the cable. See [47] for a more complete description and additional references.

We end this section with a system which is not known to be flat for generic parameter value but still enjoys the weaker property of being *orbitally flat* [11].

Example 15 (Satellite with two controls) *Consider with [4] a satellite with two control inputs u_1, u_2 described by*

$$\dot{\omega}_1 = u_1$$
$$\dot{\omega}_2 = u_2$$
$$\dot{\omega}_3 = a\omega_1\omega_2$$
$$\dot{\varphi} = \omega_1 \cos\theta + \omega_3 \sin\theta \qquad\qquad (6)$$
$$\dot{\theta} = (\omega_1 \sin\theta - \omega_3 \cos\theta)\tan\varphi + \omega_2$$
$$\dot{\psi} = \frac{(\omega_3 \cos\theta - \omega_1 \sin\theta)}{\cos\varphi},$$

where $a = (J_1 - J_2)/J_3$ (J_i are the principal moments of inertia); physical sense imposes $|a| \leq 1$. Eliminating u_1, u_2 and ω_1, ω_2 by

$$\omega_1 = \frac{\dot{\varphi} - \omega_3 \sin\theta}{\cos\theta} \quad and \quad \omega_2 = \dot{\theta} + \dot{\psi}\sin\varphi$$

yields the equivalent system

$$\dot{\omega}_3 = a(\dot{\theta} + \dot{\psi}\sin\varphi)\frac{\dot{\varphi} - \omega_3\sin\theta}{\cos\theta} \tag{7}$$

$$\dot{\psi} = \frac{\omega_3 - \dot{\varphi}\sin\theta}{\cos\varphi\cos\theta}. \tag{8}$$

But this system is in turn equivalent to

$$\cos\theta\left(\ddot{\psi}\cos\varphi - (1+a)\dot{\psi}\dot{\varphi}\sin\varphi\right) + \sin\theta\left(\ddot{\varphi} + a\dot{\psi}^2\sin\varphi\cos\varphi\right)$$
$$+ \dot{\theta}(1-a)(\dot{\varphi}\cos\theta - \dot{\psi}\sin\theta\cos\varphi) = 0$$

by substituting $\omega_3 = \dot{\psi}\cos\varphi\cos\theta + \dot{\varphi}\sin\theta$ in (7).

When $a = 1$, θ can clearly be expressed in function of φ, ψ and their derivatives. We have proved that (6) is flat with (φ, ψ) as a flat output. A similar calculation can be performed when $a = -1$.

When $|a| < 1$, whether (6) is flat is unknown. Yet, it is orbitally flat [64]. To see that, rescale time by $\dot{\sigma} = \omega_3$; by the chain rule $\dot{x} = \dot{\sigma}x'$ whatever the variable x, where ' denotes the derivation with respect to σ. Setting then

$$\bar{\omega}_1 := \omega_1/\omega_3, \quad \bar{\omega}_2 := \omega_2/\omega_3, \quad \bar{\omega}_3 := -1/a\omega_3,$$

and eliminating the controls transforms (6) into

$$\omega_3' = \bar{\omega}_1\bar{\omega}_2$$
$$\varphi' = \bar{\omega}_1\cos\theta + \sin\theta$$
$$\theta' = (\bar{\omega}_1\sin\theta - \cos\theta)\tan\varphi + \bar{\omega}_2$$
$$\psi' = \frac{(\cos\theta - \bar{\omega}_1\sin\theta)}{\cos\varphi}.$$

The equations are now independent of a. This implies the satellite with $a \neq 1$ is orbitally equivalent to the satellite with $a = 1$. Since it is flat when $a = 1$ it is orbitally flat when $a \neq 1$, with (φ, ψ) as an orbitally flat output.

5.2   Nonholonomic mechanical systems

Example 16 (Kinematics generated by *two* nonholonomic constraints)
Such systems are flat by theorem 2 since they correspond to driftless systems with n states and $n-2$ inputs. For instance the rolling disc (p. 4), the rolling sphere (p. 96) and the bicycle (p. 330) considered in the classical treatise on nonholonomic mechanics [51] are flat.

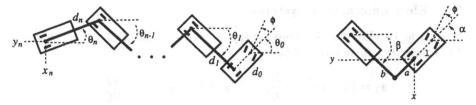

Fig. 6. n-trailer system (left) and 1-trailer system with kingpin hitch (right).

Example 17 (Mobile robots) *Many mobile robots modeled by rolling without sliding constraints, such as those considered in [5,50,76] are flat. In particular, the n-trailer system (figure 6) has for flat output the mid-point P_n of the last trailer axle [68,12]. The 1-trailer system with kingpin hitch is also flat, with a rather complicated flat output involving elliptic integrals [67,10], but by theorem 1 the system is not flat when there is more than one trailer.*

Example 18 (The rolling penny) *The dynamics of this Lagrangian system submitted to a nonholonomic constraint is described by*

$$\ddot{x} = \lambda \sin \varphi + u_1 \cos \varphi$$
$$\ddot{y} = -\lambda \cos \varphi + u_1 \sin \varphi$$
$$\ddot{\varphi} = u_2$$
$$\dot{x} \sin \varphi = \dot{y} \cos \varphi$$

where x, y, φ are the configuration variables, λ is the Lagrange multiplier of the constraint and u_1, u_2 are the control inputs. A flat output is (x, y): indeed, parameterizing time by the arclength s of the curve $t \mapsto (x(t), y(t))$ we find

$$\cos \varphi = \frac{dx}{ds}, \quad \sin \varphi = \frac{dy}{ds}, \quad u_1 = \ddot{s}, \quad u_2 = \kappa(s) \, \ddot{s} + \frac{d\kappa}{ds} \, \dot{s}^2,$$

where κ is the curvature. These formulas remain valid even if $u_1 = u_2 = 0$.

This example can be generalized to any mechanical system subject to m flat nonholonomic constraints, provided there are $n - m$ control forces independent of the constraint forces (n the number of configuration variables), i.e., a "fully-actuated" nonholonomic system as in [5].

All these flat nonholonomic systems have a controllability singularity at rest. Yet, it is possible to "blow up" the singularity by reparameterizing time with the arclength of the curve described by the flat output, hence to plan and track trajectories starting from and stopping at rest as explained in [12,68,10].

5.3   Electromechanical systems

Example 19 (DC-to-DC converter) *A Pulse Width Modulation DC-to-DC converter can be modeled by*

$$\dot{x}_1 = (u-1)\frac{x_2}{L} + \frac{E}{L}, \qquad \dot{x}_2 = (1-u)\frac{x_1}{LC} - \frac{x_2}{RC},$$

where the duty ratio $u \in [0,1]$ is the control input. The electrical stored energy $y := \dfrac{x_1^2}{2C} + \dfrac{x_2^2}{2L}$ is a flat output [71,27].

Example 20 (Magnetic bearings) *A simple flatness-based solution to motion planning and tracking is proposed in [34]. The control law ensures that only one electromagnet in each actuator works at a time and permits to reduce the number of electromagnets by a better placement of actuators.*

Example 21 (Induction motor) *The standard two-phase model of the induction motor reads in complex notation (see [32] for a complete derivation)*

$$R_s i_s + \dot{\psi}_s = u_s \qquad\qquad \psi_s = L_s i_s + M e^{jn\theta} i_r$$
$$R_r i_r + \dot{\psi}_r = 0 \qquad\qquad \psi_r = M e^{-jn\theta} i_s + L_r i_r,$$

where ψ_s and i_s (resp. ψ_r and i_r) are the complex stator (resp. rotor) flux and current, θ is the rotor position and $j = \sqrt{-1}$. The control input is the voltage u_s applied to the stator. Setting $\psi_r = \rho e^{j\alpha}$, the rotor motion is described by

$$J\frac{d^2\theta}{dt^2} = \frac{n}{R_r}\rho^2\dot{\alpha} - \tau_L(\theta,\dot{\theta}),$$

where τ_L is the load torque.

This system is flat with the two angles (θ,α) as a flat output [42] (see [7] also for a related result).

5.4   Chemical systems

Example 22 (CSTRs) *Many simple models of Continuous Stirred Tank Reactors (CSTRs) admit flats outputs with a direct physical interpretation in terms of temperatures or product concentrations [24,1], as do closely related biochemical processes [2,9]. In [65] flatness is used to steer a reactor model from a steady state to another one while respecting some physical constraints. In[44], flatness based control of nonlinear delay chemical reactors is proposed.*

A basic model of a CSTR with two chemical species and any number of exothermic or endothermic reactions is

$$\dot{x}_1 = f_1(x_1, x_2) + g_1(x_1, x_2)u$$
$$\dot{x}_2 = f_2(x_1, x_2) + g_2(x_1, x_2)u,$$

where x_1 is a concentration, x_2 a temperature and u the control input (feed-flow or heat exchange). It is obviously linearizable by static feedback, hence flat.

When more chemical species are involved, a single-input CSTR is in general not flat, see [28]. Yet, the addition of another manipulated variable often renders it flat, see [1] for an example on a free-radical polymerization CSTR. For instance basic model of a CSTR with three chemical species, any number of exothermic or and two control inputs is

$$\dot{x}_1 = f_1(x) + g_1^1(x)u_1 + g_1^2(x)u_2$$
$$\dot{x}_2 = f_2(x) + g_2^1(x)u_1 + g_2^2(x)u_2$$
$$\dot{x}_3 = f_3(x) + g_3^1(x)u_1 + g_3^2(x)u_2,$$

where x_1, x_2 are concentrations and x_3 is a temperature temperature and u_1, u_2 are the control inputs (feed-flow, heat exchange, feed-composition, ...). Such a system is always flat, see section 1.2.

Example 23 (Polymerization reactor) Consider with [74] the reactor

$$\dot{C}_m = \frac{C_{m_{m_s}}}{\tau} - \left(1 + \bar{\varepsilon}\frac{\mu_1}{\mu_1 + M_m C_m}\right)\frac{C_m}{\tau} + R_m(C_m, C_i, C_s, T)$$

$$\dot{C}_i = -k_i(T)C_i + u_2\frac{C_{i_{i_s}}}{V} - \left(1 + \bar{\varepsilon}\frac{\mu_1}{\mu_1 + M_m C_m}\right)\frac{C_i}{\tau}$$

$$\dot{C}_s = u_2\frac{C_{s_{i_s}}}{V} + \frac{C_{s_{m_s}}}{\tau} - \left(1 + \bar{\varepsilon}\frac{\mu_1}{\mu_1 + M_m C_m}\right)\frac{C_s}{\tau}$$

$$\dot{\mu}_1 = -M_m R_m(C_m, C_i, C_s, T) - \left(1 + \bar{\varepsilon}\frac{\mu_1}{\mu_1 + M_m C_m}\right)\frac{\mu_1}{\tau}$$

$$\dot{T} = \phi(C_m, C_i, C_s, \mu_1, T) + \alpha_1 T_j$$

$$\dot{T}_j = f_6(T, T_j) + \alpha_4 u_1,$$

where u_1, u_2 are the control inputs and $C_{m_{m_s}}$, M_m, $\bar{\varepsilon}$, τ, $C_{i_{i_s}}$, $C_{s_{m_s}}$, $C_{s_{i_s}}$, V, α_1, α_4 are constant parameters. The functions R_m, k_i, ϕ and f_6 are not well-known and derive from experimental data and semi-empirical considerations, involving kinetic laws, heat transfer coefficients and reaction enthalpies.

The polymerization reactor is flat whatever the functions R_m, k_i, ϕ, f_6 and admits $(C_{s_{i_s}} C_i - C_{i_{i_s}} C_s, \ M_m C_m + \mu_1)$ as a flat output [66].

References

1. J. Alvarez, R. Suarez, and A. Sanchez. Nonlinear decoupling control of free-radical polymerization continuous stirred tank reactors. *Chem. Engng. Sci*, 45:3341–3357, 1990.
2. G. Bastin and Dochain. *On-Line Estimation and Adaptive Control of Bioreactors*. Elsevier Science Publishing Co, 1990.

3. R.L. Bryant, S.S. Chern, R.B. Gardner, H.L. Goldschmidt, and P.A. Griffiths. *Exterior Differential Systems*. Springer, 1991.
4. C.I. Byrnes and A. Isidori. On the attitude stabilization of rigid spacecraft. *Automatica*, 27:87–95, 1991.
5. G. Campion, B. d'Andrea Novel, and G. Bastin. Structural properties and classification of kinematic and dynamic models of wheeled mobile robots. *IEEE Trans. Robotics Automation*, 12(1):47–62, 1996.
6. B. Charlet, J. Lévine, and R. Marino. On dynamic feedback linearization. *Systems Control Letters*, 13:143–151, 1989.
7. J. Chiasson. Dynamic feedback linearization of the induction motor. *IEEE Trans. Automat. Control*, 38:1588–1594, 1993.
8. F. Dubois, N. Petit, and P. Rouchon. Motion planing and nonlinear simulations for a tank containing a fluid. In *European Control Conference, Karlsruhe*, 1999.
9. J. El Moubaraki, G. Bastin, and J. Lévine. Nonlinear control of biological processes with growth/production decoupling. *Mathematical Biosciences*, 116:21–44, 1993.
10. M. Fliess, J. levine, P. Martin, F. Ollivier, and P. Rouchon. Controlling nonlinear systems by flatness. In C.I. Byrnes, B.N. Datta, D.S. Gilliam, and C.F. Martin, editors, *Systems and control in the Twenty-First Century*, Progress in Systems and Control Theory. Birkhauser, 1997.
11. M. Fliess, J. Lévine, Ph. Martin, and P. Rouchon. Linéarisation par bouclage dynamique et transformations de Lie-Bäcklund. *C.R. Acad. Sci. Paris*, I-317:981–986, 1993.
12. M. Fliess, J. Lévine, Ph. Martin, and P. Rouchon. Flatness and defect of nonlinear systems: introductory theory and examples. *Int. J. Control*, 61(6):1327–1361, 1995.
13. M. Fliess and R. Marquez. Continuous time linear predictive control and flatness: a module theoretic setting with examples. *Int. Journal of Control*, 73:606–623, 2000.
14. M. Fliess, Ph. Martin, N. Petit, and P. Rouchon. Active signal restoration for the telegraph equation. In *CDC 99*, Phenix, december 1999.
15. M. Fliess, H. Mounier, P. Rouchon, and J. Rudolph. Controllability and motion planning for linear delay systems with an application to a flexible rod. In *Proc. of the 34th IEEE Conf. on Decision and Control*, pages 2046–2051, New Orleans, 1995.
16. M. Fliess, H. Mounier, P. Rouchon, and J. Rudolph. Systèmes linéaires sur les opérateurs de Mikusiński et commande d'une poutre flexible. In *ESAIM Proc. "Élasticité, viscolélasticité et contrôle optimal", 8ème entretiens du centre Jacques Cartier, Lyon*, pages 157–168, 1996.
17. M. Fliess, H. Mounier, P. Rouchon, and J. Rudolph. A distributed parameter approach to the control of a tubular reactor: A multi- variable case. In *Control and Decision Conference, Tampa*, pages 439– 442, 1998.
18. M. Fliess, H. Mounier, P. Rouchon, and J. Rudolph. Tracking control of a vibrating string with an interior mass viewed as delay system. *ESAIM: COCV(www.eamth.fr/cocv*, 3:315–321, 1998.
19. M. Gromov. *Partial Differential Relations*. Springer-Verlag, 1986.
20. I.M. Guelfand and G.E. Chilov. *Les Distributions, tome 3*. Dunod, Paris, 1964.
21. J. Hauser, S. Sastry, and P. Kokotović. Nonlinear control via approximated input-output linearization: the ball and beam example. *IEEE Trans. Automat. Contr.*, 37:392–398, 1992.

22. J. Hauser, S. Sastry, and G. Meyer. Nonlinear control design for slightly nonminimum phase systems: Application to V/STOL aircraft. *Automatica*, 28(4):665–679, 1992.

23. D. Hilbert. Über den Begriff der Klasse von Differentialgleichungen. *Math. Ann.*, 73:95–108, 1912. also in Gesammelte Abhandlungen, vol. III, pp. 81–93, Chelsea, New York, 1965.

24. K.A. Hoo and J.C. Kantor. An exothermic continuous stirred tank reactor is feedback equivalent to a linear system. *Chem. Eng. Commun.*, 37:1–10, 1985.

25. L.R. Hunt, R. Su, and G. Meyer. Global transformations of nonlinear systems. *IEEE Trans. Automat. Control*, 28:24–31, 1983.

26. B. Jakubczyk and W. Respondek. On linearization of control systems. *Bull. Acad. Pol. Sci. Ser. Sci. Math.*, 28:517–522, 1980.

27. L. Karsenti and P. Rouchon. A tracking controller-observer scheme for DC-to-DC converters. In *ECC'97*, 1997.

28. C. Kravaris and C.B. Chung. Nonlinear state feedback synthesis by global input/output linearization. *AIChE J.*, 33:592–603, 1987.

29. H. Laousy, C.Z. Xu, and G. Sallet. Boundary feedback stabilization of rotation body-beam system. *IEEE Autom. Control*, 41:1 5, 1996.

30. B. Laroche, Ph. Martin, and P. Rouchon. Motion planing for the heat equation. *Int. Journal of Robust and Nonlinear Control*, 10:629–643, 2000.

31. Y. Lenoir, , Ph. Martin, and P. Rouchon. $2k\pi$, the juggling robot. In *Control and Decision Conference, Tampa*, pages 1995–2000, 1998.

32. W. Leonhard. *Control of Electrical Drives*. Elsevier, 1985.

33. J. Lévine. Are there new industrial perspectives in the control of mechanical systems? In P. Frank, editor, *In Issues in Control, ECC99*. Springer, 1999.

34. J. Lévine, J. Lottin, and J.-C. Ponsart. A nonlinear approach to the control of magnetic bearings. *IEEE Trans. Control Systems Technology*, 4:524–544, 1996.

35. J. Lévine, P. Rouchon, G. Yuan, C. Grebogi, B.R. Hunt, E. Kostelich, E. Ott, and J.A. Yorke. On the control of US Navy cranes. In *ECC97*, 1997.

36. Ph. Martin. *Contribution à l'étude des systèmes différentiellement plats*. PhD thesis, École des Mines de Paris, 1992.

37. Ph. Martin. A geometric sufficient conditions for flatness of systems with m inputs and $m + 1$ states. In *Proc. of the 32nd IEEE Conf. on Decision and Control*, pages 3431–3436, San Antonio, 1993.

38. Ph. Martin, S. Devasia, and B Paden. A different look at output feedback: control of a VTOL aircraft. *Automatica*, 32(1):101–108, 1996.

39. Ph. Martin and P. Rouchon. Feedback linearization and driftless systems. *Math. Control Signal Syst.*, 7:235–254, 1994.

40. Ph. Martin and P. Rouchon. Any (controllable) driftless system with 3 inputs and 5 states is flat. *Systems Control Letters*, 25:167–173, 1995.

41. Ph. Martin and P. Rouchon. Any (controllable) driftless system with m inputs and m+2 states is flat. In *Proc. of the 34th IEEE Conf. on Decision and Control*, pages 2886–2891, New Orleans, 1995.

42. Ph. Martin and P. Rouchon. Flatness and sampling control of induction motors. In *Proc. IFAC World Congress*, pages 389–394, San Francisco, 1996.

43. H. Mounier. *Propriétés structurelles des systèmes linéaires à retards: aspects théoriques et pratiques*. PhD thesis, Université Paris Sud, Orsay, 1995.

44. H. Mounier and J. Rudolph. Flatness based control of nonlinear delay systems: A chemical reactor example. *Int. Journal of Control*, 71:871–890, 1998.

45. H. Mounier, J. Rudolph, M. Fliess, and P. Rouchon. Tracking control of a vibrating string with an interior mass viewed as delay system. *ESAIM: COCV(www.eamth.fr/cocv*, 3:315–321, 1998.

46. H. Mounier, J. Rudolph, M. Petitot, and M. Fliess. A flexible rod as a linear delay system. In *Proc. of the 3rd European Control Conf.*, pages 3676–3681, Rome, 1995.

47. R. M. Murray. Trajectory generation for a towed cable flight control system. In *Proc. IFAC World Congress*, pages 395–400, San Francisco, 1996.

48. R. M. Murray, M. Rathinam, and W. Sluis. Differential flatness of mechanical control systems: A catalog of prototype systems. In *asmeIMECE*, San Francisco, November 1995.

49. R.M. Murray. Nilpotent bases for a class on nonintegrable distributions with applications to trajectory generation for nonholonomic systems. *Math. Control Signal Syst.*, 7:58–75, 1994.

50. R.M. Murray and S.S. Sastry. Nonholonomic motion planning: Steering using sinusoids. *IEEE Trans. Automat. Control*, 38:700–716, 1993.

51. Iu I. Neimark and N.A. Fufaev. *Dynamics of Nonholonomic Systems*, volume 33 of *Translations of Mathematical Monographs*. American Mathematical Society, Providence, Rhode Island, 1972.

52. M. van Nieuwstadt and R. M. Murray. Approximate trajectory generation for differentially flat systems with zero dynamics. In *Proc. of the 34th IEEE Conf. on Decision and Control*, pages 4224–4230, New Orleans, 1995.

53. M. van Nieuwstadt and R.M. Murray. Real time trajectory generation for differentially flat systems. *Int. Journal of Robust and Nonlinear Control*, 8(11):995–1020, 1998.

54. P.J. Olver. *Applications of Lie groups to differential equations*, volume 107 of *Graduate Texts in Mathematics*. Springer-Verlag, 2nd edition, 1993.

55. N Petit, Y. Creff, L. Lemaire, and P. Rouchon. Minimum time constrained control of acid strength on a sulfuric acid alkylation unit. *Chemical Engineering Science*, 2000. to appear.

56. N. Petit, Y. Creff, and P. Rouchon. δ-freeness of a class of linear delayed systems. In *European Control Conference, Brussels*, 1997.

57. N. Petit, Y. Creff, and P. Rouchon. Motion planning for two classes of nonlinear systems with delays depending on the control. In *Control and Decision Conference, Tampa*, pages 1007–1011, 1998.

58. J.B. Pomet. On dynamic feedback linearization of four-dimensional affine control systems with two inputs. *ESAIM-COCV*, 1997. http://www.emath.fr/Maths/Cocv/Articles/articleEng.html.

59. J.F. Pommaret. *Systems of Partial Differential Equations and Lie Pseudogroups*. Gordon & Breach, N.Y., 1978.

60. J.F. Pommaret. Dualité différentielle et applications. *C.R. Acad. Sci. Paris, Série I*, 320:1225–1230, 1995.

61. C. Raczy. *Commandes optimales en temps pour les systèmes différentiellement plats*. PhD thesis, Université des Sciences et Technologies de Lille, 1997.

62. J.P. Ramis. Dévissage Gevrey. *Astérisque*, 59-60:173–204, 1979.

63. M. Rathinam and R. Murray. Configuration flatness of Lagrangian systems underactuated by one control. *SIAM J. Control Optimization*, 36(1):164–179, 1998.

64. A. Reghai. Satellite à deux commandes. Technical report, Ecole Polytechnique, Palaiseau, France, 1995. Mémoire de fin d'études.

65. R. Rothfuß, J. Rudolph, and M. Zeitz. Flatness based control of a nonlinear chemical reactor model. *Automatica*, 32:1433–1439, 1996.
66. P. Rouchon. Necessary condition and genericity of dynamic feedback linearization. *J. Math. Systems Estim. Control*, 5(3):345–358, 1995.
67. P. Rouchon, M. Fliess, J. Lévine, and Ph. Martin. Flatness and motion planning: the car with n-trailers. In *Proc. ECC'93, Groningen*, pages 1518–1522, 1993.
68. P. Rouchon, M. Fliess, J. Lévine, and Ph. Martin. Flatness, motion planning and trailer systems. In *Proc. of the 32nd IEEE Conf. on Decision and Control*, pages 2700–2705, San Antonio, 1993.
69. P. Rouchon and J. Rudolph. *Invariant tracking and stabilization: problem formulation and examples*, pages 261–273. Lecture Notes in Control and Information Sciences 246. Springer, 1999.
70. S. Sekhavat. *Planification de Mouvements sans Collision pour Systèmes non Holonomes*. PhD thesis, LAAS-CNRS, Toulouse, 1996.
71. H. Sira-Ramirez and M. Ilic-Spong. Exact linearzation in switched-mode DC-to-DC power converters. *Int. J. Control*, 50:511–524, 1989.
72. W M Sluis. *Absolute Equivalence and its Application to Control Theory*. PhD thesis, University of Waterloo, Ontario, 1992.
73. W.M. Sluis. A necessary condition for dynamic feedback linearization. *Systems Control Letters*, 21:277–283, 1993.
74. M. Soroush and C. Kravaris. Multivariable nonlinear control of a continous polymerization reactor. In *American Control Conference*, pages 607–614, 1992.
75. D. Tilbury, O. Sørdalen, L. Bushnell, and S. Sastry. A multisteering trailer system: conversion into chained form using dynamic feedback. *IEEE Trans. Robotics Automation*, 11(6):807, 1995.
76. D.M. Tilbury. *Exterior differential systems and nonholonomic motion planning*. PhD thesis, University of California, Berkeley, 1994. Memorandum No. UCB/ERL M94/90.
77. G. Valiron. *Equations Fonctionnelles*. Masson et Cie, Editeurs, Paris, 2nd edition, 1950.
78. M. J. van Nieuwstadt. Rapid hover to forward flight transitions for a thrust vectored aircraft. *J. Guidance Control Dynam.*, 21(1):93–100, 1998.

Part II

Nonlinear Quantitative Feedback Theory

Introduction

Quantitative Feedback Theory (QFT) has been developed since the sixties around the work of Isaac Horowitz. Today it is one of the recognized techniques for designing practical control systems in many technological areas. Its most important properties are:

1. *Design to Specifications.* The plant parameters and the disturbace uncertainties to be combatted by feedback design, and the performance specifications to be achieved despite these uncertainties, are basic ingredients of the control problem.
2. *Rigourous and Systematic Design.* Relatively simple step by step design, mainly in the frequency domain, easile do–able by ordinary, practical designers.
3. *Cost of Feedback.* Great emphasis on this issue, specially in terms of loop bandwith and sensor noise effects, and their minimization.
4. *Design transparency.* Early in the design process, and at each step, the principal trade-offs are highly visible. The designer can choose between them as he proceeds, such as bandwith vs. compensator complexity, or competing sensor in multiloop design.

This Part of the book, consisting of a unique chapter, will give an introduction to the fundamentals of QFT overall, and then concentrate on uncertain nonlinear control systems. The basic idea of one technique is to convert the uncertain nonlinear (time-varying) control problem in an *equivalent* uncertain linear time invariant control problem. The solution for the resulting LTI design problem is guaranteed to solve the original nonlinear problem. Thus, relatively simple LTI design is used in most of the design. In a second technique, specially suited for plant disturbance attenuation, the uncertain nonlinear (time-varying) plant is converted into a combination of a simple LTI plant and *equivalent disturbances*. Schauder fixed point theorem provides a mathematically rigorous foundation of the different design methods.

For simplicity, the chapter will be mostly devote to the single–input single–output case. The reader is directed to [14], [16], [29], and references therein for further study of the topic, including the multivariable case.

Fundamentals of Nonlinear Quantitative Feedback Theory

Isaac Horowitz[1] and Alfonso Baños[2]

[1] Dept. Electrical Engineering, University of California at Davis, USA
[2] Dept. Infomática y Sistemas, Universidad de Murcia, Murcia, Spain

1 The Concept of Control in Technology

Examples: (1) Construct a system for production of sulfuric acid of $X \pm$ 0.1% purity, at the rate of $Y \pm 1\%$ units per hour, with a single coordinated command input to the plant and a single plant output; denoted as SISO system.

(2) A system for production of sulfuric acid simultaneously with 5 by-products, with purities of $X_1 \pm z_1, X_2 \pm z_2, ..., X_6 \pm z_6$, at rates of $r_1 \pm f_1, r_2 \pm f2, ..., r_6 \pm f_6$ units per hour. Both the desired purities and the rate commands are subject to change. It is required that after any such command change is made, the entire system should be operating at the new commands within H hours. This is a multiple input, multiple plant output (MIMO) System.

1.1 The Plant and its Problems

Assume excellent Chemical Engineers design the above plants, tune them carefully to achieve the defined objectives. Next, quoting from a Trade Journal, "We step out for a coffee break. When we return the system is in disorder, not operating properly. The reason is: Uncertainty, Disturbances, Variations in raw material and catalyst purities, etc. We don't know with sufficient precision the chemical formulae, the reaction rates, the sensitivities to temperature, pressure, etc." These can all be lumped together under *plant uncertainty*.

But we can *quantify* the uncertainty: the plant input-output relations $y = P(u)$, known only as a member of a set $\mathbf{P} = \{P\}$, the disturbance d known only as a member of a set $\mathbf{D} = \{d\}$. The greater the uncertainty, the larger are these sets.

1.2 Feedback Control is a Solution

Assume there exist sensors with accuracy at least as great as the tolerances on the output. For the SISO system, two degrees of freedom (loop compensator G, prefilter F in Fig. 1, which is one of many possible canonical structures), must be available to the designer.

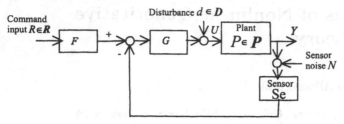

Fig. 1. A SISO Two Degree of Freedom (TDF) Structure, for a SISO Plant

Assume P is Linear Time Invariant (LTI), so using transforms, and for $n = 0$ in Fig. 1,

$$Y(s) = \frac{P(s)D(s) + P(s)G(s)F(s)R(s)}{1 + P(s)G(s)S_e(s)} \tag{1}$$

We show how feedback can solve the uncertainty problem. In Eq.(1), choose G large enough so that $1 + PGS_e \approx PGS_e$, where the argument s has been dropped by simplicity, giving

$$Y \approx \frac{PD + PGFR}{PGS_e} = \frac{D}{GS_e} + \frac{FR}{S_e} \tag{2}$$

Choose G large enough to make the disturbance term D/GS_e as small as desired. Then the only significant part of the output Y is FR/S_e. The prefilter F is available to obtain the desired Y. Henceforth, we assume $S_e = 1$, because G and F can be otherwise modified to compensate for S_e. Thus, the desired plant outputs can be obtained, despite large plant uncertainties and disturbances.

In Fig. 1, the system transfer function is $T = Y/R = FGP/(1 + GP)$. Define the sensitivity of T to the plant transfer function P as $S = \frac{dT/T}{dP/P}$, with the result

$$S = 1/(1+L) \tag{3}$$

where $L = GP$ is called the *loop transmission*, which is the fundamental feedback synthesis tool. By making it large enough, over a large enough frequency range, the closed-loop feedback system can be made as insensitive as desired to plant uncertainty (up to the sensor accuracy). But it is impossible to make the system more accurate at any frequency than the sensor accuracy at that frequency. In fact, infinite loop transmission is needed at any frequency to achieve sensor accuracy (prove this, [16], ch. 1). But there is a price to pay for these benefits of Feedback, to be studied in Sec. 5.

1.3 History of Feedback Control

Feedback Control was practiced in ancient cultures, for example in the Babylonian irrigation system. Maxwell considered the Stability problem, which is present in all practical feedback systems. But a scientific, engineering design theory for linear time invariant (LTI) Feedback Amplifiers, is due to H. W. Bode [11]. Much of it is applicable to LTI feedback control. The earliest systematic feedback control design theory seems to be that developed for the Radar Gun Control problem in World War 2. In the U.S. it was soon being taught as a graduate servomechanisms course. But it was quite qualitative, and ignored the sensitivity problem to plant uncertainty . The reason is that the radar system has inherently only *one degree of fredom* (ODF). Thus in Fig. 1, the loop compensator G acts directly on the error between the transmitter (r) and the target (y), so $F = 1$, and the system transfer function $T = L(1+L)$. The sensitivity function of Eq.(3) is still $S = 1/(1+L)$, giving $T = 1 - S$.

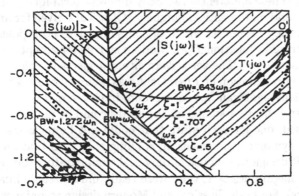

Fig. 2. In ODF system, $|S| > 1$ (worse than in open loop system, for a significant frequency range)

Thus, a desired system sensitivity (to plant uncertainty) cannot be achieved, unless it is fortuitously equal to $1 - T(s)$. The consequence is illustrated in Fig. 2, where a second order $T(s) = \omega_n^2/(s^2 + 2\xi\omega_n s + \omega_n^2)$ is plotted for $s = j\omega$ (i.e. frequency response), in the complex plane, for 3 values of damping factor ξ. The same plot displays the sensitivity $S(j\omega)$ to the plant, by using the point $O' = (1,0)$ as the origin, (because $S = 1 - T$ for ODF systems, so S is the vector from T to the point $0' = (1,0)$. $|S(j\omega)|$ increases from zero at $\omega = 0$, until it is 1.0 at ω_x (a type I system is assumed here, for which $L(0) = \infty$). It is more than 1 for $\omega > \omega_x$. Note that the system (half-power) bandwidth (BW) is > 1 for all $T(s)$ whose $\xi < 1$. For the general, any order $T(s)$, *the sensitivity is > 1 for all ω whose $T(j\omega)$ lies outside the unit circle centered at $0' = (1,0)$. All practical feedback systems certainly have an* *excess* ≥ 2 *of poles over zeros, so must therefore have an ω range outside this* $|S(j\omega)| = 1$ *circle, in which their sensitivity S to the plant is* > 1. *The smaller the damping factor, the larger is this* $|S(j\omega)| > 1$ *range, where the systems transfer function is more sensitive to plant uncertainty than that of an open loop (no feedback) system* (step response overshoot is sensitive in this frequency range).

Classical (Frequency Response) and Modern (State Space, LQR, Observers) Control have same Highly Inadequate Objectives For many years (to 1985), most feedback control research, whether Classical (Frequency Response), or "Modern" (State Space, LQR, Observer theory, etc.) was devoted to One Degree of Freedom(ODF) systems, even though the ODF constraint did not necessarily exist for their specific system. This was likely due to the heritage of the Radar Gun Control problem. No design theory was therefore developed by any of these design methods to exploit the additional degree of freedom available in Fig. 1, which permits independent design of $T(s)$ and sensitivity $S(s)$. These apparently radically different techniques differed only in terminology and mathematical language. Classical strived for an acceptable System Frequency Response, which of course involves both poles and zeros of the system transfer function $T(s)$. Modern strived for acceptable system eigenvalues (poles of $T(s)$). The zeros were either neglected, which of course is an inadequate representation of $T(s)$. Or it was assumed there were no finite zeros, which means significant loss of freedom in choice of $T(s)$. For some years Modern Control theory assumed all the states could be measured. Later, observers were used, one for each state that could not be measured. Of course, this involved dynamic compensation, spoiling the claim that one of the advantages of LQR design was its use of only real compensators (but if these real numbers had magnitudes > 1, they were really unrealizeable infinite bandwidth amplifiers). For both Classical and Modern design techniques, the objectives would be obtained only at fixed *nominal plant parameter values*. Neither Classical nor Modern developed any theory for coping with Plant Uncertainty. *The objectives of both were thus essentially the same, except that Modern tended to neglect system zeros.* Also, Modern is much less realistic in terms of System Constraints. But Modern was much more "Mathematical" oriented [22]. Both assumed that adaptive, i.e. nonlinear loop compensation G was needed to cope with nontrivial plant uncertainty. But even the adaptive structures that were considered tended to be ODF. The following statement by R. Kalman, Father of Modern Control theory, was typical (-1956):

" it is generally taken for granted that the dynamic characteristics of the process will change only slightly under any operating conditions encountered during the lifetime of the control system. Such slight cbanges are foreseen and are usually counteracted by using feedback. Should the changes become large, the control equipment as originally designed may fail to meet performance specifications. "

But even the adaptive theory was flawed. There was no engineering design, i.e. no statements of uncertainty and of specifications, followed by systematic design procedure for their achievement. Nonlinear feedback structures were presented with Qualitative Sensitivity reduction properties, but design was by cut and try [21].

1.4   Quantitative Feedback Design Theory (QFT)

First QFT design paper was in 1959 [23]. It emphasized that Feedback in Control was required principally because of Plant Uncertainty. *Therefore, Quantitative formulation of Plant Uncertainty and of Performance Specifications was essential*, and it presented systematic design procedures for their achievement.

To counteract Kalman's statement of inability of ordinary (LTI) feedback to cope with significant Plant Uncertainty, a design example with over 100: 1 Plant Uncertainty, was presented. A design for the X15 pitch control with 1000 to 1 Uncertainty over the entire frequency range (up .to Mach 6 and 100,000 feet) was published in 1964. The Flight Dynamics Lab of the Wright Patterson Air Force Base had categorically stated that LTI Feedback could not cope with such huge Plant Uncertainty. The Author of the above design was labeled as Anti-Adaptive = Personna Non Grata. QFT advanced into SISO Multiple-Loop Systems, Digital Systems, Non-minimum Phase and Open loop Unstable systems, even to Plants with both right half-plane zeros and poles, as well as to Multiple Input -Multiple Output Systems.

QFT especially emphasized the Cost of Feedback, in terms of Loop Bandwidth and Sensor Noise effects. A very important advance was made in 1975 to *rigorous design for uncertain linear time-varying systems [17], and in 1976 to nonlinear systems [18]* . An important feature of these design techniques, is that an ordinary control engineer, with hardly any knowledge of nonlinear mathematics, but with knowledge of LTI feedback design, can readily execute quantitative design for highly uncertain and complex nonlinear plants, to achieve exact system specifications.

An important point is that it is not necessary to have a mathematical model of the nonlinear plant [9,25]. Numerous applications were made, manyby Master Students, some with no previous knowledge of Feedback theory, to SISO and MIMO designs for LTI and *nonlinear plants* in advanced Flight Control,in automatic self-adjustment of aircraft damaged in flight, in Robotics, design of several types of automatic welding machines [9], in Forest regulation, in highly ill-conditioned 2 by 2 distillation column [16], in nonlinear process control, and recently to its first attempt at Irrigation Canal Flow control, with 5 to 1 improvement over the best design heretofore by PID. One Master student succeeded in longtitudinal 2 by 2 flight control stability design for the X29 Forward Swept Aircraft, for which the Plant was both *nonminimum-phase and unstable*. This problem had been abandoned as impossible by both Grummann Aircraft and Minneapolis-Honeywell, both with considerable Government funding.

2   Rigorous QFT Techniques for Design of Uncertain Nonlinear and Time–Varying Feedback Systems

The essential idea in first technique [18], is to replace the Nonlinear/Time-Varying Plant set $\mathbf{W} = \{W\}$ (a set due to plant uncertainty), by some *equivalent* LTI plant set $\mathbf{P_e} = \{P\}$, for which LTI design is applicable.

For example, $\mathbf{W}$ is a steering system to be used for a specified range of car models, and specific range of road types, which gives a nonlinear plant set $\mathbf{W} = \{W\}$. The cars will be driven by a variety of drivers, who will apply a range of command inputs $\mathbf{R} = \{r\}$ to the steering system. There is specified a set of acceptable transfer functions $\mathbf{T} = \{T\}$. Thus, an acceptable output Y due to any command R in $\mathbf{R}$, is $Y(s) = T(s)R(s)$, for T in $\mathbf{T}$. Consider any pair (R, W) in $\mathbf{R} \times \mathbf{W}$. It is required that the system output $Y(s)$ be equal to $T(s)R(s)$, for some T in $\mathbf{T}$. This is depicted in Fig. 3–a.

A crucial step is the derivation of an equivalent LTI plant set $\mathbf{P_e} = \{P\}$, as follows. The technique is applicable to problems for which the following steps can be executed: i)Choose any pair (R, T) in $\mathbf{R} \times \mathbf{T}$ giving an acceptable output $Y(s) = T(s)R(s)$, with inverse transform $y(t)$, ii)Choose any W in $\mathbf{W}$ and solve for $u(t) = W^{-1}(y(t))$, i. e. the input of W which gives the output $y(t)$ (it is assume that every W in $\mathbf{W}$ is invertible so $u(t)$ is unique), iii)Define the LTI-equivalent plant $P(s) = Y(s)/U(s)$, being $U(s)$ the Laplace transform of $u(t)$. Repeating over all R in $\mathbf{R}$, T in $\mathbf{T}$, and W in $\mathbf{W}$, the result is a LTI-equivalent plant set $\mathbf{P_e} = \{P\}$. Thus if $\mathbf{R}$, $\mathbf{T}$, and $\mathbf{W}$ have n_1, n_2, and n_3 members, then $\mathbf{P_e}$ has $n_1 \times n_2 \times n_3$ members. In practice, at least $\mathbf{T}$ has uncountable members, and so have realistic $\mathbf{R}$ and $\mathbf{W}$.

Next consider the following LTI problem (Fig. 3-b). There is given the command input set $\mathbf{R}$, the LTI plant set $\mathbf{P_e}$, and the OK tranfer function set $\mathbf{T}$. For each R in $\mathbf{R}$, and P in $\mathbf{P_e}$, the closed loop output must be $Y(s) = T(s)R(s)$ for some T in $\mathbf{T}$. This is analogous to the original problem in Fig. 3-a, except that nonlinear set $\mathbf{W}$ is replaced by the LTI set $\mathbf{P_e}$. For a large class of such LTI problems, a design in QFT is executable, i. e. a pair of controllers $F(s)$ and $G(s)$ can be found for that purpose. And it can be proven that the same pair $F(s)$, $G(s)$ solves the original nonlinear design problem, that is for each W in $\mathbf{W}$ and each R in $\mathbf{R}$, the output is guaranteed to be $Y(s) = T(s)R(s)$ fos some T in $\mathbf{T}$.

Outline of proof (See [18,20] for details). Schauder fixed point theorem is used: a continuous mapping of a convex compact set of a Banach space into itself has a fixed point. There are several choices of Banach spaces, but one in the transform domain is convenient. Choose the set $\mathbf{T}$ compact and convex in it. Pick any (R, W) pair and define the mapping over $\mathbf{T}$

$$\Phi(T) = \frac{FP(T, R)G}{1 + P(T)G} \tag{4}$$

where $P(T, R)$ is the mapping from $\mathbf{T} \times \mathbf{R}$ to $\mathbf{P_e}$, given by $P(T, R) = TR/U$, where U is the Laplace transform of $u = W^{-1}y$, and $Y = TR$ is the Laplace transform of y. In the mapping Φ, since R is fixed, U is a function of T. It can be rewritten as

$$\Phi(T) = \frac{FGTR/U(T)}{1 + GTR/U(T)} \tag{5}$$

We must prove that $\Phi(T)$ is continuous and maps $\mathbf{T}$ into itself. Continuity is a rather technical condition and it is satisfied under mild asumptions (see [18,20] and [1]). To prove that $\Phi(T)$ maps $\mathbf{T}$ into itself, note that $TR/U = Y/U$, which is precisely the definition of P in $\mathbf{P_e}$. But F and G have been designed so that $Y = TR$, T in $\mathbf{T}$ for all P in $\mathbf{P_o}$. Hence $\Phi(T)$ maps every T into $\mathbf{T}$, and thus Φ has a fixed point T^*, with

$$T^* = \frac{FGT^*R/U^*}{1 + GT^*R/U^*} \tag{6}$$

giving as a result that $U^* = G(FR - T^*R)$. Here $U^* = U(T^*)$ is the input $u^* = W^{-1}(y^*)$ in the time domain, being y^* the inverse transform of $Y^* = T^*R$. Thus, Y^* is also the output of the nonlinear original closed loop system in Fig. 3-a, and since the fixed point T^* is in the set $\mathbf{T}$, the output is in the OK output set for this (R, W) pair. The same Φ mapping can be made for all (R, W) pairs. It is therefore essential that the same (F, G) pair be a solution of the LTI-equivalent problem, for all P in $\mathbf{P_e}$. Recall that $\mathbf{P_e}$ was generated by performing the steps: i) $Y = RT$ for each R in $\mathbf{R}$, T in $\mathbf{T}$, ii)for each such y (inverse transform of Y), $u(t) = W^{-1}y(t)$ is found for each W in $\mathbf{W}$, and iii) $P(s) = Y(s)/U(s)$, being U the Laplace transform of u.

A simplication to the above treatment is to consider only a impulse reference. It is not difficult to show by means of block transformations (Fig. 3-c,d) that a valid design for the *augmented* nonlinear plant $\mathbf{R^{-1}WR}$, for an impulse reference and for the OK output set $\mathbf{T}$, is also valid for a nonlinear plant $\mathbf{W}$, for a reference set $\mathbf{R}$ and an OK output set $\mathbf{TR}$. Note that due to the fact that W is non linear, in general $R^{-1}WR \neq W$. The Schauder mapping is defined on the set $\mathbf{T}$, and a similar reasoning applies but now the set of commands is part of the set of nonlinear plants.

Schauder theorem does not guarantee a unique solution, but if each of the elements W, F, G is one to one, then the solution must be unique. The above must be applicable for all the nonlinear plants in set $\mathbf{W}$. In response to any nonlinear plant w in $\mathbf{W}$, the output must be $Y = T$, with T a member of the OK set $\mathbf{T}$.

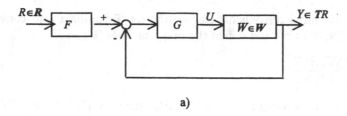

a)

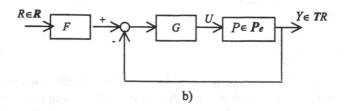

b)

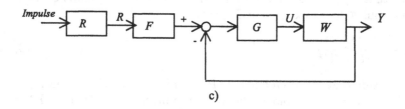

c)

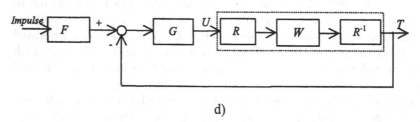

d)

Fig. 3. The SISO nonlinear problem: a)find $F(s)$, $G(s)$ so that y is in $\mathbf{T}R$ for all w in $\mathbf{W}$, R in $\mathbf{R}$; b) quivalent LTI problem; c)-d) equivalent problem for an impulse reference and plant set $\mathbf{R}^{-1}\mathbf{W}\mathbf{R}$.

In practice, it is therefore important for the designer to include all conceivable command inputs which may be applied to the system in its lifetime. This can include perturbations on normal command inputs (Sec. 6). It is important to emphasize that this is *not* a technique for *approximating* any specific nonlinear system by a linear one. The representation of the nonlinear plant set $\mathbf{W}$, by $\mathbf{P_e}$, is *exact set equivalence with respect to sets* $\mathbf{R}$, $\mathbf{T}$

(or alternatively the representation of $\mathbf{P_{eR^{-1}WR,T}}$ by its $\mathbf{P_e}$, is exact set equivalence with respect to $\mathbf{T}$).

Set equivalence suffices, because the Quantitative Feedback design problem is inherently a problem of sets: guarantee that the output belongs to a specific set, for all members of a command input set and a nonlinear plant set. It is important to include the qualification that the equivalence is only *with respect to sets* $\mathbf{R}$, $\mathbf{T}$. Our design is not guaranteed for command inputs not in set $\mathbf{R}$. But this is good Engineering. It would be silly to try to design a car steering system to respond to video signal commands.

It is not compulsory that QFT be used to solve the LTI-Equivalent problem. Any technique may be used, even cut and try , but it is essential that it be properly solved. The QFT technique can do so for a large problem class, in fact for any LTI problem which is solvable by LTI compensation. Also, and very important, the designer can tell, just from examination of $\mathbf{P_e}$ whether the LTI equivalent problem is solvable for the given $\mathbf{T}$ set, by means of LTI compensation.

In the following, the nonlinear QFT technique is further formalize, starting with the study of asymptotic values for pedagogical purposes. A general validation result is also developed for a class of nonlinear systems, based on restrictions on the set of acceptable outputs, in such a way the the resulting equivalent problem is solvable in the linear QFT framework.

2.1 Asymptotic Tracking

Although the asymptotic behavior of the closed loop system can be viewed as a particular case, its detailed analysis in this Section can illustrate the basic ideas to be used in the more general case developed above. As it will be seen, the asymptotic behavior of nonlinear QFT designs explains to a great extent the practical validity of the technique. A single but illustrative example is used here to develop the main points of the technique.

Consider the electrical circuit of Fig. 4-a, where the nonlinear resistor G has a characteristic given by Fig. 4-b, and C is assumed to be the unity.

The input-output dynamics is given by the ordinary differential equation

$$\dot{y} + (1 + ay^2(t))y(t) = u(t), y(0) = 0 \tag{7}$$

where $a \in [0, 2]$. When the TDF feedback structure is used to control this system, the control law is given by

$$\hat{u}(s) = G(s)(F(s)\hat{r}(s) - \hat{y}(s)) \tag{8}$$

where ˆ stands for Laplace transform. Assuming that the closed loop system is stable, the asymptotic value of the output $y(\infty)$ is given by

$$(1 + ay^2(\infty))y(\infty) = G(0)(F(0)r(\infty) - y(\infty)) \tag{9}$$

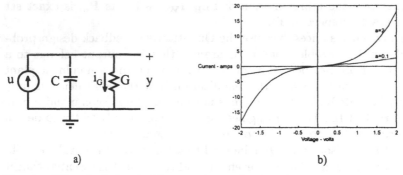

a)   b)

Fig. 4. a)Electrical circuit, b)uncertain nonlinear resistor

that is, $y(\infty)$ can be obtained as the real solution to the third order algebraic equation

$$ay^3(\infty) + (1 + G(0))y(\infty)) - G(0)F(0)r(\infty) = 0 \qquad (10)$$

By simplicity of notation $G(0)$ and $F(0)$ are substituted by G_0 and F_0, while $y(\infty)$ and $r(\infty)$ are substituted by y_∞ and r_∞. Since the equation is polynomial, the solution can be computed using the root locus technique, being $1/a$ the parameter; that is, the solutions can be obtained as the root locus of

$$1 + \frac{1}{a}\frac{(1 + G_0)y(\infty) - G_0 F_0 r_\infty}{y_\infty^3}, \frac{1}{a} \in [0.5, \infty] \qquad (11)$$

The result is given in Fig. 5 (for $G_0 = 12$, $F_0 = 1.18$, $r_\infty = 1$), where the real-valued branch, going from the pole at the origin to the zero, will be the output asymptotic value for the different values of the uncertain parameter a. Then, it can be seen that the asymptotic value of the output is relatively close to the reference for those dc-gain values of the compensators. This is an analysis result.

From a control point of view, the question would be how to use the controller, that is G_0 and F_0, to bound the values of $y(\infty)$ according to the specifications for any value of $a \in [0, 2]$. y_∞ will be upper bounded by $G_0 F_0 r_\infty / (1 + G_0)$, and lower bounded by the real solution of the algebraic equation (3.4) for $a = 2$, which in general may be hard to compute in closed form.

Given closed loop output steady-state specifications such as $y_\infty \in [y_{l,\infty}, y_{u,\infty}]$, or equivalently dc-gain closed loop transfer functions specifications in the form of $T_0 \in [T_{l0}, T_{u0}]$, the design problem can be reformulated as the computation of G_0 and F_0 such that

$$\frac{G_0 F_0}{1 + G_0} \leq T_{u,0} := \frac{y_{u,\infty}}{r_\infty} \qquad (12)$$

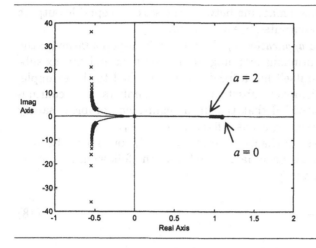

Fig. 5.   Asymptotic
output values as a
root locus

and the real solution of $2y_\infty^3 + (1 + G_0)y_\infty - G_0 F_0 r_\infty = 0$ verify

$$\frac{y_\infty}{r_\infty} \geq T_{l,0} := \frac{y_{l,\infty}}{r_\infty} \tag{13}$$

Nonlinear QFT Solution The problem given by (12) and (13) consists of
the resolution of two nonlinear algebraic equations for the two unknowns G_0
and F_0. QFT approachs the problem using Schauder's fixed-point theorem.
The complete solution follows the following steps:

1. *Definition of the set of acceptable outputs*: asymptotic values of closed
loop acceptable outputs for references with $r_\infty = 1$, are assumed to be in the
interval

$$A_1 = [0.9, 1.1] \tag{14}$$

2. *Computation of the equivalent linear family (ELF)*: a set of linear systems is defined using each acceptable output and its corresponding input. In
the example,

$$ELF_0 = \left\{ \frac{y_\infty}{u_\infty} \middle| u_\infty = (1 + ay_\infty^2)y_\infty, y_\infty \in A_1, a \in [0,2] \right\} = [0.29, 1] \tag{15}$$

3. *Definition of the closed loop mapping Φ*: substituting the nonlinear
system by the set ELF_0, the linear closed loop output $y_{L,\infty}$ is given by

$$y_{L,\infty} = \frac{P_0 G_0 F_0}{1 + P_0 G_0} r_\infty, P_0 \in ELF_0 \tag{16}$$

that is

$$y_{L,\infty} = \frac{\frac{y_\infty}{(1+ay_\infty^2)y_\infty}G_0 F_0}{1 + \frac{y_\infty}{(1+ay_\infty^2)y_\infty}G_0} = \frac{G_0 F_0}{1 + ay_\infty^2 + G_0} =: \phi(y_\infty) \tag{17}$$

Thus, $y_{L,\infty} = \phi(y_\infty)$ defines a mapping between the set of acceptable outputs and the set of closed loop outputs for every G_0 and F_0.

4. *Design based on the application of Schauder's fixed-point theorem*: the theorem states that "a continuous mapping from a compact and convex subset of a Banach space into itself has a fixed point". Applied to the example, since A_1 is convex and compact subset of the real numbers, ϕ is continuous, and $\phi(A_1) \subset A_1$, provided that the linear equivalent problem satisfy specifications, then the mapping ϕ has a fixed point in A_1.

5. Validation of the design: the fixed point will be the output of the nonlinear closed loop system. This can be deduced from the following reasoning: the fixed point $y\star_\infty$ must verify

$$y\star_\infty = \frac{G_0 F_0}{1 + ay\star_\infty^2 + G_0} \tag{18}$$

but this is exactly the identity that must satisfy the closed loop output corresponding to the nonlinear control problem. Then by construction, a solution to the equivalent linear problem is also a solution to the original nonlinear problem. Note that the fixed point will depend, in general, on uncertain parameters and the reference.

TDF solution to the equivalent linear problem QFT translates a nonlinear robust control problem to a linear robust control problem, which may be solved using any robust control technique. In particular, linear QFT is a good candidate for solving the equivalent linear problem. Using again the example, the (asymptotic) linear control problem can be stated in the following way: compute G_0 and F_0 such that

$$\frac{P_0 G_0 F_0}{1 + P_0 G_0} \in [0.9, 11] \tag{19}$$

for each $P_0 \in [0.29, 1]$. Here, it is clear the utility of the two degrees of freedom structure, that is F_0 and G_0. A key step is the use of logarithms to solve the problem, that is

$$\log\left(\frac{P_0 G_0 F_0}{1 + P_0 G_0}\right) \in [-0.046, 0.041] \tag{20}$$

In terms of logarithms, the problem is to find values of F_0 and G_0 such that

$$\max_{P_0 \in [0.29, 1]} \log\left(\frac{P_0 G_0 F_0}{1 + P_0 G_0}\right) = \log\left(\frac{G_0 F_0}{1 + G_0}\right) < 0.041 \tag{21}$$

$$\min_{P_0 \in [0.29, 1]} \log\left(\frac{P_0 G_0 F_0}{1 + P_0 G_0}\right) = \log\left(\frac{0.29 G_0 F_0}{1 + 0.29 G_0}\right) > -0.046 \tag{22}$$

Since

$$
\Delta \log \left(\frac{P_0 G_0 F_0}{1+P_0 G_0} \right) := \log \left(\frac{G_0 F_0}{1+G_0} \right) - \log \left(\frac{0.29 G_0 F_0}{1+0.29 G_0} \right) =
$$
$$
= \log \left(\frac{G_0}{1+G_0} \right) - \log \left(\frac{0.29 G_0}{1+0.29 G_0} \right)
$$
(23)

the design problem (for the steady-state) is finally given by the inequality

$$
\begin{cases}
\log \left(\frac{G_0}{1+G_0} \right) - \log \left(\frac{0.29 G_0}{1+0.29 G_0} \right) < 0.087 \\
0.9 \frac{1+0.29 G_0}{0.29 G_0} < F_0 < 1.1 \frac{1+G_0}{G_0}
\end{cases}
$$
(24)

A "economic" solution to the problem is $G_0 = 10.1$ and $F_0 = 1.2$.

SDF Solution to the equivalent linear problem . A single degree of freedom structure can also be used to solve the equivalent linear control problem. In this case the problem, using a similar development to the TDF structure, is to find G_0 such that

$$
\frac{P_0 G_0}{1+P_0 G_0} \in [0.9, 11]
$$
(25)

for every $P_0 \in [0.29, 1]$. A single analysis shows that G_0 must satisfy the inequalities

$$
\begin{cases}
\frac{0.29 G_0}{1+0.29 G_0} > 0.9 \frac{G_0}{1+G_0} < 1.1
\end{cases}
$$
(26)

from where we finally obtain that $G_0 > 31$. Note that the use of a single structure leads to a less economic controller than the linear QFT (TDF) solution.

Validation of the nonlinear QFT design A first property of QFT designs is that they are "economic" solutions to the linear equivalent problem. The two key points have been the use of the TDF structure and the logarithmic analysis of specifications. Now the question is whether the linear QFT design is also an economic design for the original nonlinear problem.

For the nonlinear control problem, Schauder's fixed-point theorem guaranties that the output, which is the fixed point, belongs to the set of acceptable outputs. However, it is not clear if the output takes values in some subset of the set of acceptable outputs or in the whole set, for different values of the uncertain parameters. In the example, it is not known a priori if $y\star_\infty$ takes values in $[0.9, 1.1]$ or in some subinterval, meaning a more conservative design. This limitation of nonlinear QFT relies upon the fact that it is a solution of the problem given by (24), while the original problem is given by (12) and (13).

The QFT solution, given by $G_0 = 10.1$ and $F_0 = 1.2$, results using (12) and (13) in that the output of the nonlinear control system is in the interval

$$y\star_\infty \in [0.95, 1.09] \subset [0.9, 1.1] \tag{27}$$

showing that in this case, the design is also "economic" for the original nonlinear system. A more detailed analysis of the fixed point values for each value of a, and its relationship with the output $y\star_\infty$ is given in Table 1 and Fig. 6. For some values of the uncertain parameter a, values of ELF_0, linear closed loop outputs $y_{L,\infty}$, and nonlinear closed loop outputs $y\star_\infty$ are shown.

a	ELF_0	$y_{L,\infty}$	$y\star_\infty$
0	$\{1\}$	$\{1.09\}$	1.09
0.5	$[0.62, 0.71]$	$[1.04, 1.06]$	1.05
1	$[0.45, 0.55]$	$[0.99, 1.02]$	1.01
1.5	$[0.35, 0.45]$	$[0.93, 0.98]$	0.97
2	$[0.29, 0.38]$	$[0.90, 0.95]$	0.95

Table 1. Fixed points for different values of the parameter a

Note that since $y\star_\infty$ must belong to the interval defined by the outputs $y\star_\infty$, the result is not overly conservative. This fact is basically related with the structure of the plant uncertainty, which in this case is given by the intervals defined by ELF_0. Since the same control effort, that is G_0, is used to map the uncertainty in ELF_0 with the uncertainty in the output $y\star_\infty$, one may expect good results for the nonlinear control problem if the uncertainty is similar to the exhibited by ELF_0 in this example.

More general reference inputs In the above design, only the case $r_\infty = 1$ is considered. In this Section we extend the design to consider more general references. The (linear) specification for the asymptotic value of the output is defined as the set

$$A_r = [0.9, 1.1]r_\infty \tag{28}$$

where in general r_∞ may take values different to 1. This specification means steady output values between the 90% and the 110% of the steady reference value. In this case, the equivalent linear family is given by

$$ELF_0 = \left\{ \frac{y_\infty}{u_\infty} \,|\, u_\infty = (1 + ay_\infty^2)y_\infty, y_\infty \in A_r, a \in [0, 2] \right\} \tag{29}$$

where the elements of ELF_0 depends on r_∞. In particular, they are intervals of the form $[\alpha_r, 1]$, being α_r dependent on r_∞ . Using the method exposed in

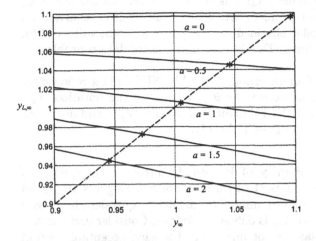

Fig. 6. The mapping ϕ and its fixed points($\star$)

previous Sections, we arrive to the problem of finding G_0 and F_0 to satisfy the following inequalities:

$$\begin{cases} \log\left(\frac{G_0}{1+G_0}\right) - \log\left(\frac{0.29G_0}{1+0.29G_0}\right) < 0.087, \forall \alpha_r \\ \frac{0.9(1+\alpha_r G_0)}{0.29\alpha_r G_0} r_\infty < F_0 < \frac{1.1(1+G_0)}{G_0} r_\infty \end{cases}$$ (30)

A solution for different values of r_∞ is given in next Table. Note that the control effort grows with increasing values of the reference, since the control system is nonlinear.

r_∞	ELF_0	G_0	F_0
$[0, 1]$	$[0.29, 1]$	10.1	1.2
$[0, 3]$	$[0.17, 1]$	21.2	1.15
$[0.5]$	$[0.016, 1]$	283.5	0.9

Table 2. dc-gains G_0 and F_0 for different asymptotic references values

2.2   Tracking

This Section considers the problem of designing a controller according to the TDF structure, such that some closed loop tracking specifications are achieved, taking into account not only asymptotic values. The design method closely follows the steps given in Section 2, but following a more formal approach. The material presented in this Section is based on the seminal works [17,18] and [1].

The core concept of nonlinear QFT is the replacement of the nonlinear robust control system design problem with a linear one, that is "equivalent", in the sense that a controller obtained for the linear problem will work for the original nonlinear problem. This is known as the validation problem in the QFT literature.

We consider the case of an uncertain, invertible, NLTV plant represented by set $\mathbf{W}$. This set is parameterized by $\theta \in \Theta$, we use the notation $W_\theta : U \to Y$, which is an invertible NLTV plant with inverse W_θ^{-1}, and define the NLTV plant set as $\mathbf{W} = \{W_\theta : U \to Y | \theta \in \Theta\}$.

The Equivalent Linear Family of Plants To introduce the concept of equivalent linear family of plants (ELF) first consider a single NLTV plant $W : U \to Y$. $\mathbf{P}(\mathbf{A_R})$, the ELF or family of linear plants equivalent to W with respect to the OK output set A_R, is defined as follows. Consider first the case in which the nonlinear plant is not uncertain. For each acceptable output $y \in A_r$ we choose (by some method) an invertible LTI mapping $P^y : U \to Y$. That is, for each $r \in R$ we define a mapping $\gamma_r : A_r \to (U \to Y)$ such that $P^y := \gamma_r(y)$ is determined for each acceptable output $y \in A_r$. The set $\mathbf{P}(A_R)$, the ELF of W with respect to A_R, is then defined as

$$\mathbf{P}(A_R) = \bigcup_{r \in R} \gamma_r A_r \tag{31}$$

For the general case in which the nonlinear plant is uncertain, for each $\theta \in \Theta$ we construct an ELF of LTI plants. That is, we define a mapping $P_\theta^y : U \to Y$ and a family of linear plants

$$\mathbf{P}_\theta(A_R) = \bigcup_{r \in R} \gamma_{r,\theta} A_r \tag{32}$$

is generated for each $\theta \in \Theta$ and for each $r \in R$. The linear equivalent family for the NLTV plant set $\mathbf{W}$ is then defined as

$$\mathbf{P}(A_R) = \bigcup_{\theta \in \Theta} P_\theta(A_R) \tag{33}$$

At this point, we also note that the substitution of a nonlinear plant W_θ by a linear equivalent P_θ^y in the TDF feedback system leads to the relation

$$y_{L,\theta} = P_\theta^y(GFr - Gy_{L,\theta}) \tag{34}$$

which defines a closed loop mapping $\phi_{r,\theta} : A_r \to Y$ such that $y_{L,\theta} = \phi_{r,\theta}(y)$, where $y_{L,\theta}$ is the output of the TDF system and r, y, and θ selects the plant as $P_\theta^y = \phi_{r,\theta}(y)$.

We now identify two conditions useful in studying a potential ELF:

i)the *equivalence condition*:

$$P_\theta^y(W_\theta^{-1}y) = y, \forall \theta \in \Theta \tag{35}$$

and

ii)the *continuity condition*: the closed loop mapping $\phi_{r,\theta}$ is continuous on A_r, $\forall \theta \in \Theta$, and $\forall r \in R$.

A General Validation Theorem. Given $P(A_R)$, the ELF of **W** defined above, a key question is the feasibility of a controller $K = (F, G)$ for the (possibly time-varying) NLTV plant set **W**, when it is feasible for the entire ELF $P(A_R)$. It is understood that a feasible controller for a set of plants is capable of achieving design specifications for any plant of the set. More formally, $K \in \Gamma(\mathbf{P}, R, A_R)$ if the output of the TDF feedback system with controller K and plant P is in the set A_r for any $r \in R$ and any $P \in \mathbf{P}$.

The question is: does $K \in \Gamma(\mathbf{P}(A_R), R, A_R)$ imply that $K \in \Gamma(\mathbf{W}, R, A_R)$? If this question can be answered in the affirmative we have a technique for translating a nonlinear control problem into a robust linear control problem. A solution to this validation problem is given in the following theorem: "Assume that for each $r \in R$ the set A_r is a convex compact subset of a Banach space Y_S. If $P(A_R)$ is an ELF of the NLTV plant set **W** with respect to A_R chosen so that conditions i) and ii) are satisfied, then $K \in \Gamma(\mathbf{P}(A_R), R, A_R)$ $\implies K \in \Gamma(\mathbf{W}, R, A_R)$.

A proof is given in [1], strongly based on early ideas of [17,18]. The proof involves an application of the Schauder fixed point theorem to the closed loop mapping $\phi_{r,\theta}$. Thus the continuity of this mapping is a key requirement. The above result reduces the validation question to one of defining an ELF and then testing conditions i) and ii) defined above. This result is very general and applies to any mapping $\phi_{r,\theta}$ with the two desired properties. Note that in addition nothing has been said about the controller $K = (F, G)$, it does not have to be necessarily linear. Also the ELF does not have to be linear.

The result is about to translate an uncertain nonlinear control problem in a uncertain (usually linear) control problem. The appoach followed by QFT has been to choose a LTI transformed control problem easily integrated in the LTI QFT technique. In the following, a possible election of $\phi_{r,\theta}$ in the framework of QFT is given.

Nonlinear QFT solution Given the general framework for validation of linear control of uncertain NLTV plants developed above, the QFT approach to robust nonlinear control can be analyzed as a particular case. Next we use the validation results developed above to demonstrate the validity of the nonlinear QFT approach. We begin by introducing some assumptions and identifying the ELF used in nonlinear QFT. Here the space $RH_{2,e} = U = Y$ is chosen as the input and output signals space. $U_S = Y_S = RH_2(\subset RH_\infty)$ is the Banach space of stable signals Y_S, given by the set of stable and strictly proper real-rational complex functions.

Allowable Nonlinear Plants. An uncertain NLTV plant **W** is a parameterized set of NLTV plants $W_\theta : U \to Y$, where $\theta \in \Theta$. The individual NLTV plants must satisfy the following assumptions:

P1) Each plant W_θ must be representable by mapping from $RH_{2,e}$ to $RH_{2,e}$

P2) Each plant W_θ must be invertible with inverse W_θ^{-1} continuous over A_R

P3) (High frequency linear behaviour) Each plant W_θ can be represented by an ordinary differential equation of the form

$$y^{(n)}(t) + f_y(y^{(n-1)}(t), .., y(t), t; \theta) = K(\theta)u^{(m)}f_u(u^{(n-1)}(t), .., u(t), t; \theta) \quad (36)$$

where

$$lim_{t \to 0+} y^{(n)}(t) + f_y(y^{(n-1)}(t), ..., y(t), t; \theta) = lim_{t \to 0+} y^{(n)}(t) \quad (37)$$

$$lim_{t \to 0+} K(\theta)u^{(m)} + f_u(u^{(n-1)}(t), ..., u(t), t; \theta) = lim_{t \to 0+} K(\theta)u^{(m)} \quad (38)$$

$\forall y \in A_R$, and $\forall u \in W_\theta^{-1} A_R$.

The invertibility condition P2 is needed in the procedure for obtaining the ELF described below. The continuity condition in P2 is needed to insure that condition ii) above can be satisfied. Condition P4 is related with the fact that the nonlinear plant must be well approximated by a linear plant at high frequencies in order to obtain a feasible LTI QFT solution of the equivalent linear problem. This restriction can be avoided in many cases by the use of a nonlinear precompensator [16,19] although this extension will not be described here.

The Set of Acceptable Outputs. For each $r \in R$, the set of acceptable outputs A_r satisfies the following conditions:

O1) The set A_r is a compact and convex subset of the Banach space $Y_S(= RH_2)$

O2) The set $W^{-1}A_r$ must be a subset of $U_S(= RH_2)$

O3) Any function y_r in A_r has a fixed relative order e, such as $n - m < e \le e_{yr}$, where $e_y r$ is a constant upper bound depending on r.

Both requirements O1 and O3 are necessary to apply Schauder fixed point theorem in the solution of the validation problem. According to O3 and P3, it is straightforward to show that all plants in ELF are strictly proper with relative order n-m. Condition O2 is included to insure that the linear design would not have to be internally unstable.

The Equivalent Linear Family. The equivalent linear plant family is obtained using transfer functions representing input-output pairs obtained from the given NLTV plant. That is, for each acceptable output $y \in A_r$ corresponding to $r \in R$, and for each $\theta \in \Theta$, P_θ^y is defined as a LTI mapping with the transfer function given by

$$P_\theta^y(s) = \frac{y(s)}{u(s)} \tag{39}$$

where $u = W_\theta^{-1} y$. Thus $P(A_r)$, the family of equivalent linear plants with respect to A_r, is defined as

$$\mathbf{P}(A_r) := \left\{ P_\theta^y(s) = \frac{y(s)}{u(s)} | u = W_\theta^{-1} y, \forall \theta \in \Theta, \forall y \in A_r \right\} \tag{40}$$

With assumptions P1-P3 and O1-O3 the NLQFT approach is valid in the sense of that the resulting LTI controller will provide satisfactory robust control of the original uncertain NLTV plant. The result can be stated as: given an uncertain NLTV plant set $\mathbf{W}$ satisfying assumptions P1-P3, let us define for each input r sets of acceptable outputs A_r satisfying assumptions O1-O3, and $P(A_r)$ as the ELF of $\mathbf{W}$ with respect to A_r; if, in addtion, $G(s)F(s)r(s)$ has relative degree greater than or equal to e_{ur}, then $K \in \Gamma(\mathbf{P}(A_R), R, A_R) \Longrightarrow K \in \Gamma(\mathbf{W}, R, A_R)$.

2.3 Stability

This question is usually defined as "what happens in a system when it suffers from small changes, in plant, in compensators, in inputs, etc?. The classical question is whether the system is infinitely sensitive to changes from an equilibrium situation. If the sensitivity is finite to any such small changes then we have Bounded input Bounded output (BIBO) stability, with respect to the particular (infinitesimally) small deviation. Since apriori the plant is uncertain, there is really no need to consider small changes of open plant set, nor of compensation G, since it is in cascade with the plant. If one is nevertheless concerned with these, simply make the plant uncertainty a bit larger. If serious plant changes during operation may occur in practice, one can include uncertain time-varying plants in the equivalent plant set with many different scenarios (see later for an uncertain time-varying plant set example). It is only necessary that one can work backwards from the desired output, to solve for the plant input u, to derive one more member of $\mathbf{Pe}$.

With respect to changes in the command or other inputs, infinite variety of departures from equilibrium can be postulated. (1) The command input R is a member of a set which has been defined for t in $[0, t_1)$, but at t_1, its character changes abruptly by addition of another signal. No matter, include the total as another member R_x of $\mathbf{R}$, providing of course that one wants the system response to R_x be in $\{T R_x, \forall T \in \mathbf{T}\}$. Of course, one may want *nonlinear* closed loop system response, i.e different **T** sets for different **R** sets. One could even want such nonlinear system response for systems with LTI Uncertain Plants. Very little (Sec. 11.4 in [16]) has been done of this nature. One can include any number of such signals. (2) One could demand "Time-Varying" type of system response functions, whether for LTI or Nonlinear/Time-Varying Plants. This can be done. It requires a Linear Time-Varying Prefilter F.

Classical stability techniques have been modified for application to the stability problem of QFT Nonlinear design. This approach has been pioneered by Baños and Barreiro [2,7].

There is a second Nonlinear QFT design technique, wherein the uncertain nonlinear time-varying plant set is replaced by an "equivalent" disturbance set $d_e = \{d_e\}$, plus a very simple LTI Plant of form k/s^n, with n the order of the nonlinear plant; d_e a function of the nonlinear plant and of the output specifications. This technique is best suited for nonlinear plants which are linear with respect to their highest derivative. It is extraordinarily well suited to handle the disturbance attenuation problem, for nonlinear or LTI Uncertain plants, because of its very simple plant uncertainty. It will be shown (Sec. 6), how this technique can be used to establish "Quantitative Bounded Input-Bounded Output Stability".

Special attention should be paid to noise. In considering a noise signal added to a command input, note that it is filtered by the prefilter F, before entering into the feedback loop and applied to the nonlinear plant. The designer will find that F tends to be a low-pass filter with bandwidth approximately that of the desired closed-loop system response. Such a noise signal N, can be treated as a deterministic signal, $NF(s)$, and a reasonable number of samples are used. It will be seen that sensor noise is of much concern, because in feedback systems dealing with troublesome plants and disturbances, such noise signals tend to be amplified over a large frequency range. This is a very important cost of feedback, which may be too high, as to force modification of the desired benefit of feedback, or more complex feedback structures, such as use of internal sensors [16].

In Section 6, we analyze two techniques for designing stabilizing controllers in nonlinear QFT. The first technique is based on the adaptation of absolute stability results, while the second uses a equivalence disturbance approach.

Relation to Describing Function The Describing Function is also a restricted equivalence representation, but unlike the **Pe** technique, it is approximate even for the restricted situation. The input (to the nonlinear w) consists of a set of sinusoid inputs, over given frequency and amplitude ranges. Only the stead-state fundamental component of the output is used. Despite this narrow restriction, and its use only in an approximate manner, it is often very useful for stability analysis, especially for hard nonlinearities such as saturation, coulomb friction, hysteresis. The dual input describing function is less restrictive, in assuming the sum of two signals as input, and their respective component outputs.

The describing function technique can be easily incorporated in the nonlinear QFT framework ([13],[4]), giving in general less conservartive (but approximate) results than absolute stability criteria. All these approaches have been unified under a common QFT framework ([4]), allowing a very transparent tool for analyzing the differents types of stability restrictions and associated control effort.

The describing function can also be very useful for special situations, specially for oscillating adaptive systems, but it can not be used for finding system responses, where the P_e-technique is exact and rigorous, providing quantitative design for highly uncertain nonlinear plants.

3   Design Examples for the P_e–LTI Plant Equivalent Method

The following two examples only describe derivation of the LTI P_e sets. The balance of the design, whereby F, G are chosen to satisfy the specs. for P_e is postponed until the detailed LTI QFT method (for LTI Plants), is presented in Sec. 4.

3.1   Design Example 1 (LTV Plant)

Plant: $y = w(u)$: $\frac{dy(t)}{dt} + m(t)y(t) = k(t)u(t)$, where $m(t) = A + Be^{-at}$, and $56/k(t) = E + Je^{-bt}$.

Uncertainty: $A, E \in [1, 4]$, $B \in [2, 5]$, $a, b \in [.2, 1]$. The parameters $A, .., b$ though uncertain are fixed, but uncertain time-variations of m, k are present. This is a means of modelling uncertain time variations of plant parameters. Some $m(t), k(t)$ are shown in Figs. 7. The plant is time varying with uncertain rates of variation.

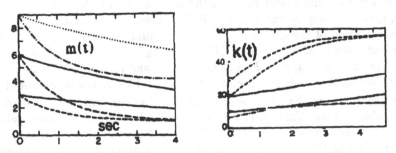

Fig. 7. Linear Time-varying (LTV) uncertain plant

Performance specifications: The closed-loop system to be designed is LTV because the plant is LTV. The differential equation relating system output to input has LTV terms. In response to a system impulse input at $t = t_x$, $r = \delta(t - t_x)$, the system output $y(t_x, \tau)$ is a function of both t_x and of $\tau = t - t_x$. $y(t_x, t) = 0$, for $t < t_x$. It is usually denoted as $y(t_x, \tau)$. In a LTI system, output y is a function of t only, so fixed $t_x = 0$ may be assumed. However, we can make the above system perform like a LTI system, by specifying an acceptable impulse response set **A**, whose elements are independent of t_x, so are "time invariant".

Six second-order models for the impulse response, $Y(s) = \omega_n/(s^2 + 2\xi\omega_n s + \omega_n^2$, were used to generate the OK set of Fig. 8: $(\omega_n, \xi) = (3, .6)$, $(4, 1)$, $(3, 1)$, $(4, 2)$, $(2.3, 1)$, $(6, 2)$. The integrals of the impulse responses are shown in Fig. 8-b, as we are accustomed to this form. The set **T** is obtained from their transforms with respect to $\tau = t - t_x$, shown in Fig. 8-c. Their envelopes are used for the upper and lower ω-domain bounds of the OK system transfer function set $\mathbf{T} = \{T(s)\}$.

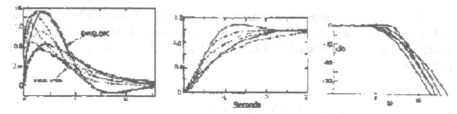

Fig. 8. Tolerances of System Response

The OK **T** set is independent of t_x, so it is easy to obtain the LTI Equivalent, $P(s) = Y(s)/U(s)$ which is a function of (a)the six plant parameters E, J, A, B, a, b, (b)the **A** parameters ξ, ω_n, and (c)the command input instant t_x. We used the six (ξ, ω_n) pairs of (b), the two extreme values of the 6 elements of (a) and eleven values of t_x : $(0, .2, .5, 1, 2, 5, 10, 20, 50, 100, 200)$, a total of $6 \times 26 \times 11 = 4224$ runs. This is easily programmed on the computer to plot directly the numerical values of the LTl equivalent $P(j\omega)$ for any desired ω, giving the equivalent LTI sets $\mathbf{P}(j\omega)$, needed for execution of QFT LTI design. Some of are shown in Fig. 9 for a number of ω values,and are called *plant templates*. A "nominal LTI plant" P_0 was chosen with nominal parameters: $A = E = 4$, $B = F = 5$, $a = b = 1$, and $t_x = 0$, $\xi = 1$, $\omega_n = 4$. We postpone further design to Sec. 4, where QFT LTI design is presented.

It is worth noting that if so desired, we could design in the performance sense, a Linear Time Varying (LTV) system, by specifying the acceptable output set **A** as a function of T: $\mathbf{A}(T)$, as in Fig. 10, with continuity of **A** with respect to T. Then, one could proceed as previously, at a sufficient number of discrete values of T.

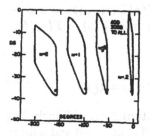
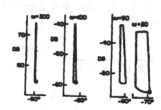

Fig. 9. LTI Equivalent Plant Sets (Templates) $\mathbf{P}(j\omega)$

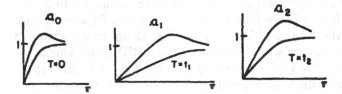

Fig. 10. LTV system type Specifications

3.2 Nonlinear Design Example with LTI Output Specifications

The plant relation is

$$W : k_1 y^2(t) + k_2 \frac{dy(t)}{dt} = u \tag{41}$$

where $k_1, k_2 \in [1, 10]$ are independently uncertain. The command input set **R** consists of steps of amplitude $M \in [1, 4]$. The bounds on the unit step response to be attained for all W are shown in Fig. 11-a. The bounds on T were obtained by using a second order model for the OK system "transfer function": $T(s) = \omega_n^2/(s^2 + 2\xi\omega_m s + \omega_n^2)$, with resulting OK (ξ, ω) range shown in Fig. 11-b, which is translated into bounds on $|T(j\omega)|$.

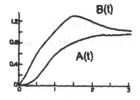
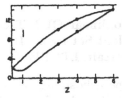

Fig. 11. Bounds on $T(s)$ parameters: ξ, ω_n

To generate equivalent LTI $\mathbf{P_e}$ use $1/P(s) = Us/Y(s)$, $u(t) = k_1 y^2 + k_2 dy/dt$, so $1/P = k_2 s + k_1 Lapl(y^2)/Y(s)$. Find y from $Y(s) = T(s)M/s$, square it to give

$$P(s) = \frac{(s + 2\xi\lambda)(s^2 + 4\xi\lambda s + 4\lambda^2)}{k_2 s(s + 2\xi\lambda)(s^2 + 4\xi\lambda s + 4\lambda^2) + 2\lambda^2 k_1 M(3s + 4\xi\lambda)} \tag{42}$$

which is a function of the plant's k_1, k_2 and of its output ξ, ω_n, M, including the sign of M (due to y^2 in W). It is easy to check $P(s)$ at $\omega = 0$ and infinity because these correspond to t at infinity and zero. As $t \to \infty$, $dy/dt \to 0$ so $u \to k_1 y^2 = k_1 M^2$, while $y \to M$, so the set $\{P(0) = Y(0)/U(0)\} = \{1/(k_1 M)\}$, and the set $P(0)$ consists of two subsets 180 apart due to $M > 0$ and $M < 0$. As $t \to 0$, $u \to k_2 y$ so $U(\infty) = k_2 s Y(\infty)$, giving $P(\infty) = 1/(k_2 s)$ with angle -90, for the entire set. The $P(\omega)$ in Fig. 12 are characteristic of a plant set, some (not all) of whose members have a RHP pole. This is the effect of $k_1 y^2$ in $k_2 dy/dt + k_1 y^2 = u$ when y and dy/dt are < 0, but $k_1 y^2$ is > 0. In a LTI system with term $k_1 y$ rather than $k_1 y^2$, both terms are then negative during the initial (small t) part of the response. Analytically, it is evident from the coefficients of s and of s^0 in the denominator of $P(s)$ above: $2\omega_n^2[(4\xi\omega_n k_2 + 3k_1 M)s + 4k_1 M\xi\omega_n]$. When $M < 0$, the last term is < 0 and when $3k_1\omega_n|M| > 4\xi\omega_n k_2$ both coefficients are < 0. It turns out that the P instability is likeliest at $M < 0$, $|M|k_1$ maximum, k_2 and ω_n minimum. The balance of the design follows usual QFT LTI principies. Thus, the LTI Plant Equivalent can be unstable, even if the nonlinear original does not appear to be so.

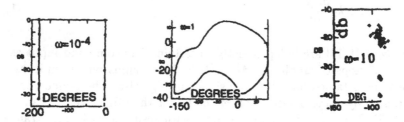

Fig. 12. LTI $\mathbf{P_{eWA}}$

3.3   A Second Nonlinear QFT Technique: Replacement of Nonlinear Plant Set by an Equivalent Disturbance Set and a Simple, Uncertain LTI Plant

This technique is especially suited for handling disturbance signals entering into any point in the plant, but for simplicity, assume it enters at the plant input as in Fig. 13.

In Fig. 13-a, $\mathbf{D} = \{d\}$ is the set of disturbances to be attenuated, to achieve output y in $\mathbf{A} = \{a\}$, an acceptable set, for all plants in set $\mathbf{W} = \{w\}$. The bounds defined on a in $\mathbf{A}$ are as follows. Let the differential equations w in $\mathbf{W}$ be of order n. There must be assigned bounds on the derivatives of $a(t)$ from zero order to order $n - 1$. These are performance bounds, and will generally be small. $\mathbf{D}$ is a set of functions of bounded variation, with magnitude bound, so certainly includes all realistic signals (see [19] for mathematical details). We work backwards from the output $y = a \in \mathbf{A}$. The input to the plant is $x = v(y)$, $v = w^{-1}$, $y = w(x)$. We divide $x(t)$ into two parts: $x_1(t), x_2(t)$. The first, x_1, is associated with a "linear" part of x, and the second, x_2, with all of the nonlinear part and possibly some linear part.

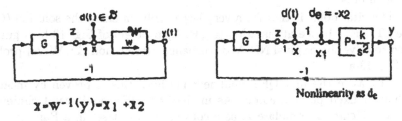

$$x = w^{-1}(y) = x_1 + x_2$$

Fig. 13. Nonlinear System and Equiv. Disturbance-LTI System

We demonstrate with an example:

$$w : \ddot{y} + A^2 \dot{y} + B y^3 = kx \qquad (43)$$

Let $kx_1 = \ddot{y}, kx_2 = A^2 \dot{y} + By^3$. Another possibility is $kx_1 = \ddot{y} + E\dot{y} + Jy, kx_2 = A^2\dot{y} + By^3 - E\dot{y} + Jy$. There is an infinitude of ways to split x between x_1, x_2. We prefer the first. Let $P(s) = k_1/s^2$ and replace Fig. 13-a by 13-b, in which P is a LTI plant and $d_e = -x_2$ is an "equivalent disturbance", in order that x, d and y are the same in Fig. 13. The two figures are then "equivalent" with respect to z, d, y; $x_1 = x + d_e$, so *in general*

$$d_e = x_1 - x = -x_2 \qquad (44)$$

here $d_e = -(A^2\dot{y} + By^3)/k$. From the viewpoint of d_e as input and y as output, there is no difference between Figs. 13. It is impossible to distinguish between them, if the only measurements allowed are of z, d, y. However, the design problems are radically different for the two. In Fig. 13-a one has a *nonlinear time-varying (NLTV) problem*, which can be horrendous, especially for the kind of uncertain plant, disturbances and performance demanded below. In 13-b, one has a LTI plant with only k uncertainty, and an additional disturbance d_e, which is a function of the plant output y and the nonlinear plant parameters k, A, B. The performance specifications demand that the resulting output $y(t)$ is a member of an acceptable set $\mathbf{A} = \{a\}$, so each combination of $(a(t), k, A, B)$ generates a different $d_e(t)$. Thus, the pair of sets $(\mathbf{A}, \mathbf{W})$ generate a set $\mathbf{d_e}$ of equivalent disturbances. This is a well-defined disturbance set if we properly define $\mathbf{A}$ and $\mathbf{W}$. The external disturbance d is in general, a member of a set $\mathbf{D}$. The two sets $\mathbf{D}$ and $\mathbf{d_e}$ are summed to give $\mathbf{d_t} = \{d_t = d + d_e\}$, the total disturbance set. So we see that Fig. 13-b constitutes an ordinary LTI disturbance attenuation problem with only gain k uncertainty in the LTI plant P, subjected to a disturbance d_t which can be any member of the set $\mathbf{d_t}$. *This is a standard QFT LTI disturbance attenuation problem*: choose G to guarantee that for all $P \in \mathbf{P}$, $d_t \in \mathbf{d_t}$, the output y is a member of the acceptable set $\mathbf{A}$. Obviously, this Fig. 13-b problem is far, far easier than the NLTV uncertainty problem of Fig. 13-a. This problem is solvable because the d_t of $\mathbf{d_t}$ have the same form of bounds as the a of $\mathbf{A}$.

The vital point is that for a very large problem class, the solution (G) of the LTI disturbance attenuation $(\mathbf{d_t}, \mathbf{P}, \mathbf{A})$ problem of Fig. 13-b, is guaranteed to solve the original NLTV disturbance attenuation $(\mathbf{D}, \mathbf{W}, \mathbf{A})$ problem of Fig. 13-a.

Just as in the first QFT nonlinear method, this is proven by means of Schauder fixed point theorem. As in the first QFT nonlinear technique, it is not difficult to formulate $\mathbf{A}$ as a convex, compact set in a Banach Space. Define, in Fig. 13-b, $H(s) = P/(1+PG) = Lapl.(h(t))$. For fixed d_t, w, define the potential Schauder mapping over $\mathbf{A}$:

$$\Phi(a) = h * d_t, \tag{45}$$

with $d_t = d - x_2$, x_2 a function of a and $*$ indicating convolution. In a properly designed system, this is a continuous mapping, and to satisfy the equivalent L TI problem, Φ must map the set $\mathbf{A}$ into itself, giving a fixed point $a_q = h * (d - x_{2q})$, which corresponds in transform language (with *Lapl.* indicating Laplace transform)to $[(1 + PG)/P]Lapl.(a_q) = d_t$. The left side is: $(1/P)Lapl.(a_q) + GLapl.(a_q)$, which in the time domain is, $x_{1q} - z$, of Fig. 13-b, giving, $x_{1q} = z + d_t = x$ of Fig. 13-b, and proves that x_q is a solution of the nonlinear Fig. 13-a. If each element w, G is 1:1, then the solution is unique. An important point is that if the nonlinear plant w is of order n, then bounds on x_2 from zero to $n-1$ order must exist, in order that d_e exists, which has been done. But one more condition must be satisfied in order that $\mathbf{A} = \{a\}$ is a Banach space, [19], the nth derivative of $a(t)$ must be bounded, the precise value unimportant. This can be guaranteed, because of the order of the plant.

It is essential that the LTI problem be solved, for only then does Φ map $\mathbf{A}$ into itself. In practice, the designer need not worry over the details of the proof. The precautions needed to make $\mathbf{A}$ a convex, compact set in a Banach space, and for Φ to map $\mathbf{A}$ into itself, are normally automatically satisfied in the QFT LTI disturbance attenuation technique.

3.4   Example

Problem Statement: There is a set of time-varying nonlinear plants $\mathbf{W} = \{w\}$, $y = w(u)$,

$$\ddot{y} + \frac{Bt+1}{t+1}\dot{y}^m y^2 + (1 + E\sin\omega_0 t)y^3 = ku \qquad (46)$$

Uncertainty: $B \in [-5, 5]$, $k \in [10, 40]$, $\omega_0 \in [1, 2]$, $m \in [.5, 1.5]$, $E \in [-2, 10]$. Clearly w is unstable to a large input class for a significant plant parameter range.

Specifications: The disturbance set $\mathbf{D} = \{d\}$ at plant input is defined by $\mathbf{D} = \{|d| \leq D_O = 100\}$. Any realistic disturbance belongs to a class of such form, i.e. of bounded magnitude.

The Acceptable output set: $\mathbf{A} = \{a\}$ is the set of continuous functions defined on t in $(0, \infty)$, with $|a(t)| < a_0 = 0.2$, and $|da(t)/dt| < a_1 = 5$. $\mathbf{A}$ may appear overdefined, as in many cases bounds on y may suffice. The reason for its inclusion will soon be seen.

The LTI Equivalent: Replace the time-varying, nonlinear Fig. 13-a by the LTI input-output equivalent of Fig. 13-b as follows. Let $x = w^{-1}y = x_l + x_n$ where $x_i = P^{-1}y$ is associated with the LTI P of Fig. 13-b, and $x_n = x - x_l$ is the balance. Here $P(s) = k/s^2$ (i.e. $kx_l = d^2y/dt^2$). Then Fig. 13-b is an input-output equivalent of Fig. 13-a if $d_e = -x_n$. Replace y by a of $\mathbf{A}$ to generate $\mathbf{d_e}$, giving d_e a function of w and a:

$$-kd_e = kx_n = \frac{Bt+1}{t+1}\dot{a}^m a^2 + (1 + E\sin\omega_0 t)a^3 \qquad (47)$$

which gives the properties of d_e. Since the relation between d and y is not known, a_0, a_1 are used for a, da/dt in (13), giving $k|d_e|_{max} = 5 \times 5^1.5 \times .2^2 + (1 + 10) \times .2^3 = 2.32$, so $|d_e|_{max} = .23 \ll |d|_{max} = 100$, and will be ignored, letting the total $|d_t| \leq 100$. *Thus, bounds on both y and dy/dt were needed in order to derive the bound on d_e.* This explains the bound on da/dt. We are confining ourselves in these notes to NLTV differential equations wherein the leading derivative in the output y and the leading derivative in input u, appear linearly, as in this example (this class can be enlarged [19]). Then P in the LTI equivalent plant of Fig. 13-b must involve the highest derivatives of y and u. For this class d_e can be, in general, a function of all the lower states. Therefore bounds on all these lower states are needed in order to derive the bound on d_e. This explains the bounds on a and da/dt in this example. The formal proof of this technique requires much more precise definition of $\mathbf{A}$ (see previous reference). But these need not concern the practical designer. He needs only to set bounds on all the states which appear in d_e, in order to be able to define d_e, even though he may not really care what the values are of the higher states. The lower the bounds, the smaller is $max(|d_e|)$, so it would appear desirable to assign small values to a_0, a_1,..., but remember that it is essential to solve the LTI problem of Fig. 13-b, i.e. guarantee that the output y satisfies the bounds assigned on **A**. The more stringent these bounds are, the larger the magnitude and bandwidth of G in Fig. 13-b, will have to be. Of course, some of the bounds are dictated -after all, disturbance attenuation is the reason for applying feedback around the NLTV plant, but for those states of which there is no concern, the designer should use the highest bounds which do not significantly affect the magnitude of the total d.

3.5   Solution of the Equivalent LTI Disturbance Attenuation Problem

It is imperative the LTI problem is solved, for only then is it guaranteed that the same G solves Fig. 13-a. Part of this solution, guaranteeing that $|y| < a_0 = .2$ is quite easy for this class of problems. In Fig. 13-b,

$$Y = \frac{D_t P}{1 + PG} = D_t H \tag{48}$$

where $H = P/(1 + PG)$, $d_t = d + d_e$, so $y = d_t * h$. Suppose $h(t) = Lapl.^{-1}(H(s))$ is as shown in Fig. 14. What input $d_x(t)$, bounded by D_0, gives an extreme output? By considering the mechanics of graphical convolution, it is obvious that $d_x(-t)$ in Fig. 14 is the one. It is clear that if $J = \int_0^\infty h(t)dt$ is fixed, then the extreme output is minimized by having $h(t) \geq 0$ for all t (or ≤ 0 for all t), i.e., the step response of $h(t)$ should have no overshoot. If so, the peak output is JD_0, with

$$J = H(0) = \frac{P(0)}{1 + P(0)G(0)} = \frac{1}{G(0)} \tag{49}$$

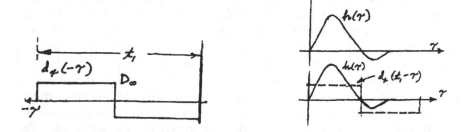

Fig. 14. The Convolution of H and d_t gives the output

Here the extreme d_x is a step function of magnitude 100, requiring $100J < .2$, so $G(0) \geq 500$ if there is no overshoot in the step response of $P/(1+PG)$. As a safety factor, we take $G(0) = 750$. Of course $L(s) = P(s)G(s)$ must be shaped for stability over all **P** (here only k uncertainty), making the problem much easier than the usual shaping of L. The technique for this purpose is presented later (Sec. 5), so only the final result is given. Nominal $P_0(s) = 10/s^2$ is used with $G = 750 \times 1700^2 \frac{.0213s+1}{s^2+2040s+1700^2}$. Nichols plots of L for $k = 10, 40$ are shown in Fig. 15. We have to also guarantee that $|dy/dt| < a_1 = 5$. Simulations (LTI) over the k uncertainty range verified that this was indeed so. If the result had been different, we could simply increase the bandwidth of G, and if necessary its magnitude, to achieve this. Or it might be simpler to increase the value of a_1, because its effect on total d is slight. Some cut and try may be required for a_1 control. In the case of a fourth order NLTV plant, we would have also bounds on the second and third derivatives of y to satisfy in the LTI design. It is guaranteed (if the LTI plant is minimum-phase and satisfies the conditions applicable to LTI plants) that such bounds can always be satisfied (see above reference).

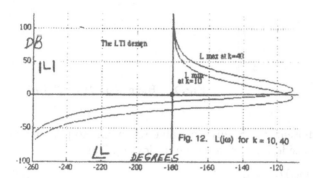

Fig. 12. L(jω) for k = 10, 40

Fig. 15. The LTI Design m the NiChols Chart

Simulations. A large number of nonlinear simulations of the original NLTV problem were done. Despite much effort we could not find inputs (in the defined class) which gave outputs that violated the performance specs. Steps were applied at different time instants. It did not make much difference in the response. Disturbances $d = M\sin(\omega_d t), (Mt)exp^{-.5t}$, with $M = 150, 500$, and the same ones applied at $t = 5$, gave very similar outputs. The reader is encouraged to experiment with all kinds of esoteric disturbances, satisfying $|d|_{max} < 100$. He should find it is impossible to find any resulting in $|y| > .2$ for any value of $t \in (0, \infty)$.

4   QFT LTI Design for the Tracking Problem

Fig. 1, redrawn here as Fig. 16, is the basic structure. The LTI Plant P is uncertain, known only to be any member of a set **P**. The designer is free to choose LTI prefilter F and loop compensator G, to ensure that the system transfer function $T = FPG/(1+PG)$ satisfies assigned specifications. Those on magnitude $M(\omega) = |T(j\omega)|$ suffice in minimum-phase systems, for example those in Fig. 17, in decibel units $(20\log_{10}(M))$. This gives an OK set **M** for $|T(j\omega)|$, to be achieved for all $P \in \mathbf{P}$, with $b(\omega), a(\omega)$ the upper and lower bounds on **M**.

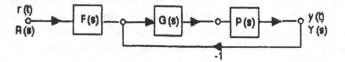

Fig. 16. LTI TDF (Two Degrees of Freedom) Structure

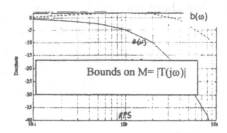

Fig. 17. Bounds on $M = |T(j\omega)|$

F and G are to be chosen. It is highly desirable to be able to pick one at a time, easily done by considering the *variation* in $\log|T|$. Thus,

$$\Delta \log|T(j\omega)| = \Delta |L(j\omega)/(1 + L(j\omega)|\tag{50}$$

thereby eliminating F. The purpose of G in $L = PG$ is to ensure that the variation $\delta(\omega) = b(\omega) - a(\omega)$, allowed in $M(\omega) = |T(j\omega)|$ in Fig. 17, is not exceeded at each ω, so it is worth making a plot of $\delta(\omega)$, in Fig. 18.

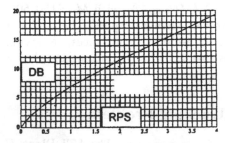

Fig. 18. $\delta(\omega) = b(\omega) - a(\omega)$

The Logarithmic Complex Plane (Nichols chart, Fig. 19) is a highly transparent, excellent medium for visualizing the design procedure for the above purpose. It consists of loci of constant $M = |T|$, and $Arg(T)$ in the logarithmic L plane: $Angle(L)$ in degrees, $Magnitude(L)$ in db ($20 \log_{10} |L|$). $Arg(T)$

is not needed. It is easy to see that in certain regions, very large changes in L cause very little change in M. For example, let $|L|$ range from 15 to 75 db (factor of 1000 arithmetic) and $Angle(L)$ from -90 to -270; the resulting maximum change in M is only -2 db. The designer soon obtains intimate understanding of the relations between L and variation of M.

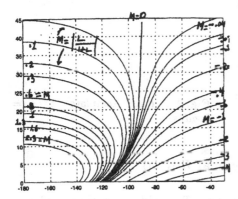

Fig. 19. Nichols Chart

4.1   STEP 1: Plant Templates (P) and Loop Templates (L = PG)

Display the Plant Uncertainty on the Nichols Chart. For example, $\mathbf{P}(s) = \{k/(As^2 + Bs + C), k \in [1,4], A \in [1,4], B \in [-2,2], C \in [1,6.25]\}$.

Detailed design is done in the frequency domain. At each ω, $P(j\omega)$ is a complex number. Because of plant uncertainty, there is a *set* of plants, so at each ω, we get a set of complex numbers, which we call the *plant template* $\mathbf{P}(j\omega)$. For example, at $\omega = 3rad/s$, the plant template is given by the set of complex numbers shown in Fig. 20: $\mathbf{P}(j3) = \{k/(-9A + j3B + C), k \in [1,4], A \in [1,4], B \in [-2,2], C \in [1,6.25]\}$.

Fig. 18 allows variation $\delta(3) = 15.3$ db. in $M(3) = |T(j3)|$. Find the bounds on $L(j3)$ to assure this. Since $L = PG$ varies with P, it is convenient to choose a nominal P_N, giving a nominal $L_N = P_N G$, for this purpose. Values used here for nominal plant are: $k = 1$, $A = 4$, $B = 2$, $C = 6.25$, so $1/P_N = 4s^2 + 2s + 6.25$, for $s = j3$.

The nominal P_N (which corresponds to the nominal $L_N = P_N G$ value), is marked N in Fig. 20. Note that the template of $\mathbf{L} = \mathbf{P}G$, is isometric to the template of $\mathbf{P}$: The template $\mathbf{L}(j\omega)$ is obtained by shifting the plant template $\mathbf{P}(j\omega)$, by $Angle(G(j\omega))$ horizontally, by $|G(j\omega)|$ vertically. Its shape and size is the same as the plant template $\mathbf{P}$.

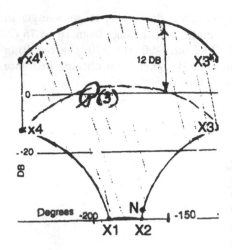

Fig. 20. Plant template at $\omega = 3rad/s$

4.2   STEP 2: Bounds on the Nominal $L_N(j\omega)$

The nominal loop $L_N(j\omega)$ is our design tool. By making it sufficiently large, a huge plant template can be forced into a very small $T(j\omega)$ template, i.e the loop transmission is the means by which large plant uncertainty can be translated into small system transfer function uncertainty. But large L is expensive (Sec. 5), so we seek the smallest possible L_N which satisfies the specifications. Simply manipulate the plant = loop template on the Nichols chart, until (at any fixed angle), the minimum $|L_N|$ is found which satisfies the specs. It is seen that this $|L_N|_{min}$ is a function of $Angle(L_N)$. The resulting curve is called the Bound $B(j\omega)$ on $L_N(j\omega)$. $B(\omega)$ for $\omega = .1, .55, 1, 2, 3, 5, 10, 50$ are shown in the Nichols Chart of Fig. 23. Of course, the computer can be programmed to find the bounds. It is a good idea for the novice to do a few by hand. We indicate one possible procedure by finding. A point on $B(3)$, whose template is shown in Fig. 20, and for which $\delta(3) = 15.3$ db in Fig. 18.

A point on $B(3)$: Pick any value of angle for the bound on $L_N(j3)$, say -20. Set the point N of the template at -20, at location *Trial 1* in Fig. 21. Check the M loci and thereby ascertain that the variation is from -8 db to -40 db $= 32$ db, which greatly exceeds the allowed 15.3 db variation of Fig. 18. So for *Trial 2*, we move the L Template much higher, as shown. Now the M variation is from 0.08 db to -2.9 db $= -3$ db, much less than the permitted 15.3 db. In the *Trial 3*, the variation in M is from -15.3 to .1, very close to that permitted, which we use: Point N of Trial 3, in the Nichols chart, is a point on $B(3)$. Thus, if $Angle(L_N(j3)) = -20$ degrees, then $|L_N(j3)|$ must be on or above this point N. We next try a different angle for $L_N(j3)$, say -50 degrees, and repeat the above search, to find the minimum $|L_N(j3)|$ at angle of -50 degrees.

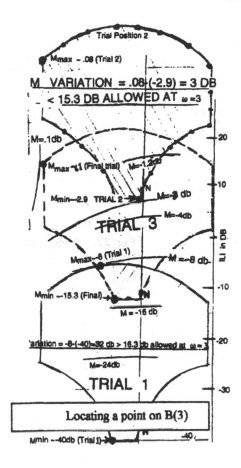

Fig. 21. Locating a point on $B(3)$

Some plant templates $\mathbf{P}(j\omega)$ are shown in Fig. 22. For $\omega = 2$ it is semi-infinite in magnitude and is 360 wide, typical of a plant whose poles wander from left-half plane into right half-plane, crossing the imaginary axis. At $\omega = 0.2$, $\mathbf{P}(j0.2)$ has two separate parts.

4.3   STEP 3: The Universal High-Frequency Bound UH(ωB)

For $\omega \geq 100$, the templates $\mathbf{P}(j\omega)$ are almost vertical lines, because $P(s)$ is close to k/s^2 at such large s, so P uncertainty is closely that of k only. *In this frequency range the allowed M variation is larger than the plant uncertainty.* It is in fact essential that this be so at large enough ω, otherwise $L(j\omega)$ is not allowed to $\to 0(-\infty db)$ as $\omega \to \infty$. In fact, Bode (the Feedback Amplifier pioneer), proved long ago, that *on the average (over the arithmetic ω range), the feedback benefit (sensitivity reduction in db),is zero in any practical feedback system (one whose L has an excess of poles over zeros $\geq$ 2).* This means that if sensitivity reduction is obtained in one frequency range, it must be

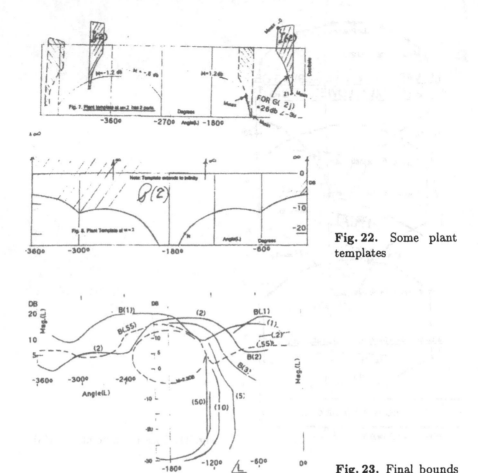

Fig. 22. Some plant templates

Fig. 23. Final bounds on L_N

balanced by some other range, in which the sensitivity > 1, i.e. worse than in an open loop system. In the Two Degrees of Freedom (TDF) system (Sec. 2), we have at least control of the frequency range in which $|S| > 1$. We can postpone it to high enough frequency, in which it is of no matter, because $|T(j\omega)|$ is so small there. Say it is from -30 db to -76 db (46 db range), even though the plant uncertainty is only 40 db. The equivalent in the time domain is that at small t the step response of the plant can vary, say between 10^{-6} and 10^{-4} due to plant uncertainty. But for the closed loop system, the variation is twice as much, say between $.5 \times 10^{-7}$ and 10^{-5}. However, in the One Degree of Freedom (ODF) system, we have seen (Fig. 2), that we have no control over the frequency range where $|Sens.| > 1$. It is determined by the system transfer function, and tends to occur in a significant frequency range.

But sensitivity to the plant is not the only matter of concern. There is also the effect of disturbances D, say entering at u in Fig. 1, the plant input, giving system output $DP/(1 + L)$, which is precisely DP multiplied by the sensitivity, which may be too large to bear; certainly so if the disturbance is at the plant output in Fig. 1, for it is then $D/(I+L)$. We have been assuming that here disturbances are not the major problem (see Sec. 7), where they are assumed to be the major problem). But we see that they cannot be totally ignored. We must therefore add an additional constraint in this "higher" frequency range, which we can recognize by the specifications allowing us to have L_N increasingly approach -1, where the sensitivity $S = 1/(1 + L)$ becomes very large. This new constraint, is called the γ constraint.

γ **constraint:** $|L/(1 + L)| \leq \gamma$ Typical γ value is 2.3 db, corresponding to a damping factor of 0.707 for a second order system. A more conservative constraint is that $|S| = |1/(1 + L)|$ be less than a chosen γ value. At high enough frequency, the γ constraint dominates and determines the bounds on nominal L_N. The combination of the γ constraint, and the fact that in the higher range the plant template is a fixed vertical line (because P approach k/s^n, so there is only the uncertainty of k), leads to *universal high frequency bound (UHωB)*: find the γ magnitude locus in the Nichols Chart, i.e. the locus for which $|L/(1+L)| = \gamma$, for example, in Fig. 24, the locus $LKJUVXW'$ for $\gamma = 2.3$ db. Project this locus downward by the amount of the k uncertainty, which is the length of the verlical line XX'. This gives the entire UHωB as $LKJW'HWXVUOL$ (proof is left to the reader). Part of this Universal Boundary is shown in Fig. 23, for our present problem. Some prefer to use the sensitivity function $|S| = |1/(1 + L)| < \gamma$ as the γ constraint.

4.4   STEP 4: Find $L_N(j\omega)$ which Satisfies its Bounds $B(\omega)$

Computer programs have been written for this purpose. But it is a good idea for the beginner to do a few problems by hand. We only present the results here for this example. There was chosen (Figs. 25-26)

$$L_N(s) = \frac{9.5 \times 8.5 \times 280^2}{14} \frac{s + 14}{s(s + 8.5)(s^2 + 1.2 \times 280s + 280^2)} \tag{51}$$

The other L_i shown in Figs. 26 were chosen later (Sec. 5) to illustrate the *Cost of Feedback*, which we will discuss later. All of them satisfy the bounds of Fig. 23, but they differ considerably in complexity of the resulting loop compensator G, and in the associated *Cost of Feedback*, which is a very important practical concept to be discussed in Section E.

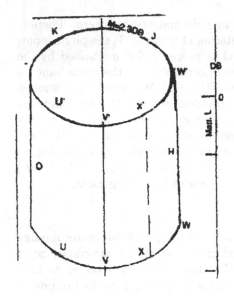

Fig. 24. Universal High-Frequency Bound

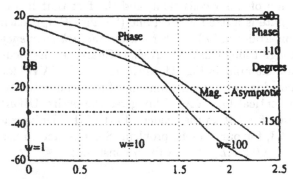

Fig. 25. Bode plots of L_N

4.5   Design of Feedback Controller $G(s)$ and Prefilter $F(s)$

$G(s) = L_N(s)/P_N(s)$ is available from L_N, giving

$$G(s) = \frac{9.5 \times 8.5 \times 280^2}{14} \frac{(s+14)(4s^2 + 2s + 6.25)}{s(s+8.5)(s^2 + 1.2 \times 280s + 280^2)} \tag{52}$$

which has an excess of one pole over zeros. Thus, we have chosen G, (associated with $L_N = P_N G$), but this is only one of the two degrees of freedom G has been chosen to assure that the allowed *variation* in $|T(j\omega)|$, $M_{max} - M_{min} = \delta(\omega)$ of Fig. 18, is satisfied for all ω. But this is insufficient. For example, suppose that at some ω_x, the specification dictate $M_{max} = 2$ db, $M_{min} = -5$ db, giving $\delta(\omega_x) = 7$ db, and suppose that $G(j\omega_x)$ is such that at this frequency, $M_{max} = -3$ db, $M_{min} = -9$ db, with variation of

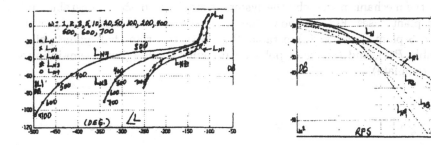

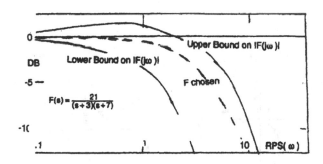

Fig. 26. Other designs to illustrate *Cost of Feedback*

$6 < 7$ db. Even though the variation specification has been satisfied, the M specification has not been satisfied. Since $|T| = |F|M$, the function of F is to fit the actual OK variation inside the M specification. This is achieved here by assigning bounds on $|F(j\omega_\tau)|$: minimum value = 4 db = $-5 - (-9)$ db, maximum value = 5 db = $2 - (-3)$ db. Note that this free range for $|F|$ of $5 - 4 = 1$ db is precisely the amount of overdesign of $G(j\omega_x)$, which gave a variation of 6 db, when 7 db variation was allowed by the specifications.

In this example, using L_N, a few of the resulting bounds of $|F(j\omega)|$ were $\omega = 0.1 : -.4$ to $.4$; $\omega = .5 : -2.3$ to 1.4; $\omega = 1 : -3.6$ to $.9$; $\omega = 2 : -9.2$ to $-.3$; $\omega = 5 : -25$ to -4. This pair of bounds on $|F(j\omega)|$, is shown in Fig. 27. $F(s) = \frac{21}{(s+3)(s+7)}$ is a simple function which satisfies these bounds.

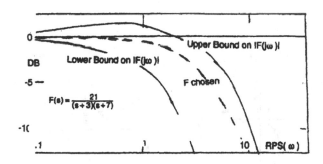

Fig. 27. Bounds on $|F(j\omega)|$

4.6 Simulations

Step responses for over 300 cases were checked. They all satisfied the specifications. Note that the plant is unstable over part of its parameter range. This has no effect on performance. Fig. 28-a shows unit step responses for 81 unstable cases, Fig. 28-b for 81 stable cases, Fig. 28-c for 27 cases of a pair of plant poles on imaginary axis. They are indistinguishable. It is interesting to see the mechanism whereby the system is stable for both the unstable and stable plant cases. Fig. 29 shows the Nyquist encirclements which prevail for the 3 plant cases of: (a)unstable $P(s)$ with two right half plane poles, (b)stable $P(s)$, and (c)pair of poles on imaginary axis (giving semi-infinite plant template).

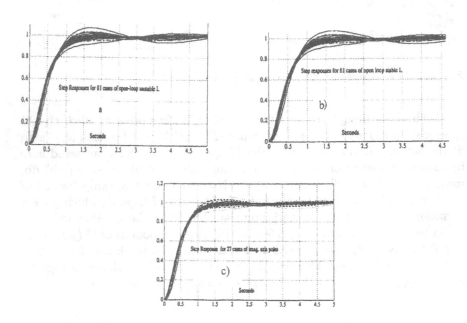

Fig. 28. Step responses

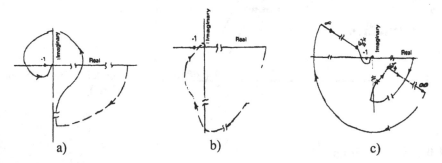

Fig. 29. Nyquist Plots

Figs. 30-a,b,c shows simulations of the same 3 classes of $L(j(\omega)$ on the Nichols Chart. Fig. 31-a presents the system output due to unit disturbances entering directly at the system output, so output $y(0) = d(0)$, because no practical feedback can be instantaneous, as it would require infinite loop bandwidth. The larger the loop bandwidth, the faster the feedback acts to attenuate the disturbance. Fig. 31-b presents the effect of unit step d entering at the plant input. Note its small scale. In this TDF structure of Fig. 16, the available feedback benefits are: (1)to decrease the sensitivity of tracking response to plant uncertainty, and (2)attenuation of disturbances entering the plant. Both of these benefits are entirely determined by the single $L(s)$, so cannot be realized independently. One or the other may dominate at any ω value. So, both problems should be quantitatively considered. We will present QFT Disturbance Attenuation in Sec. 7; only note here that it is fairly simple when done in ω domain.

5 Cost of Feedback: Sensor Noise Effect, Loop Bandwidth, ω cut-off

The tremendous reduction of the effect of plant uncertainty, achievable by large L_N, may tempt the designer to be extravagant in choosing L_N. But there is a price to pay. In Fig. 32, consider the effect of sensor (at plant output) noise N, at plant input U_n. Although noise is a random process, we can treat it adequately for our purpose as a deterministic signal. The effect is

$$T_n = \frac{U_n}{N} = \frac{G}{1+L} \approx \frac{G}{L} = \frac{G}{PG} = \frac{1}{P} \tag{53}$$

in range in which $|L| \gg 1$. So in this ω range, we have no control over T_n. This is generally the low frequency range contained within the bandwidth of $T(j\omega)$. It is not the ω range of importance for this sensor noise problem. The problem is in the ω range in which $|L|$ is small, so $1 + L \approx 1$, for then

$$T_n \approx G = \frac{L}{P} \tag{54}$$

In this range, the sensor noise is amplified by the amount in which $|L| > |P|$. Both L and P tend to be small in this range, but $G = L/P$ tends to be large. We illustrate the importance of the Sensor Noise Amplification by means of various design examples.

Fig. 33 presents the results of 4 different designs for the same problem, satisfying the same bounds. G_1 has one zero, one pole (1,1) -impractical. $G_2 = (1,2)$. $G_3 = (5,8)$. $G_c = (12,15)$. Which is the better design? If complexity of G is the only criterion, then $G2$ is the best design. But over nearly all of the frequency range shown in Fig. 33, the sensor noise is highly amplified, with maximum amplification of 130 db (arithmetic $10^{6.4}$, over a million). To find rms effect of the sensor noise, if it is white and Gaussian, this must be squared, and integrated over the *arithmetic* ω range, and its square root taken. The result is enormous. The design would be impractical, unless the sensor noise

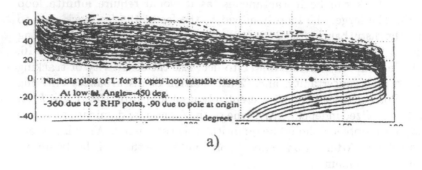

a)

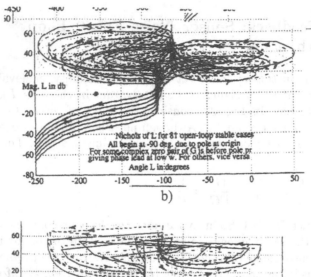

b)

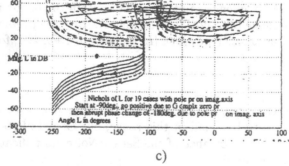

c)

Fig. 30. Nichols plots

is very small. By adding 4 more zeros, and 6 more poles to L, one obtains the very much smaller G_3, whose maximum arithmetic value is $\approx 10^3$, a very great improvement over G_2, in terms of sensor noise effect. If we use the much more complex *conditionally stable* design with additional 7 zeros and 7 poles, there is much more sensor noise reduction, maximum < 30, and over frequency range of 100 rad/s, instead of 10,000 rad/s. The difference is enormous. So the

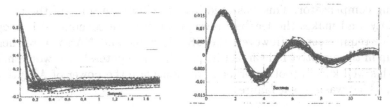

Fig. 31. Effect at output of plant disturbance:a)at output, b)at input

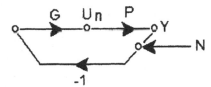

Fig. 32. Sensor noise effect at plant input

answer to the question "which is the better design?" is, "that it depends on the sensor noise". QFT is very transparent and highly visible, in this tradeoff available to the designer between compensator complexity and sensor noise effect.

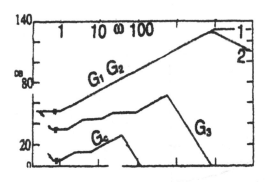

Fig. 33. Comparison of more economic but more complex designs

Fig. 34 ([16], p. 188) presents another example, involving the X15, first explorer of upper space, in which the speed could vary between mach 0.1 and mach 6, and altitude from ground to 100,000 feet. The uncertainty is 60 db (1000) over the entire frequency range. The design is quite complex. In the first QFT design G_1 has 2 zeros, 5 poles. In the second G_2 has 7 zeros, 10 poles. The third and most economical design with respect to sensor noise effect has G_3 with only 3 zeros, 6 poles, but this is due to use of *scheduling*, explained as follows: The factors causing the plant uncertainty can be measured by air data measurements which can be related to the plant parameters, and much of their variation cancelled by adjustments of a scheduling compensator. This was used here to eliminate some of the plant uncertainty, and makes the feedback design much more economical in loop bandwidth requirements. However, the US Air Force and NASA were then caught up in the adaptive fad (in fact its principal promoters), and would not exploit this well known tool. in order to demonstrate the power of Adaptive systems, which was a great failure.

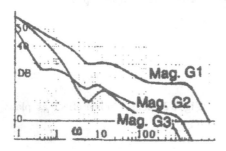

Fig. 34. X-15 designs

Finally, we return to Fig. 35, in which the Trade-Off between design complexity and sensor noise effect is again displayed. The first design L_N has 1 zero, 4 poles; L_{N2} has 4 zeros, 7 poles; L_{N3} has 5 zeros, 9 poles; and L_{N4} has 5 zeros, 11 poles.

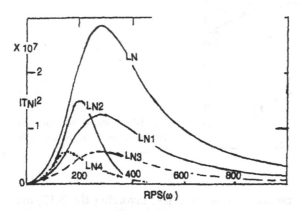

Fig. 35. Trade-off between design complexity and bandwidth economy

5.1   Graphical Ddisplay of the Sensor Noise Effect. Loop Bandwidth and cut–off Frequency

A second important cost of feedback, is its bandwidth multiplication effect, which is illustrated by the above LTI problem of Figs. 16-31. We use here the concept of bandwidth to indicate the ω range over which the function is effective, and over which we must be certain that it truly has its theoretical value. If we use the -5 db point, then from Fig. 17, the bandwidth range is from 0 to ~ 8 rad/s. One is not much concerned with the behavior of a system a few octaves past its bandwidth, if we know it is decreasing in magnitude.

Let us now consider the "bandwidth" of the loop transmission L, which is the basic function which controls the system sensitivity to plant uncertainty. Over what frequency range must we be certain of the validity of our theoretic paper design value of L? For we must not forget the fragility of theoretical models. A resistor R is the model of a piece of metal wire, but at high enough frequency, its model becomes more complex, eventually the partial differential equations of Maxwell must be used. We might happily concentrate on mathematical optimization, leading us to operate at high frequencies at which the presumed model R is far from valid.

To answer this vital question, we should first seek the optimum L function, in order to be fair. The optimum is defined as follows (Quantitative Feedback Design has led us to bounds on L_N, the nominal loop function): suppose we choose a specific excess e_L of poles over zeros for $L_N = P_N G$, $e_L = e_P + e_G$, the excess of poles over zeros of the plant P and of the compensator G. The value of e_G should at least be 1, because any practical transfer function must go to zero at infinity. Trade-off considerations between complexity of G and sensor noise effect (previously discussed) determine our choice of e_G. The larger its value, the smaller the sensor noise effect. Assume e_G has been accordingly chosen, giving e_L. Hence, at high enough ω, $L_N(s) \to k/s^{e_L}$ at large s.

L_{opt}, the optimum L_N, is defined as that which satisfies the Bounds on L_N with minimum value of k. It has been shown ([16], Sec. 10.5), that for a large class of bounds, L_{opt} lies on its Bounds at all ω values, that such an L_{opt} exists (if necessary, only in the limit), and is unique. This information is valuable to the designer, as it tells him what to aim for, and how far he is from optimum design. Furthermore, thanks to the Hilbert-Bode integrals relating magnitude and phase of analytic functions, the designer can at any point in his Loop Shaping, find the improvement possible by further sharpening of his design, and judge whether is worth the effort and the resulting greater Compensator (G) complexity. As deduced by QFT, the typical form of $|L_{opt}(j\omega)|$ vs ω, in the crucial high frequency range (which is the important range for cost of feedback), is shown in Fig. 36 as $|L|$.

It is encouraging that *H. W. Bode [11], the Master of Feedback Amplifier theory, who was deeply concerned with the loop bandwidth and cut-off Frequency problem, has the same high ω characteristic.* This figure is very revealing. The crossover frequency ω_c, is defined as that at which $|L|$ is zero db, so cuts the horizontal log ω axis. For (approximately) $\omega > \omega_c$, the sensor noise is amplified at all frequencies for which $|L_{opt}| > |P|$, so we have labeled this region 'cost of feedback'. The cut-off frequency (ω_{cut}) is defined as the last *corner* (or break) frequency of $L_{opt}(j\omega)$. Its great importance lies in that *design effort* can be *relaxed* for $\omega > \omega_{cut}$. We need not pay much attention to L for $\omega > \omega_{cut}$, just as we need not pay attention to $T(j\omega)$ beyond its bandwidth, if we know their magnitudes are decreasing thereafter. *The great importance of ω_{cut} lies in that we must be sure of the numbers upon which our paper design is based, and upon which practical hardware design with all its trouble and expense will be built, at least up to ω_{cut}, and preferably for several octaves beyond it.* It is easy to be careless and optimistic when one considers the great sensitivity reduction and disturbance attenuation achievable by feedback, that one forgets the high price that must be paid for its benefits. In Fig. 36, note the typical, often huge difference between ω_T (bandwidth of T), and ω_{cut}. *In an open-loop (no feedback) design, the designer need concern himself only for a few octaves beyond ω_T. In a demanding feedback design, it is best he does so for a few octaves beyond ω_{cut}. The latter may be several decades larger, say 1000 Hertz for ω_{cut}, instead of 10 Hertz for ω_T. Modern Control theory has been notoriously indifferent to this vitally important Cost of Feedback (see specially p. 290 of [20]).*

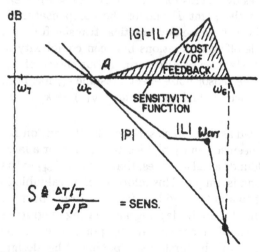

Fig. 36. Sensitivity function is very insensitive to cost of feedback

The following might be worth pursuing. Consider the important factors and ratios involved in the cost and benefits of feedback: such as: (1)the ratio of amount of plant uncertainty $\mathbf{U}(P)$, which can vary greatly vs. ω, (2)the permissible uncertainty in closed-loop response, possibly the allowed variation in the latter's bandwidth $\mathbf{U}(T)$. Another factor: the ratio ω_{cut} to ω_T, as discussed in last paragraph. It would be valuable to the designer, if such a *universal* ratio could be found between benefits and costs. There do exist many Quantitative Feedback Designs, whose data could be used for such an investigation. But we should warn the potential investigator, that he should first learn OFT thoroughly, in order to obtain deep understanding of the trade-offs in the Feedback Control Problem.

5.2   The Sensitivity Function (Design Tool of H_∞) is Highly Insensitive to Cost of Feedback

It is interesting that Bode [11] invented the sensitivity function, which is $1/(I+L)$ in SISO systems, and derived for it some important theorems, such as the *Equality of Positive and Negative Feedback Areas* (i.e. in practical feedback systems, there is zero average feedback benefit, Fig. 2). However, as a design tool, he used exclusively the loop transmission L, because that is what must be actually built in practice, and because it directly gives the cost of feedback in terms of the highly important cut-off frequency ω_{cut}. Both Classical Feedback Control theory, and Modern Control theory (State Space, LQR, Observers) which followed, ignored the uncertainty and sensitivity problem.

In the early 1980s, there emerged H_∞, which was a significant positive advance in Modern Control theory, because it did seriously concern itself with the sensitivity of the feedback system to plant uncertainty. However, it used the sensitivity function, $1/(1+L)$, as its principal design tool, and continues to do. It would not have done so, if it had paid any attention to the cost of feedback, because the sensitivity function is highly insensitive to the cost of feedback, so is an exceedingly poor tool for its minimization. In fact H_∞ designs (including at least one which received a Best Practical Paper Award from the IEEE Control Society) tend to have *infinite cost of feedback*. The poor insensitivity of the sensitivity function is seen as follows: In Fig. 36, note that in the frequency region with high cost of feedback, $|L|$ is very small, so $1 + L$ is very insensitive to L. Thus, suppose we have a choice of $|L| = .05$, or .005 (with $|P|$ less than either, as is very often so in practical design. This makes a big difference (factor of 10) in the sensor noise effect at that frequency, so the designer using L as the synthesis tool (Bode, QFT) recognizes its importance. He can't miss it. But the difference for the sensitivity function between these choices is at most only between 111.05 and 111.005, approximately 4.5% (hardly noticeable in the sensitivity locus in Fig. 36), instead 1000% for QFT.

Most of the more recent Control techniques (H_∞, Fuzzy design, neural networks) continue to ignore the cost of feedback. This is unfortunate for the advancement of genuine (Quantitative) Feedback theory. The reason is that awareness of the high cost of feedback, would motivate study of ways of decreasing the cost. Such awareness has inspired QFT to develop practical Quantitative design techniques for Multiloop Systems, which often permit fantastic reduction in sensor noise effect, at the price of additional sensors. It has led QFT, for the same purpose of cost of feedback reduction, to a special nonlinear device (First Order Reset Element-FORE), to Oscillating Adaptive systems (used by Minn. Honeywell for the X15, but unaccompanied by any Quantitative Design theory), and to other cost of feedback reduction techniques.

6   Stability of QFT Nonlinear Synthesis Technique

The robust stability of feedback systems is a hard problem even for the case of (finite dimensional) linear time-invariant systems, where there are no known tests for the general case ([20], [10]). As it can be expected, for the nonlinear case the situation is still more complicated. Here we consider two approches: i) the adaptation of classical absolute stability results to QFT synthesis ([2,7]), and ii) the use of an equivalent disturbance method ([6]).

Adaptation of absolute stability results Consider a Lur'e-type nonlinear feedback system, Fig. 37, given by the feedback interconnection of a linear uncertain system $H(s)$, including the feedback compensator G, and a nonlinear system N satisfying a sector condition. Since the goal is to obtain frequency domain restrictions for G, there exists a clear connection with the classical absolute stability work (see for example [30] for a clear exposition of the absolute stability problem). Closed-loop stability is defined as I/O stability from any input entering additively to the feedback system to the rest of signals. Our sense of I/O stability is finite gain L_2 stability. More formally, a mapping $H : L_{2,e} \rightarrow L_{2,e}$ is finite gain L_2 stable if for every input $x \in L_2$, the output $y = Hx \in L_2$ and, in addition, $\|y\|_2 < \alpha\|x\|_2 + \beta$ for some finite constants α, β.

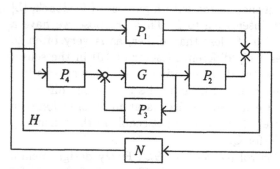

Fig. 37. The Lur'e type nonlinear system: G is the feedback compensator, P_i, $i = 1, .., 4$ are (possibly uncertain) LTI blocks of the plant, and N is the nonlinear block of the plant

The nonlinear plant is supposed to be given by (possibly uncertain) linear and nonlinear subsystems, where, in addition, the nonlinear subsystem satisfied a sector condition. For the sake of completness some definitions follows. Suppose a mapping $\Phi : R_+ \times R \rightarrow R$, and $a, b \in R$ being $a < b$, then $\Phi \in$

Conic Sector $[a, b]$ or simply $\Phi \in sector[a, b]$ if i) $\Phi(t, 0) = 0$, $\forall t \in R$, and ii) $a \le \Phi(t, x)/x \le b$, $\forall x \ne 0$, $t \in R$; let $D[a, b]$ be the disk in the complex plane which is centred on the real axis and whose circumference passes through the two points $-1/a$ and $-1/b$.

In the following, the adaptation of the Circle Criterion is given to cope with plant uncertainty. It will be referred to as the robust Circle Criterion. Related work about robust absolute stability can be found in [8] and references therein. Most of the material of this Section is taken from [2], where in addition some other criteria such as the Popov Criterion are also explored. Only the SISO case is considered here. For extensions of this work to the multivariable case see [7]. Conditions given by the robust Circle Criterion take a rather simple form, given as a set of linear inequalities from which a set of boundaries can be computed using, for example, the ideas given in [12].

If the nonlinear plant P of Fig. 1 is given by a combination of series, parallel or feedback interconnections of linear and nonlinear blocks, closed loop system stability can be always inferred from the stability of the equivalent Lur'e system of Fig. 37, where the linear subsystem is a Linear Fractional Transformation (LFT) of the compensator G, that is

$$\mathbf{H}(s) = \mathbf{P_1}(s) + \frac{\mathbf{P_2}(s)\mathbf{G}(s)\mathbf{P_4}(s)}{1 + \mathbf{P_3}(s)\mathbf{G}(s)} \tag{55}$$

However, since all transfer functions are scalar, after simple manipulations the linear part H reduces to

$$\mathbf{H}(s) = \frac{\mathbf{P_1}(s) + \mathbf{P_2}(s)G(s)}{1 + \mathbf{P_3}(s)G(s)} \tag{56}$$

where $\mathbf{P_2}$ replaces $\mathbf{P_1P_3} + \mathbf{P_2P_4}$. Here the nonlinear part is supposed to be contained in a sector, $N \in sector[a, b]$, while the linear systems $\mathbf{P_i}$, $i = 1, 2, 3$, are in general uncertain and represented as sets of transfer functions $P_i(s)$, respectively. As a result, $\mathbf{H}$ will be also uncertain in general. Thus (54) and (55) must be read as the definition of a set of transfer functions, for all the possible combinations of elements $P_i \in \mathbf{P_i}$, i =1,2,3.

Without further restrictions, the derivation of conditions on G for feedback system stability is still a difficult problem. Note that in general H does not need to be stable for the nonlinear feedback system to be stable. For obtaining a robust Circle Criterion in this case, a natural restriction would be that the number of unstable poles of H be an invariant for all the values of the uncertain linear components P_i, $i = 1, 2, 3$. Now, since G may appears as a feedback loop around the linear components of the plant, this number can be theoretically fixed by the designer. In other situations, if G does not appear in the feedback loop, that is $P_3 = 0$, the number of unstable poles of H is more influenced by the linear dynamic of the plant. This is a more tractable case, and can be treated separately [2]. However, in the more general case, it is not clear what conditions must satisfy G in order that $H = LFT(G)$ have a predetermined and fixed number of unstable poles for any value of the uncertain systems P_i, $i = 1, 2, 3$. Even more, it is not clear which number of unstable poles would give the less conservative condition for the compensator G. The solution adopted here has been to consider H stable, which gives a first set of conditions for G, and then use the Circle Criterion adapted to this case, for coping with the uncertainty.

Robust Circle Criterion: Consider the feedback system of Fig. 37, where $N \in sector[a, b]$, and $\mathbf{P_i}$ are uncertain linear systems with single-connected templates, for $i = 1, 2, 3$. Let G be such that H is stable for any $P_i \in \mathbf{P_i}$, i =1,2,3. Then the feedback system is stable (L_2 stable with finite gain), if some of the following conditions, as appropriate, are satisfied (the $j\omega$ argument is dropped by symplicity):

(1) $ab > 0$: The Nichols diagram of G does not intersect, for any $\omega > 0$, the NC region defined by

$$\left\{ (arg(G), |G|) \in NC \left\| \frac{P_1 + P_2 G}{1 + P_3 G} - c \right\| \geq r, \forall P_i \in \mathbf{P_i}, i = 1 - 3 \right\} \qquad (57)$$

where c is the center and r is the radii of the disk $D[a, b]$. In addition, there is no net crossings of the nominal $H_0(j\omega)$ with the ray $-180 \times [(1/a)_{db}, \infty)$ in the case $0 < a < b$, or with $0 \times [(1/b)_{db}, \infty)$ in the case $a < b < 0$.

(2) $0 = a < b$: for any frequency $\omega > 0$ the Nichols diagram of $G(s)$ is out of the NC region defined by

$$\left\{ (arg(G), |G|) \in NC \left| Re \left\{ \frac{P_1 + P_2 G}{1 + P_3 G} \right\} \geq -\frac{1}{b}, \forall P_i \in \mathbf{P_i}, i = 1 - 3 \right\} \right. \quad (58)$$

(3) $a < 0 < b$: for any frequency $\omega > 0$ the Nichols diagram of $G(s)$ is out of the NC region defined by

$$\left\{ (arg(G), |G|) \in NC \left\| \frac{P_1 + P_2 G}{1 + P_3 G} - c \right\| \leq r, \forall P_i \in \mathbf{P_i}, i = 1 - 3 \right\} \qquad (59)$$

where c is the center and r is the radii of the disk $D[a, b]$.

In the three above Cases, application of Circle Criterion results in two sets of conditions: i) a stability linear condition, $H(s)$ must be stable, that may be solved by using linear QFT, given as result a set of boundaries on $G(s)$, and ii) a restriction over $G(s)$ given by Eq. (2.3), (2.4) or (2.5) depending on the case. This last condition is a rather straigthforward adaptation of Circle Criterion to cope with uncertainty. The good news is that it can be treated as a new boundary, using standard algorithms previously used in QFT, the procedure given in [12] is just one possibility. Next Section gives an illustrative example.

Example: The nonlinear system in this example (Fig. 38) is borrowed from Example 1 in [27] and has been also developed in [2].

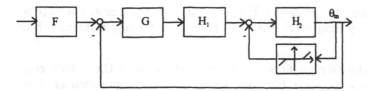

Fig. 38. Electric motor

It represents an electric motor driving a load through a gear, that is embedded in a two-degrees of fredom structure.

$$H_1(s) = \frac{k_m}{k_l} \frac{J_m s + B_m}{(J_l + J_m)s + B_l + B_m} \tag{60}$$

$$H_2(s) = k_l \frac{(J_l + J_m)s + B_l + B_m}{s(J_m s + B_m)(J_l s + B_l)} \tag{61}$$

and the parameters are known to be in intervals:

$$k_m \in 0.041[1, 1.2], k_l \in 4.8[0.85, 1] \tag{62}$$

$$B_m \in 0.0032[1, 20], B_l \in 0.00275[0.4, 1] \tag{63}$$

$$J_m = 6.39 \times 10^{-6}, J_l = 0.0015 \tag{64}$$

The goal is to analyze the stability of the two-degrees of feedback systems represented in Fig. 38, representing the control system in which the motor is embedded. F is supposed to be stable. The nontrivial part of the problem is related with finding what conditions on the compensator G guaranties the stability of the feedback system. Eliminating F from the block diagram, we again obtain a Lure's system (Fig. 37) after some simple manipulations, where

$$H(s) = \frac{H_2(s)}{1 + G(s)H_1(s)H_2(s)} \tag{65}$$

and N is given by the deadzone nonlinearity.

This stability problem corresponds with Case 2 of the Robust Circle Criterion given above, since the nonlinearity clearly belongs to the sector $[0,1]$. The first condition, that is the stability of H, can be treated in a linear QFT framework. The stability requirement gives a set of boundaries for the function $L_0(j\omega) = k_{m,0}G(j\omega)P_{1,0}(j\omega)$ (nominal values are the left extremal interval values). In this example, the frequencies $\omega = 1, 10, 100, 1,000$ y 10,000 rad/s are chosen for shown these boundaries (Fig. 39-a, where in addition, a typical 3 db stability contour has been used to provide an extra degree of robustness in the design. If $L_0(j\omega)$ satisfies these restrictions (theoretically for every frequency), then $H(s)$ is stable for any combination of plant parameters.

The second stability condition imposed by the Robust Circle Criterion, given by (57), is a bit more involved. However, the computation of these boundaries can be done by using quadratic inequalities similar to those used in [12]. Results are shown in 39-b. Finally, the two sets of boundaries need to be regrouped to obtain a final worst case boundary. Regrouping boundaries of 39-a and 39-b, the final result is given in 39-c. With these boundaries, the designer can now shape properly the nominal open loop gain L_0 to obtain a stable design: $L_0(j\omega)$ must be above its boundary for every frequency ω.

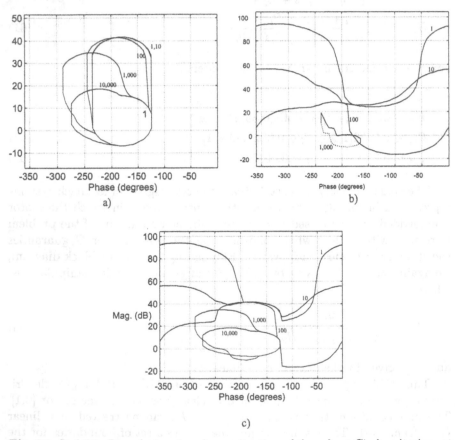

Fig. 39. Stability Boundaries given by application of the robust Circle criterion, at frequencies $\omega = 1, 10, 100, 1,000, 10000$

Use of equivalent disturbance method Our objective here is to add a stability guarantee to the QFT nonlinear **Pe** technique, by means of an equivalent disturbance method. We concentrate on the effect of bounded changes in the command inputs. In order to accommodate an almost universal class of deviations, they are formulated as signals with assigned bounds on derivatives from zero order to order $n-1$, where n is the order (highest derivative) of the nonlinear differential equations in W (it is assumed the order is the same for all its members, although some deviation in order may be possible ([16], Sec. 7.8). Bounds of the same form, are assigned on the derivatives of the deviations in the command inputs. Denote this as *Quantitative Stability Bounds*.

Using the "Nonlinear Equivalent Disturbance Technique" (Sec. 3.3), we shall show how the Nonlinear Pequiv. Method of Sec. 2, needs only slight modification, in order to achieve Quantitative Stability Bounds, at the same time that it does its primary task of acceptable signal tracking.

In Fig. 40-a, let $y = y_0$ be the output due to command input r_0, and in Fig. 40-b, $y = y_0 + \Delta$ the output due to $r_0 + \delta(r)$, $c = \delta(r) : u_0 + \delta = G[F(r_0 + c) - (y_0 + \Delta)]$. Also, $u_0 = G[Fr_0 - y_0]$. Subtracting gives $\delta = G[Fc - \Delta]$. Also, $y_0 = W(u_0)$, $y_0 + \Delta = W(u_0 + \delta)$. Combining all of these, gives Fig. 40-c, with nonlinear plant $W_x : \Delta = W(u_0 + \delta) - W(u_0)$. For example, if

$$W : \frac{dy}{dt} + Ay^3 = ku \tag{66}$$

then

$$W_x : \frac{d\Delta}{dt} + A(3y_0^2\Delta + 3y_0\Delta^2 + \Delta^3) = k\delta \tag{67}$$

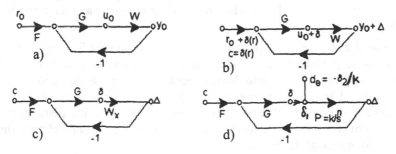

Fig. 40. Derivation of equivalent system for perturbed command input

Next step is to use the QFT Equivalent disturbance technique of Sec. 3.3, to let $\delta = \delta_1 + \delta_2$, with $k\delta_1 = d(\Delta)/dt$, and so replace the W_x nonlinear Plant relating (δ, Δ), by its equivalent LTI plant of Fig. 40-d, with $P = \{k/s\}$, plus plant disturbance $d_e = -\delta_2/k$. Fig. 40-d is an LTI Structure. Is it stable?

The original LTI equiv. technique of Sec. 2 was used only to obtain OK Performance for the **R, W, T** sets. For the example of (F2) its $\mathbf{P_e} = k/[s+X]$, with $X = A + Lapl.(y^3)/Lapl.(y)$, which is larger and more complex than the **P** of 40-d (for nonlinear plant set W of order n, **P** would consist of k/s^n. The two sets $\mathbf{P, P_e}$ are very closely the same in the high frequency range, where both plant templates approach vertical lines (if no bending modes are present), and wherein shaping for stability is the major design effort. If we add **P** to $\mathbf{P_e}$ to give $\mathbf{P_{e,k}}$ and do the Sec. 2 design for $\mathbf{P_{ek}}$, then Fig. 40-d is definitely stable, because $\mathbf{P_k}$,is a subset of $\mathbf{P_{ek}}$. The templates of $\mathbf{P_{ek}}$ will at most be larger in the low and possibly mid-frequency range, where "free uncertainty" ([16], Sec. 12.3.4) is such that the bounds on the nominal loop transmission will hardly be affected. Note that it is very easy, by simply looking at the templates of $\mathbf{P_e}$ to see if there is need for the augmentation by **P**. But even if it should be necessary to do this, we can differentiate between performance type bounds for the **R, W, T** triple, to satisfy $\delta(\omega)$ performance bounds of Fig. 18, and the Gamma Constraint of Sec. 4. $\mathbf{P_{ek}}$ need be used only for the latter bounds, and $\mathbf{P_e}$ for the Performance bounds.

But we can do more than just assure Stability for Fig. 40-d. Consider the added effort needed to add Quantitative Stability to the design effort of this Section. Fig. 40-d is LTI, so linear superposition may be used. The output Δ has two components: Δ_c due to command c, Δ_d due to d_e. The Δ_c component will be closely $Lapl.(c)F(s)$, because of the large bandwidth of the $L = GP$ loop of Fig. 40-d. This is expected, because in such large loop bandwidth systems, $L/(1 + L)$ is closely unity over the smaller bandwidth of F. Quantitative Stability requires that we assign bounds on Δ(i.e. on its derivatives from zero to $n - 1$ order, for plant order n (here $n = 1$), just as we assigned similar order bounds on the command input deviation c. The bounds $B(\Delta)$ on Δ are preferably of the form $\lambda B(c)$, $\lambda > 1$. We deduce d_e from the bounds on Δ, and make sure that the PG loop is "strong" enough, so that the sum $\Delta = \Delta_c + \Delta_d$ and their derivatives to $n-1$ have their bounds satisfied.

It is essential to prove that the above can be done, so that Schauder's theorem is applicable. One's first inclination might be to choose $\lambda \ll 1$, because of Feedback's ability to attenuate disturbances. However the Δ_c component, due to command input c is a tracking component, and the larger G is the closer it is to $F(s)Lapl.(c)$. But the Δ_d component due to d_e, is a disturbance component which can be attenuated. Therefore choose $\lambda > 1$, and choose the GP loop strong enough to achieve it.

Hence, the addition of Quantitative BIBO Stability for **R, W, T** problem of Sec. 2, needs at most the addition of set $\mathbf{P_k} = k/s^n$ to the $\mathbf{P_e}$ set. There is no difference in the high frequency where the major stability shaping is concentrated. The bounds on the output due to the addition of c, with its bounded derivatives from zero to $n - 1$, can be made very close to those on c. *The above treatment explains why all our past QFT designs, which totally* *neglected even Classical Stability considerations, emerged highly "robust" to deviations of the command input signals, even to highly different types of signals, and to Plant disturbances.*

7   Feedback Techniques for Disturbance Attenuation

A major reason for using Feedback is for attenuation of plant disturbances, generally assumed inaccessible, as otherwise they might be blocked form entering the plant. They are also assumed uncertain, otherwise the plant input may be programmed to cancel them, without having to measure the plant output. The disturbances may act at any or all D_i in Fig. 41, and are countered by measuring the output y (which is not the plant output z if $D_1 \neq 0$), and fed back via G and v in Fig. 41.

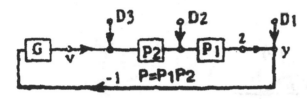

Fig. 41.   Different types of disturbances

The system output due to D_i is

$$Y_D = \frac{D_1 + D_2 P_1 + D_3 P}{1 + GP} \tag{68}$$

Arbitrarily large disturbance attenuation can be achieved by having $|GP| \gg 1$ over the desired ω range. So, again it is a matter of having the loop transmission large enough over a large enough ω range, with the attendant "cost of feedback." Two kinds of problems are considered. In the first a very broad class of disturbances is formulated with specifications directly on the tolerable output. In the second, bounds are specified on the transfer function relating system output to disturbances, $T_{D_i} = Y_{D_i}/D_i$. *Specifications on output due to bounded disturbances.* There is a comparatively simple design technique for this problem class which comprises all practical $d(t) : |d(t)| \leq D_0, |\dot{d}(t)| \leq D_1$. The tolerances can be on the maximum output magnitude y_x or perhaps on its maximum area $\int_0^\infty |y(t)| dt$. The design technique is presented by means of examples. *Example 1*: In Fig. 41, the disturbance set $\mathbf{D_3} = \{d_3(t) : |d_3| \leq 30, |\dot{d}_3| \leq 50\}$, $D_1 = D_2 = 0$. The plant is uncertain: $\mathbf{P} = \{P = P_1 P_2 = \frac{k}{s+a}\frac{b}{s+b}, k \in [1,5], a \in [2,5], b \in [1,2]\}$. The specifications are: $|y| \leq .5$, for all $P \in \mathbf{P}$ and all $d_3 \in \mathbf{D_3}$. *Design:*

$Y(s) = \frac{D_3(s)P(s)}{1+l(s)} = D_3(s)T_d(s)$. In the time domain $y(t) = d_3(t) * h(t)$, with $T_d(s) = Lapl.(h(t))$ (* indicates convolution). Use graphical convolution to study the problem: Suppose d_3 is a truncated ramp, $d_3(-t)$ as shown in Fig. 42-a, and an assumed $h(\tau)$. To find $y(t_1)$, pull $d_3(-\tau)$ to the right by t_1, to give $d_3(t_1 - \tau)$ in Fig. 42-b; multiply it by $h(t)$, which is zero for $t < 0$, and negligible for $t > t_0$. The area of this product is $\int_0^{t_1} h(\tau)d_3(t_1 - \tau)d\tau = y(t_1)$. Clearly, for $h(t)$ of Fig. 42 with area A, the "maximum $y(t)$" $= y_x$, is obtained soonest if d_3 has its maximum value of 50 for 0.6 seconds, until d_3 reaches its maximum of 30. Then $y_x = 30A$, which is reached at $t = .6 + t_0$ in Fig. 42-c.

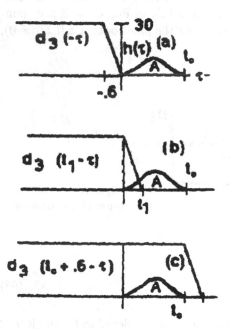

Fig. 42. Use of convolution

However, for $h(t)$ of Fig. 43-a, the "extreme" $d_3(-\tau)$ is as shown (infinite $\dot{d}_3$ is assumed to simplify the calculation, at the price of some overdesign), and $y_x = D_x(A + B)$ in Fig. 43-b, $D_x = 30$.

Let $B = \alpha A$, so $y_x = 30A(1+\alpha)$. However, $H(0) = \int_0^\infty h(\tau)d\tau = A - B = A(1 - \alpha)$. Since $y_x < 0.5$ is required, this means

$$H(0) \leq \frac{y_x(1 - \alpha)}{D_x(1 + \alpha)} \tag{69}$$

is needed, where $H = P/(1 + PG) \approx 1/G(0)$ at $s = 0$, so

$$G(0) \geq \frac{60(1 + \alpha)}{(1 - \alpha)} \tag{70}$$

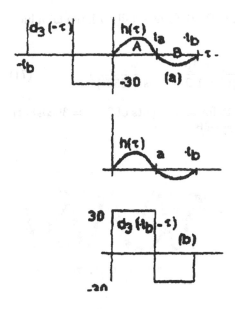

Fig. 43. Use of convolution

is required. If $P_N(s) = \frac{1}{(s+1)(s+2)}$ is the nominal plant, $L_N(0) \geq 30$ if $\alpha = 0$. We shall allow 1.33 for $(1+\alpha)/(1-\alpha)$, corresponding to 14% overshoot in the step response of $T_{D_3}(s)$, giving $L_N(0) \approx 32$ db needed, for this primary performance specification. Of course, stability is necessary ($\gamma = 3$ is used) plus stability at the nominal plant. This guarantees stability for all $\mathbf{P}$. In a more careful design, one relates the 14% allowed overshoot to constraints on $|T_{D_3}(j\omega)|$. Stability bounds $B(\omega)$ are needed. This can be done entirely on the computer. Let $L = L_N(P/P_N) = m(cos\theta + jsin\theta)P/P_N$. Set $|1+L|^2 = 2$ for $\gamma = 3$ db, solve the quadratic equation for m at given θ, ω over set $\mathbf{P}$, giving $max(m)$ and $min(m)$ which are the bounds on L_N for that e value. Repeat over e, etc. The results are shown in Fig. 44 for nominal $k = 1$, $a = 2$, $b = 1$.

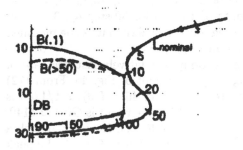

Fig. 44. Use of convolution

There is not much difference in the $B(\omega)$ from small to large ω values. The chosen L_N is (Fig. 44)

$$L_N(s) = \frac{80 \times 350^4}{15} \frac{(s+15)}{(s+1)(s+2)(s^2+1.2 \times 350s + 350^2)^2} \qquad (71)$$

Simulation: results are shown in Figs. 45 for step inputs of 30, $d = 30cos(\omega t)$, $\omega = 2, 5, 18, 30$, which all satisfy the specifications.

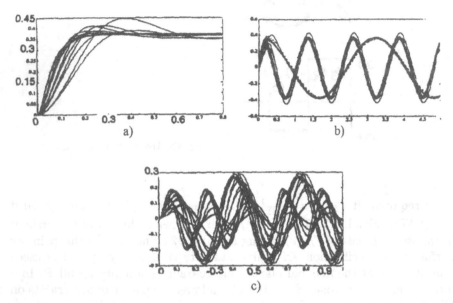

Fig. 45. Simulations

Disturbances at input to P_1 (Fig. 41). Suppose the same disturbances occur at the input to P1 in Fig. 41 instead of to P. The bound on $L_N(0)$ is the same because $P_2(0) = 1$, so presumably the same $L_N(s)$ of (74) can be used. However, the results in Fig. 46 for a step of 30 exhibit considerable overshoot, with max. of 4, even though the final $y(\infty) < 0.5$ as predicted. Furthermore, the"extreme" input has the form of d_3 in Fig. 43, giving $y_x > 4$. In the notation of Fig. 43 and the data of Fig. 46 for 30-step d_2, $30A \approx 4$, $30(A - B) \approx 0.4$, giving $B/A = 0.9$ in Fig. 43-a. The design must be modified to secure $y_x = 0.5$, which we do for infinite $\dot{d}_2, d_2 = 30$.

The idea is to use the previous philosophy of relatively fast decrease of $L_N(j\omega)$ of Fig. 44 with the resulting large overshoot of Fig. 46. From (72) for $D_x = 30$, $y_x = .5$, $\alpha = .9$ (estimated from Fig. 46), $H(0) \approx 1/1140$ is needed. Since $H(s) = P_1/(1+L)$ with $L = GP_1P_2$, the result is $G(0) \approx 1140$, giving $L_N(0) \approx 570$ (55 db instead of the 32 db of Ex. 1). However, if such a design were implemented with poles of L_N at $-1, -2$ as in Fig. 44, the peak

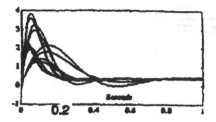

Fig. 46. Disturbances at P_1 input

values would be significantly $\geq .5$, even for 30-step inputs. The step-response overshoot of T_{D_2} can be decreased if the first negative real axis T_{D_2} pole precedes its first zero. This is achieved by

$$L_N(s) = \frac{562(1 + \frac{s}{32})(1 + \frac{s}{450})}{(1 + \frac{s}{4})^2(1 + \frac{s}{120})(1 + \frac{1.2s}{3500} + \frac{s^2}{3500^2})} \tag{72}$$

with the resulting outputs of Fig. 47.

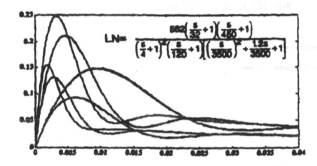

Fig. 47. Disturbances at P_1 input

For Fig. 43 inputs, the peak output is $-.32$. It is possible to go to the extreme of an effectively first order loop with only -6 db/octave decrease. In that case there is little or no overshoot and $L_N(0) = 32$ db suffices. The design

$$L_N(s) = \frac{40}{(1 + \frac{s}{4})(1 + \frac{1.3s}{12000} + \frac{s^2}{12000^2})} \tag{73}$$

gives the results shown in Fig. 48. But note the large ω cutoff of 12000, because of the slow (6 db/octave) decrease of L_N from 30 db to -30 db.

Disturbances at Plant output. In Fig. 41, $Y_D(s) = 1/(s(1 + L(s))$ for $D_1 = 1/s$. As $s \to \infty$, $Y_D \to 1$, so $y_d(0) = 1$. This is reasonable because D_1 acts instantaneously on the output, but the correcting feedback to v, from y_d via G, does not. It is convenient to write

$$Y_D = \frac{D_1}{1 + L} = D_1(1 - \frac{L}{1 + L}) = D_1(1 - T') = D_1 - D_1 T' \tag{74}$$

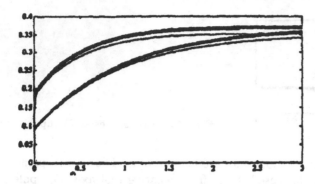

Fig. 48. Disturbances at P_1 input

Here $D_1 T'$ is the step response $\dot{y}$ of $T' = L/(1+L)$, shown in Fig. 49 for a typical $\dot{y}$. The faster $\dot{y}$, the faster the attenuation, with the price paid in the bandwidth of L. Large overshoot of $\dot{y}$ is undesirable if D_1 can change quickly as in previous examples, with the price paid in larger stability margins and consequent slower decrease of $L(j\omega)$). For this disturbance class, it is easier to convert the problem into specifications on $T'(s)$.

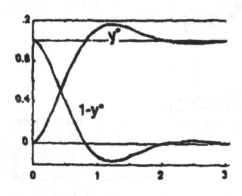

Fig. 49. Disturbances at plant output

7.1   Specifications on Disturbance Transmission Functions

In this second disturbance class, specifications are assigned on the transfer functions $T_{D_i} = Y/D_i$, assumed in the ω-domain, in the form of upper bounds only, $B_i(\omega)$. The reason is that T_{D_i} design is one-degree-of-freedom (ODF, Sec. 1) because G is the only free function. It is always possible to relate any T_{D_i} to $T' = L(1+L)$, as in (73) for D_1. An excellent perspective of the problem is easily obtained. Suppose $P = P_2 = 5/(s+2)$ with disturbance D at plant input, so $T_D = Y_D/D = P_2/(1 + L)$. If there is no feedback, $T_D = P_2$, in Fig. 50, so this T_D(open loop) must be the high-ω asymptote of the closed loop T_D, because as $\omega \to \infty$, $L \to 0$ so $T_D \to P_2 = T_{D_2}$(open loop). Hence, the desired closed-loop T_D must end up on P_2 in Fig. 50. The question is "where on P_2?".

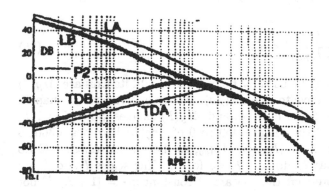

Fig. 50. Realistic ω-tolerances

Two T_D alternatives (A, B) are shown. The difference between T_D^{QL} (Open loop) and T_D must be made up by $(1 + L)$. The demands on $(1 + L)$ are very easily seen: at lower ω where $|T_D^{QL}| - |T_D|$ is large, $1 + L \approx L$ so the required L is simply the difference between them, e.g. at $\omega = 0.1$, $|L_A| = 54$ db $= |P_2| - |T_D^A|$, $|L_B| = 50$ db $= |P_2| - |T_D^A|$. Feedback benefits are no longer needed when $|T_D| = |P_2|$, so crossover ω_c is near $\omega = 6$ for B, near 20 for A. The careful observer will note that near ω_c, there is an ω range where $|T_D| > |T_D^{QL}|$, i.e., in this range feedback is making things worse. This must be so in every practical system, as previously explained. It happens automatically if excess of L, $e \geq 2$. One decides on the desired benefits (disturbance attenuation) and pays the price. Fig. 50 is easily drawn and presents a clear perspective of the benefits of feedback vs. its "costs". The design procedure is obviously quite simple if there is small plant uncertainty.

7.2   The Inverse Nichols Chart (INC)

Consider the case of $T_{D_1} = 1/(1 + L)$ in Fig. 41, for disturbances at plant output, and a significant plant uncertainty set **P**. Suppose it is required that $|T_{D_1}(j\omega) \leq b(\omega)$ for all **P** with $b(\omega)$ given. It is important that as $\omega \to \infty$ $b(\omega) \to 0$ db because $T_{D_1} = (1 + L)^{-1} \to 1$ as $\omega \to \infty$. In practice, there is no need to specify $b(\omega)$ for $\omega > \omega_c$. Thus, in Case A of Fig. 50, $b(\omega)$ is not needed for $\omega > 15$, or for $\omega > 7$ for Case B. The stability bound suffices. No special effort is needed to ensure $T_d \approx P_2$ as $\omega \to \infty$. It will automatically be so. This greatly simplifies design for disturbance attenuation. The first design step is to obtain plant templates $\mathbf{P}(\omega)$ for a sufficient number of ω values, exactly as in the command response problem. Next, one finds the bounds $B(\omega)$ on a nominal loop $L_N(j\omega) = P_N G(j\omega)$, such that

$$\left| \frac{1}{1 + L(j\omega)} \right| \leq b(\omega), \ \forall P \in \mathbf{P} \tag{75}$$

One could prepare a chart with loci of constant $|1 + L(j\omega)|$ values, just as the Nichols chart contains loci of constant $|L/(1+L)|$. Instead, note that the Nichols chart can be used by rotating it 180 and some relabeling. Thus, let $I = 1/L$, so

$$\left|\frac{1}{1 + L(j\omega)}\right| = \left|\frac{1}{1 + I(j\omega)}\right| \leq b(\omega) \tag{76}$$

The Nichols chart may be used for bounds on $I(j\omega)$ if its axes are $Arg(I)$ and $|I|$. We would then have to shape $I(j\omega)$. We are experienced in shaping $L(j\omega)$, not $I(j\omega)$, so prefer to have $|L|$ and $Arg(L)$ as axes. Since $I = 1/L$, $Arg(I) = -Arg(L)$, and the range -360 to 0 of $Arg(L)$ becomes $+360$ to 0 for $Arg(L)$ (Fig. 51). Also, $|I|_{db} = -|L|_{db}$, so the vertical range -20 db to 20 db for $|I|$ becomes 20 db to -20 db for $|L|$. The scales I and L are shown in Fig. 51-a. Note that the loci are of constant $m\star = |1 + L|^{-1}|$ not of $M = |L(1 + L)^{-1}|$. We prefer the L scales, but are accustomed to values increasing vertically upward and horizontally to the right. This is achieved by rotating Fig. 51-a by 180(or -180), giving Fig. 51-b, called the "Inverse Nichols chart" (INC). Then horizontal axis is from 0 to 360 for $Arg(L)$, as shown in brackets, but one can add any multiple of 360, so -360 is added, giving the usual angle scale from -360 to 0 in Fig. 51-b. Fig. 51-b makes sense, because if one wants small $m\star = |1 + L|^{-1}$, then large $|L|$ is needed.

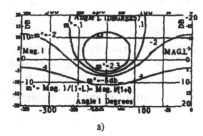

Fig. 51. Inverse Nichols Chart

Design Procedure. The INC is convenient for obtaining the bounds $B(\omega)$ on the nominal $L_N(j\omega)$ to satisfy (74). One maneuvers the template $\mathbf{P}(\omega)$ in precisely the same manner as on the ordinary Nichols Chart (for $T(j\omega)$ bounds) until (74) is satisfied, etc. Then $L_N(j\omega)$ is shaped to satisfy the $B(\omega)$.

Disturbances at plant input For $D = D_3$ in Fig. 41, the equivalent of (74) is $|P/(1+L)| \leq b(\omega)$. Of course, $b(\omega)$ must $\to |P(j\omega)|$ as $\omega \to \infty$. Again there is no need to specify $b(\omega)$ beyond the crossover ω_c of $L(j\omega)$. For $\omega < \omega_c$, simplification is possible. At low-ω, where $1 + L \approx L$, the above constraint become closely $|G(j\omega)| \approx 1/b(\omega)$ or $|L_N(j\omega)| \approx P_N(j\omega)/b(\omega)$. It is a good idea to take the nominal plant at the lowest point on the template, for then if $|1+L_N| \approx |L_N|$ is satisfied at P_N, it is certainly satisfied for the balance of **P**. This approximation is usually satisfactory up to $B(\omega) \approx 6$ db. And for larger $\omega > \omega_c$, let the stability bounds determine $B(\omega)$. These, in most cases (in the form $|L/(1 + L)| \le \gamma$ or $|1/(1 + L)| \le \gamma$ all **P**) are more stringent than the performance bounds in this higher ω range, especially if there is significant uncertainty in the high-ω gain of P. This simple approach is adequate for most problems. or precise design, write $|P/(1 + L)| \le b(\omega)$in the form

$$|\rho(j\omega) + L_N(j\omega)| \ge \frac{P_N(j\omega)}{b(\omega)} \tag{77}$$

where $\rho = P_N/P$. The NC or INC are no longer appropriate design tools. The arithmetic complex plane is more convenient. At any fixed ω, find **P**(ρ) in arithmetic units. If P_N is in **P**, **P**(ρ) includes the point 1. At any $\omega = \omega_1$, draw the boundary of -**P**$(\rho(j\omega_1))$. It is convenient to choose as P_N a low-gain P , for then $|\rho|$ tends to be < 1, or not much > 1. In Fig. 52, A is a tentative value of $L_N(j\omega_1)$, so $(\rho + L_N) = BA$, with B any point in -**P**(ρ). Obviously AC tangent to -**P**(ρ) is the smallest value of $|\rho + L_N|$, which from (76) must be $\ge |(P_N(j\omega)/b(\omega_1))$, determining a point on $B(\omega_1)$. In this way the bounds $B(\omega)$ on $|L_N(j\omega)|$ are obtained. The shaping of $L_N(j\omega)$ can be done in the arithmetic complex plane, or the $B(\omega)$ can be copied into the Nichols chart. The latter is also more convenient for the stability bounds, which usually dominate in the higher ω range. The above is conceptually useful for understanding the nature of the bounds $B(\omega)$. But they are usually easier to obtain via the computer. The above approaches (approximate and exact) can also be used for $T_{D_2} = P_1/(1 + L)$ for disturbance D_2.

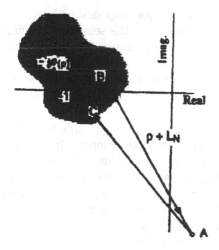

Fig. 52. Use of arithmetic complex plane

Simultaneous command and disturbance inputs: Overdesign The command response and disturbance attenuation problems have been considered separately, which is acceptable because linear systems are being considered, so the individual effects add. Each results in bounds B_T, B_D on the nominal L_N. Obviously the final bound B chosen must satisfy both. At any ω value, part of the final B may be due to B_T and part due to B_D. If B_T dominates over B_D, better disturbance attenuation than specified will be achieved. And if the plant capacity has been determined by the specified disturbance attenuation, there is then the possibility of plant saturation.

In Fig. 41, all three disturbances can be present and may or may not be correlated. If they are, assign tolerances on their combined effect and do a single design for all three. If they are independent, one might have tolerances on each separately. Of course, the final $B(\omega)$ must satisfy all the separate bounds and so achieve better attenuation for one or more of the D_i and/or smaller $T(j\omega)$ variation than required. But this is unavoidable in a linear design in which only one $L(j\omega)$ is available to satisfy multiple specifications.

8   Design of Multiloop Nonlinear Control Systems

As well as in the linear case, a fundamental limitation of feedback for highly nonlinear and uncertain plants subject to demanding specifications is due to sensor noise effect and the bandwidth of the controller. These effects are significant in many cases and, if ignored, it can invalidate many designs in practice. In general, as presented in the above Sections, QFT offers the designer a very convenient and transparent technique for accommodating the noise and bandwidth problems, both in the linear and the nonlinear single-loop cases.

There are practical situations in which a single-loop design is not affordable because the controller is very demanding, or the sensor noise effect affects considerably the design. In these cases, an alternative is to use internal feedback loops, by means of internal variable sensors. The uncertainty of the plant at high frequencies, where noise effect is important, is somehow transferred from the outer loop to the internal loops, allowing a balance between all the controllers bandwidths.

With some sligth modifications the technique is also applicable to nonlinear systems. It is important to mention that in the single loop nonlinear case, it is not surprising to obtain controllers with very high bandwith, since nonlinear dynamic is transformed in uncertainty over a equivalent linear family. This is a basic limitation of the feedback structure. The multi-loop technique can alleviate this problem, and it can be expected to be a more appropiate feedback structure in many nonlinear problems. Alternatively, use of nonlinear compensation can often alleviate the problem [16].

In the following we develop a multiloop design procedure for nonlinear plants with high frequency linear behavior. The method can be extended to more general types of nonlinear plants by means of nonlinear compensation ([5].

8.1 Nonlinear Plants with High Frequency Linear Behavior and Single Loop Design

We will consider a nonlinear system W given by the differential equation (36)-(38). Also serial connections of systems of this type will be allowed. More general nonlinear systems will be considered in the next subsection.

This type of nonlinear systems has the property of having leading linear derivatives, which result in a high frequency linear behavior. For the above nonlinear system, its high frequency linear dynamics P^∞ is defined by

$$y^{(n)}(t) = K(\theta)u^{(m)}(t) \tag{78}$$

having the transfer function $P^\infty(s) = K/s^{n-m}$. In the following, a specific nonlinear control problem is defined to help in the presentation of the proposed nonlinear multiloop technique. Consider a nonlinear plant given by the following differential equation:

$$W : u_1 \to y, \begin{cases} \frac{du_2(t)}{dt} + Au_2^3(t) = Ku_1(t) \\ \frac{d^2y(t)}{dt} + B|y(t)|^m sgn(y(t)) = Cu_2(t) \end{cases} \tag{79}$$

where the parameters have interval uncertainty given by $A \in [-5,5]$, $B \in [-3,3]$, $K \in [1,10]$, $m \in [1.5,3.5]$, and $C \in [2,40]$.

Note that for $\omega \to \infty$ or $t \to 0^+$, W can be approximated by the linear system $P^\infty(s) = CK/s^3$, with a poles-zeros excess of 3, and a vertical line template of approximately 46 db. The tracking specification is to track steps with amplitudes $Q \in [0.5, 1.5]$. In the frequency domain, this specification is given using the upper and lower bounds in Fig. 53.

Acceptable outputs sets A_r are defined for each value of the reference $r(s) = Q/s$, for $Q \in [0.5, 1.5]$. A_r must be a convex and compact subset of a Banach space. Since in response to the step, the output must contain a pole at the origin, it is not possible to directly define appropriate acceptable outputs sets. But we can easily can avoid the problem by using the idea of blocks transformation developed previously. Tracking steps with the plant W is equivalent to unit impulses with the plant $R^{-1}WR$, where R is a mapping with transfer function Q/s. For the impulse reference, we define the set of acceptable outputs A_δ as the set of signals y given by strictly proper, minimum phase and stable rational functions bounded in magnitude by the bounds of Fig. 53, and in addition with magnitude derivatives bounded by a given funcion $K(\omega)$, defined over the ω axis.

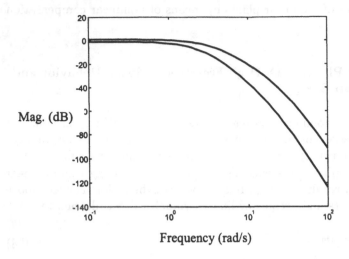

Fig. 53. Upper and lower magnitude bounds for closed loo acceptable outputs

Single–loop Design No details design are given here for a possible TDF single-loop desing, but a solution is given in Fig. 54-a, where the Bode diagram of the compensator G is given in Fig. 54-b. The Bode plot of the transfer function function from the sensor noise to the control input is given in Fig. 54-c. It can been seen that the effect of sensor noise is very significant for frequencies until 10^{14} rad/s.

8.2   Multiloop Design

For a given specification, single loop designs can exhibit a large demand in terms of control bandwidth as it is shown in the previous single-loop design. It should be noticed that this is a basic limitation of the single loop feedback structure. Thus, the effect of realistic sensor noise can invalidate the design in many nonlinear cases. This problem can be encountered in highly uncertain linear designs with demanding specifications, but is still more important in highly nonlinear designs, even with slightly uncertainty, where nonlinearity is transformed to uncertainty of an equivalent linear problem.

For the linear case, [16] gives a technique for reduction of the effect of the sensor noise based in a multi-loop technique. In this Section, this multi-loop is adapted to the nonlinear case. In particular, a two-loop design technique is used for the example, but the results can be generalized to more general situations. A two-loop feedback structure is given in Fig. 55, where, assuming that the designer has access to the internal variable $u_2(t)$, the nonlinear plant is divided into two blocks, W_1 and W_2, given by the differential equations

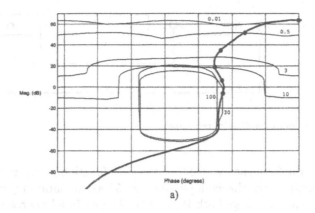

a)

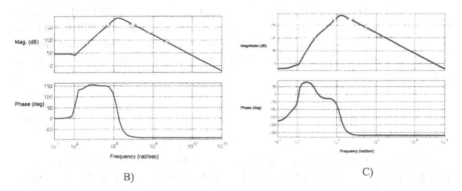

B)   C)

Fig. 54. Single-loop design: a)tracking and stability boundaries, and loop shaping, b)Bode plot of $G(s)$, c)Bode plot of the nominal transfer function from the noise input to the control signal.

$$W_2 : u_1 \rightarrow u_2 \ , \ \frac{du_2(t)}{dt} + Au_2^3(t) = Ku_1(t) \tag{80}$$

$$W_1 : u_2 \rightarrow y \ , \ \frac{d^2y(t)}{dt} + B|y(t)|^m sgn(y(t)) = Cu_2(t) \tag{81}$$

Here P_1 and P_2 stand for the equivalent linear families of W_1 and W_2, respectively.

Outer-loop design For designing the compensator G_1, only the high frequency template of the nonlinear plant W_2 is used, since it is only in the high frequency region where benefits can be expected, in terms of reducing the sensor noise effect [16]. In general, high frequencies may be considered to

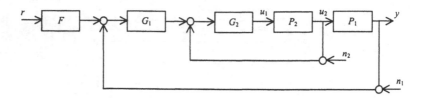

Fig. 55. Two-loops feedback structure

start at w_x, the frequency where the loop gain function contacts the high frequency boundary. For the example (see Fig. 54-a), we obtain $w_x = 300$ rad/s. Over this frequency range both W_1 and W_2 behave in a linear way. In particular, W_1 behaves like a linear system with transfer function $P_1^\infty(s) = C/s^2$, corresponding to a vertical line template of 26 db. Using this approximation for templates of P_1, for frequencies higher than $w_x = 300$ rad/s, a high frequency boundary is computed using the same stability specification than in the single-loop case, since in this frequency region stability specifications dominates tracking specifications. In the example, the specification is given as a γ-constraining

$$\left| \frac{L_1(j\omega)}{1 + L_1(j\omega)} \right| \leq \gamma \tag{82}$$

where $\gamma = 1.2$ is used. The resulting boundary is shown in Fig. 56(thick line) jointly with the boundaries of the single loop design. For the shaping of the nominal L_1, that is L_{10}, the same boundaries are considered, except for the high frequency boundary, where the new boundary is used instead. As it can be seen, the result is a reduction of 20 db with respect to the old one. A shaping of L_{10} is also shown in Fig. 56.

Note that in general the transfer function $L_1(s)$ is given by

$$L_{10} = G_1(s) P_{10}(s) \frac{G_2(s) P_{20}(s)}{1 + G_2(s) P_{20}(s)} \tag{83}$$

where $P_{10}(s)$ and $P_{20}(s)$ are the nominal values of the equivalent linear family of the nonlinear systems W_1 and W_2, respectively. Thus, for the computation of the compensator G_1, first the inner loop design must be completed. On the other hand, a single analysis of the above equation shows that if a realistic design is wanted, in terms of having controllers $G_1(s)$ and $G_2(s)$ with strictly proper transfer functions, $L_{10}(s)$ should have minimum poles-zeros excess. In the example, since the poles-zeros excess of $P_{10}(s)$ and $P_{20}(s)$ is 2 and 1, respectively, the pole-zeros excess of $L_{10}(s)$ should be at least of 5. This is a major criterion for the shaping of $L_{10}(s)$.

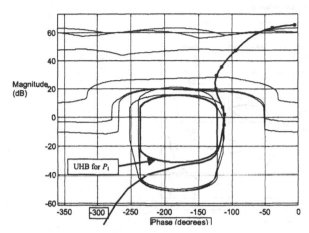

Fig. 56. Two-loops design

Inner loop design Since the main interest is in the reduction of the outer loop gain at high frequencies, for reducing the effect of the sensor noise, only the contribution of the compensator G_2 over the range $\omega > \omega_x$ is considered. For $\omega > \omega_x$, the stability specification dominates. Thus G_2 must verify the γ-constraint

$$\left| \frac{L_{10}P + L_{10}P_{20}PG_2}{P_0 + L_{10}P + (P_0 + L_{10}PP_{20})P_2G_2} \right| \leq \gamma \tag{84}$$

where P and P_2 are any member of the equivalents linear families of the nonlinear systems W (as a whole) and W_2, respectively, P_0 and P_{20} are nominal values of those families, and L_{10} is the outer loop gain. For the example, L_{10} is already computed when closing the outer loop, the templates of P were obtained in the single loop design, as well as those of P_2. Templates of P_2 reduce to the high frequency template which can be computed analytically. Finally, the computation of P_{20} is only a bit more involved. First, for a nominal acceptable output y, and a nominal parameters combination of the nonlinear plants W_1 and W_2, the signal u_2 is computed. Since P_0 is already computed in the single-loop design, it is only necessary to compute $P_{10}(s) = y(s)/u_2(s)$.

Using all this information, the γ-constraint can be transformed to boundaries for the frequency response $G_2(j\omega)$ of the inner compensator, for frequencies $\omega > \omega_x$. The result is given in Fig. 57, where feasible shaping is given simply by

$$G_2(s) = \frac{55}{1 + s/100} \tag{85}$$

Once the inner feedback compensator in designed, G_1 can be easily computed by direct substitution in (82). It will be not explicitly shown here.

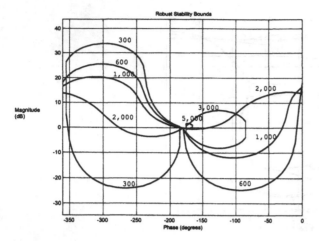

Fig. 57. Stability boundaries for G_2

To analyze the sensor noise effect, see Fig. 58, where the single loop design is compared with the two-loop design. It can be seen how the bandwidth of the closed loop transfer function, from the sensor noise signal to the control signal, is reduced from 10^{14} to 10^8 rad/s, a reduction of six orders of magnitude. This is the main justification for introducing the multi-looping technique. Note, in addition, how this can be achieved by the very single inner feedback controller given by (85).

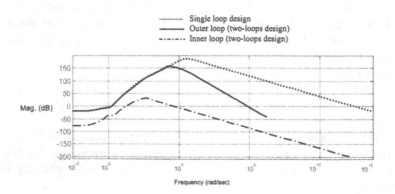

Fig. 58. Bode plots of nominal closed loop transfer functions, from the sensor noise to the control signal, in the single loop and the two-loops cases

References

1. Baños, A., and Bailey, F. N., 1998, "Design and validation of linear robust controllers for nonlinear plants", *Int. Journal Robust and Nonlinear Control*, 8, 803-816.

2. Baños, A., and Barreiro, A., 2000, "Stability of nonlinear QFT designs based on robust absolute stability criteria", *Int. J. Control*, 73, 1,74-88.

3. Baños, A., Bailey, F. N., and Montoya, F. J., "Some results in nonlinear QFT", *to appear in Int. Journal Robust and Nonlinear Control*.

4. Baños, A., Barreiro, A., Gordillo, F., and Aracil, J., "Nonlinear QFT synthesis based on harmonic balance and multiplier theory", in A. Isidori, F. Lamnabhi-Lagarrigue, and V. Respondek (eds.), *Nonlinear Control in the year 200*, Lectures Notes in Control and Information Sciences, Springer Verlag, 2000.

5. Baños, A., and Horowitz, I., "QFT design of multiloop nonlinear control systems", *to appear in Int. Journal Robust and Nonlinear Control*.

6. Baños, A., and Horowitz, I., "A method for stability of nonlinear control systems", in preparation.

7. Barreiro, A., and Baños, A., 2000, "Nonlinear robust stabilisation by conicity and QFT techniques", *Automatica*, 36,9.

8. Battachariya, S. P., Chapellat, H., and Keel, L. H., 1995, *Robust control: the parametric approach*, Prentice Hall, Upper Saddle River, NJ.

9. Bentley, A. E., 1994, "QFT with Applications in Welding", *Int.J. Robust and Nonlinear Control*, 4, 2.

10. Blondel, V., 1994, *Simultaneous stabilization of linear systems*, LNCIS no. 191, Springer–Verlag, London.

11. Bode, H. W., *Network Analysis and Feedback Amplifier Design*, Van Nostrand, 1946.

12. Chait, Y., and Yaniv, O., 1993, "MISO computer–aided control design using the Quantitative Feedback Theory", *International Journal of Robust and Nonlinear Control*, 3, 47-54.

13. Eitelberg, E., and Boje, E., 1989, "Some practical low frequency bounds in Quantitative Feedback Design", Proc. ICCON'89, WP–2–1, 1-5.

14. Eitelberg, E., 2000, *Control Engineering*, NOYB Press, 58 Baines Road, Durban 4001, South Africa.

15. Golubev, B., and Horowitz, I., 1982, "Plant Rational Trasnfer Function Approximation from Input–Output Data", *Int. J. Control*, 36, 711-23.

16. Horowitz, I., 1993, *Quantitative Feedback Design Theory (QFT, Vol. 1)*, QFT Publishers, 1470 Grinnel Ave., Boulder, CO 80305, USA.

17. Horowitz, I., 1975, "A Synthesis theory for linear time–varying feedback systems with plant uncertainty", *IEEE Trans. Automat. Control*, AC–20, 4, 454-64.

18. Horowitz, I., 1976, "Synthesis of feedback systems with Nonlinear Time-varying uncertain plants to satisfy quantitative performance specifications", *Proc. IEEE*, 64, 123-130.

19. Horowitz, I., 1982, "Feedback Systems with nonlinear uncertain plants", *Int. J. Control*, 36, 1, 155-171.

20. Horowitz, I., 1991, "Survey of QFT", *Int. J. Control* 53, 255-91, 1991.

21. Horowitz, I.,1952, "Plant adaptive systems vs ordinary feedback systems", IRE AC–i7i, 48-56.

22. Horowitz,I., and U. Shaked, 1975, "Superiority of transfer function over state variable methods in linear time invariant feedback system design", *IEEE Trans Auto Control*, AC–20, 84–97.

23. Horowitz, I., 1959, "Fundamental Theory of Linear Feedback Control Systems", *Trans. IRE on Auto. Control*, AC–4.

24. Kobylarz, T., and Barfield, F., 1990, "Flight controller design with nonlinear aerodynamics, large uncertainty and pilot compensation", *Proc. AIAA*, Portland Oregon, 1–15.

25. Miller, R., et al, 1994, "MIMO Flight Control Design for YF–16 using nonlinear QFT and Pilot compensation, *Int. J. Robust and Nonlinear Control*, 4, 211–30.

26. Nataraj, P. S. V., 1992, "QFT and Robust Process Control", *Proc. First QFT Symposium*, WPAFB, 275–841.

27. Oldak, S., Baril, C., and Gutman, P. O., "Quantitative design of a class of nonlinear systems with parametric uncertainty", *QFT Symposium*, Wright Laboratory, 1992.

28. Yaniv, O., 1987, and Horowitz, I., "QFT-reply to criticisms", *Int. J. Control*, 945–62.

29. Yaniv, O., 1999, *Quantitative Feedback Design of linear and nonlinear control systems*, Kluwer Academic Publishers.

30. Vidyasagar, M., 1993, *Nonlinear systems analysis*, Prentice–Hall International, London.

Part III

Hybrid Systems

Part III

Hybrid Systems

Introduction

The concept of "hybrid system" (HS for short) is used to refer to often complex dynamic systems. Typically a HS model consists of several components interacting via common dynamics (either continuous or discrete time) or via common event driven dynamics. During the last decade this concept has attracted the attention of many researchers who have seen this approach as a suitable way to model and analyze complex systems that would be rather difficult to describe otherwise. Although, in general, the modeling and analysis of HS is a quite difficult topic of study due to the potential complexity of such systems, for some particular though important classes of HS' (like piecewise-linear or switched linear systems), some promising results have been obtained recently.

This part begins with a brief introduction to the fundamentals of discrete event systems (DES'). Then, new features are introduced to these systems to include some temporal (though simple) aspects, like continuous time progress. Finally several formalisms for modeling some specific classes of HS' are introduced.

Introduction to Discrete Event Systems

Francisco J. Montoya[1] and René K. Boel[2]

[1] Departamento de Informática y Sistemas, Universidad de Murcia, Murcia, Spain
[2] SYSTeMS Group, Electrical Engineering Department, Universiteit Gent,
Belgium

1 Introduction

As mentioned in the introduction of this Part, a Discrete Events System
(DES for short) working together and sharing dynamics with a Continuous
System (CS) results in a Hybrid System (HS). Since the reader is supposed
to be more familiar with concepts from CS than with concepts from DES,
this chapter briefly introduces the fundamentals of DES. The main goal of
this chapter is to clarify the need for particular formalisms and tools in order
to properly model and analyse DES.

This chapter is organized as follows: in Section 2 the most important
issues that distinguish a DES dynamic systems from a CS are introduced.
In Section 3 we describe how DES may interact with CS and what should
be considered as a HS. Section 4 describes the levels of abstraction that one
can consider when modeling DES, while Section 5 describes the important
concepts of *languages* and *automata*. Finally, Section 6 presents a brief review
of *supervisory control* of DES.

2 Motivation of DES study

First of all, we will discuss the main differences between DES and more
traditional dynamic systems. This will motivate the need for specific tools
to model and analyze DES. The mentioned differences are related to two
important issues in the modelling of any dynamic system:

- State space
- State evolution

2.1 State space

The *state* of a dynamic system at time τ_0 is defined as the minimum set of
variables, usually denoted as $\mathbf{x}(\tau_0)$, that defines the future evolution of the
system, provided that the system inputs (if any) $\mathbf{u}(\tau)$ are also known for
$\tau \geq \tau_0$. The *state space* is the set of all possible values that $\mathbf{x}(\tau)$ may take at
different time instants.

In CS the state space is a subset of $\mathbb{R}^n$, where the dimension n is either
finite or infinite.

$$\mathbf{x}(\tau) = [x_1(\tau), x_2(\tau), \ldots x_n(\tau)]^T \in SS \subseteq \mathbb{R}^n \tag{1}$$

Example 1. In the example shown in Fig. 1 a wagon with mass M moves along a track. Its motion is due to the force $F(\tau)$ applied to the wagon, and is also influenced by the friction between the wagon and the track rail. The system output is the position, denoted by $x(\tau)$, of the wagon with respect to some reference point. This is a MISO system (*multiple-input, single-output*) which could be seen as a *black box* with two inputs and one output, as shown in Fig. 2.

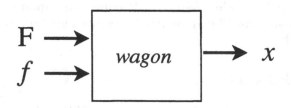

Fig. 1. Example of *continuous state* system

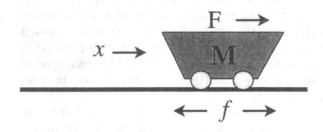

Fig. 2. System of Fig. 1 seen as a black box

The state of the system of Fig. 1 could be defined as Eq. 2 shows.

$$\mathbf{x}(\tau) = \begin{bmatrix} x(\tau) \\ \dot{x}(\tau) \end{bmatrix} \in SS \subseteq \mathbb{R}^2 \tag{2}$$

SS may be bounded, or not, depending on physical cosntraints of the system, and depending on whether system inputs are bounded or not. In either case, one can easily see that both state variables, $x(\tau)$ and $\dot{x}(\tau)$ take values from a continuous subset of $\mathbb{R}$.

In contrast, in a DES the state variables take values in a countable (possibly finite) set of values. In general, the state of a DES can be defined as

$$\mathbf{q} = [q_1, q_2, \ldots, q_n]^T \in Q = Q_1 \times Q_2 \times \cdots \times Q_n \tag{3}$$
$$q_i \in Q_i = \{v_1^i, v_2^i, v_3^i, \ldots\}$$

Note that, although in Eq. 3 sets Q_i are written so that an implicit order relation exists among its elements, in general however there is no such ordering relation, as will be seen in the following examples.

Example 2 (A simple DES). Consider a part of a manufacturing system as shown in Fig. 3. In this subsystem there exist two conveyors that transport pieces towards machines M_1 and M_2. The input queue for machine M_1 is denoted as "queue#1", and the queue for machine M_2 as "queue#2". Pieces are supposed to enter the system at random time instants through queue#1.

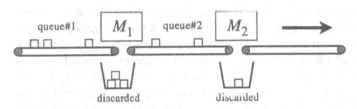

Fig. 3. Example of a simple DES

After processing a piece, machine M_1 can either pass the piece to queue#2 or discard it to a waste basket. Machine M_2 works similarly, except that pieces that are not discarded are assumed to leave the system through the third conveyor, whose state is not relevant for further evolution of the subsystem under consideration in this example.

Assume that the number of pieces that queue#1 and queue#2 can hold is bounded by some integer N, and waste baskets are emptied often (so that their states are not actually relevant). Then the state space of this system could be defined as

$$\mathbf{q} = [q_1, q_2, q_3, q_4]^T \in SS = Q_1 \times Q_2 \times Q_3 \times Q_4 \tag{4a}$$
$$q_1, q_3 \in Q_1 = Q_3 = \{0, 1, \ldots, N\} \tag{4b}$$
$$q_2, q_4 \in Q_2 = Q_4 = \{\text{busy}, \text{idle}\} \tag{4c}$$

The state variables q_1 and q_3 represent the number of pieces in queue#1 and queue#2, respectively. Variables q_2 and q_4 represent the state of machines M_1 and M_2 respectively.

The following examples illustrate the important point that the state defined for the system of Example 2 could be either more complex, or simpler, depending on how many details are relevant for purposes of the model. This is similar to the simplification of CS models depending on whether some dynamics may be considered as negligible for the modeling/control purposes of the specific problem.

Example 3 (A simpler state space). Consider again the system shown in Fig. 3, with some further assumptions which lead to a simpler model:

- A machine is never idle if its queue is not empty (the state where a machine has finished working on a piece and has not begun to work with the next one, is considered as an irrelevant *vanishing* state).
- It is highly unlike that the maximum lengths of queues can be reached in practice, so we can model queues with an unbounded length.

Under these conditions, we can forget about the state of machines, and make this component of the state implicit in the state of the respective queues: for instance, a value of $q_1 = 1$ in Eqs. 4a–4c represents the state where machine M_1 is working on a piece and no other piece is waiting to be processed in queue#1; $q_1 = 2$ represents the state where machine M_1 is working on a piece and there exist one piece waiting to be processed in queue#1; $q_1 = 0$ may represent the state where no piece is waiting, and machine M_1 is idle. Since machine M_1 (resp. M_2) is never idle if there is a piece in queue#1 (resp. queue#2), the state q_2 of machine M_1 is implicit in q_1 (resp. q_4 is implicit in q_3).

$$q_2 = \begin{cases} \text{busy : if} & q_1 \geq 1 \\ \text{idle : if} & q_1 = 0 \end{cases} \tag{5}$$

$$q_4 = \begin{cases} \text{busy : if} & q_3 \geq 1 \\ \text{idle : if} & q_3 = 0 \end{cases} \tag{6}$$

These considerations lead to the following state space for the model of Fig. 3:

$$\mathbf{q} = [q_1, q_2]^T \in SS = \mathbb{N}^2 \tag{7}$$

Example 4 (A more complex state space). Suppose now that we take again the system of Fig. 3, but now we the performance analysis for which this model is to be used requires the modeller to take into consideration the following additional aspects:

- The states where a machine is moving pieces need to be considered.
- The state of the conveyors are relevant too (*running* or *stopped*).
- Discarded pieces are buffered into the baskets. The storage space of these baskets is limited to N' pieces.
- A machine cannot continue working if it has to put a piece in the next queue and this one is full, or it has to discard one piece and its basket is full.

Under these assumptions, we could obtain the model shown in Eqs. 8a–8e.

$$\mathbf{q} = [q_1, q_2, q_3, q_4, q_5, q_6, q_7, q_8]^T \tag{8a}$$

$$q_1, q_3 \in Q_1 = Q_3 = \{0, 1, \ldots, N\} \tag{8b}$$

$$q_2, q_4 \in Q_2 = Q_4 = \{\text{busy, idle, moving–out, discarding}\} \tag{8c}$$

$$q_5, q_6 \in Q_5 = Q_6 = \{0, 1, \ldots, N'\} \tag{8d}$$

$$q_7, q_8 \in Q_7 = Q_8 = \{\text{running, stopped}\} \tag{8e}$$

Note that state variables q_2 and q_4 have now a larger domain. The new state variables q_5, q_6 represent the state of the waste baskets of machine M_1 and M_2 respectively, while q_7, q_8 represent the state of the two conveyors in the subsystem under consideration.

Looking at examples 2–4, it is easy to realize that there does not exist a *unique* model for the same DES plant.

As we noted at the beginning of this subsection, in any case we can identify important differences between the state space of a DES, and the state space of a CS:

- The state variables of a DES do not take values from continuous sets, but from countable (and often finite) sets.
- Moreover, these discrete sets are often defined by enumeration, because no arithmetical or order relation exists among their elements; often the state sets of components are totally unstructured (for instance, Eqs. 8c and 8e).

2.2   State evolution

The behaviour of a dynamical system – whether CS or DES – is determined by the (set of) possible trajectories of the state as time evolves. We need a suitable mathematical formalism for representing this evolution. In any CS the state $\mathbf{x}(t)$ evolves continuously along time. Ordinary differential equations (ODE's) and all the mathematical tools related to them provide suitable methods to model, analyze and design controllers for these systems.

The CS of example 1 could be modeled by means of an ODE, whose most general expression is shown in Eq. 9a. If a *linear* approximation of the system dynamic may be sufficient for the specific problem purposes, we would take a model like Eq. 9b. If in addition to the linear property we can consider the system to be *time invariant*, we could take the well known expression of the *Linear Time Invariant* systems dynamic equation shown in Eq. 9c.

$$\dot{\mathbf{x}}(\tau) = f(\mathbf{x}(\tau), \mathbf{u}(\tau), \tau) \tag{9a}$$

$$\dot{\mathbf{x}}(\tau) = \mathbf{A}(\tau) \cdot \mathbf{x}(\tau) + \mathbf{B}(\tau) \cdot \mathbf{u}(\tau) \tag{9b}$$

$$\dot{\mathbf{x}}(\tau) = \mathbf{A} \cdot \mathbf{x}(\tau) + \mathbf{B} \cdot \mathbf{u}(\tau) \tag{9c}$$

If we compare the state evolution of a CS described by any of the above models to the evolution of the state of a DES, we find important differences:

- In a DES, the state does not evolve continuously along time; it remains constant except for abrupt jumps at the occurrence of *events*. Some of the events that cause the state to jump, in the example 2 with state representation 4a, are as follows:
  - A new piece arrives to queue#1
  - Machine M_1 finishes working on a piece and discards it
  - Machine M_1 finishes working on a piece and lets it go ahead
  - Machine M_1 begins to work on a piece
  - ... etc.

  At any given time point τ, the current system state $\mathbf{q}(\tau)$ uniquely determines the set of events that possibly may happen. Hence this also uniquely determines the set of all possible future trajectories of the system state.
- The state of a DES is a piecewise constant function of time, since events do not happen *continuously* along time, but just in a countable set of time instants. Between two consecutive occurrences of events, the state remains unchanged.
- There exist no synchronization method for the state evolution, that is, there does not exist some device like a *clock* that provides *ticks* or specific *time instants* where state changes are allowed to happen (events happen in an *asynchronous* way). The model only must describe the order in which the events occur.

One always could try to model and analyze a DES in a by using an ODE. Consider the queue length $q(t)$ in example Fig. 3, and define impulse–like inputs signals $u_1(t)$ and $u_2(t)$ whose occurrence denote respectively the arrival and departure of a piece to/from the queue. Then

$$q(\tau) = \int_{-\infty}^{\tau} u_1(t) \cdot dt - \int_{-\infty}^{\tau} u_2(t) \cdot dt \tag{10}$$

Fig. 4 shows an example of inputs $u_1(\tau)$, $u_2(\tau)$ and the corresponding sample path for state variable $q(\tau)$. In this formalism there is unfortunately no obvious way to explicitly express the requirement that the n-th departure must occur after the n-th arrival (implicitly expressed by the condition that $q(\tau) \geq 0$). This limitation of the formalism of Eq. 10 becomes even more cumbersome for large systems with many different events and with complicated precedence relations between them.

- Modelling and analysis of systems that involve asynchronous impulse–like (and often non–deterministic) signals is rather difficult within the ODE framework. These signals often do not come from outside the system, but are generated by internal dynamics.

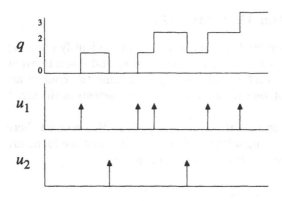

Fig. 4. Example of inputs signals and sample path for the model of Eq. 10

- In the above queue model example (Fig. 4, Eq. 10) it is clear that input signals may not take any arbitrary value at any arbitrary time instant: it is not possible that $u_2(\tau) = 1$ if $q(\tau) = 0$ (that is, no piece can leave an empty storage space).

- Some state spaces like the ones of examples 2–4 show no arithmetical or order relation among their values. How could we model this into the ODE's framework? One could think of assigning an integer value to each of these elements, and modeling the changing inputs also as impulse–like signals (either unitary or not).

  In that case we would be introducing in the model some semantics and relationships that are not contained in the system by itself.

- We actually do not need a continuous time model, because we only have state changes in a countable number of time instants.

- Another possibility could be the construction of a discrete time model, sampling the state at instants kT, and defining difference equations for relating the state $\mathbf{q}_k = \mathbf{q}(kT)$ to the state $\mathbf{q}_{k-1} = \mathbf{q}((k-1)T)$ at previous sampling instant(s). The problem in this case is that events may happen too close together, so that between two consecutive sampling instants more than one event may happen. Since the state evolution depends not only on what events happened, but also on the order in which they happened in, such a model is unable to represent the actual state evolution of the system. To fix this problem, one could try to take as sampling interval (if possible) a pessimistic estimate (lower bound) of the minimal time between two consecutive occurrences of events. In both cases, continuous or discrete time, one would get models where *nothing* would happen *most* of time.

The conclusion of this section is that ordinary differential equations or difference equations are not suitable to model these discrete event dynamic systems.

3   Interactions between DES and CS

In the control of complex automated systems, one can often identify dynamic subsystems of different kinds (continuous, discrete time, and discrete event systems). How these systems interact will greatly determine the control architecture, and whether or not we can talk about *hybrid systems* in the strict sense.

One possible control architecture is shown in Fig. 5. In this schema there exists (at least) one low level control layer, where continuous–time loops are controlled by either continuous or discrete time controllers.

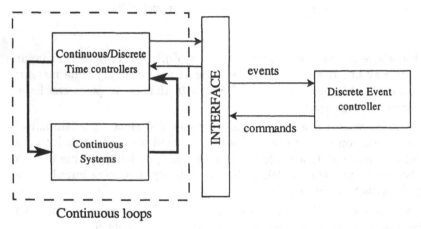

Continuous loops

Fig. 5. Hierarchical architecture of monitoring and control

This control layer accepts *commands* from a higher level controller. These commands may be, for example, changing the reference for some loop(s), modifying some controller(s) parameter(s), etc. On the other hand, the highest level controller *observes* some *events* from the lowest levels. Examples of these events may be, for example, that some process has reached its maximum allowed time to remain in a given state.

The highest level controller may be seen as a *discrete event system*, because it works with the discrete event nature of input/output signals that are received/issued in an asynchronous way. This controller is usually responsible for supervising the lower levels layers, monitoring the system startup and shutdown, changing operation modes, detecting system failures, etc. This is why this controller is usually known as *supervisory controller*, or just *supervisor*. The supervisor requires a global model of the whole plant for taking its decisions. This global model is usually obtained on a modular compositional approach basis, that allows the systematic construction of large complex models starting from very simple submodels, as we will see in this chapter.

Between the DES controller and the continuous plant, there must exist an *interface*, responsible for translating commands coming from the supervisor into the adequate actions over the lower level controllers, and similarly, observing the continuous states, detecting the occurrence of events, and issuing the adequate event to the supervisor.

Somehow, the control schema of Fig. 5 could be seen as a hybrid system, in the sense that it involves both continuous/discrete time dynamics, and event driven dynamics. Whether this should be considered as a hybrid system (HS) depends on how closely the DES dynamics and the CS dynamics are coupled. If CS problems can be solved just by using continuous time techniques without caring about the higher control level, then the system should not be treated as a HS. Similarly, the DE controller does not need to keep the track of the behaviour in the continuous loops, and can perform its task with just event–like information, then again the plant should not be treated as a HS. If the dynamics cannot be decoupled and events affect directly the evolution of the CS, and vice versa, then we need to model the system as a true HS (Fig. 6).

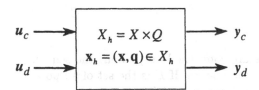

Fig. 6. Hybrid system

In a HS, the state space X_h is composed of a continuous subspace which contains all continuous dynamics state variables (X) and a discrete subspace (Q). The state at a given time instant $x_h(\tau) = (x(\tau), q(\tau))$ represents the continuous state variables $x(\tau)$ and the discrete state variables $q(\tau)$.

In general, we have both continuous time inputs and event driven inputs $u(\tau) = (u_c(\tau), u_d(\tau))$, The output $y(\tau) = (y_c(\tau), y_d(\tau))$ can be separated into 2 classes in the same way. The dynamics of the continuous state variables $x(\tau)$ are expressed by an ODE with right hand side depending on the *hybrid state* $x_h(\tau)$, on the inputs $u_h(\tau)$, at time τ.

$$\dot{x}(\tau) = f(x_h(\tau), u(\tau), \tau) \tag{11a}$$

$$q(\tau^+) = g(x_h(\tau), u(\tau), \tau) \tag{11b}$$

$$y(\tau) = h(x_h(\tau), u(\tau), \tau) \tag{11c}$$

Similar considerations apply for the discrete state variables q. We use the notation τ^+ in Eq. 11b to emphasize the fact that changes in q happen in an asynchronous event driven way. Note that $q(\tau^+) = q(\tau)$ whenever no event takes place. Which events are allowed at time τ, and what is the next state $q(\tau^+)$ is a function of $q(\tau)$, the value of the state just prior to the time when the event changing the state takes place. It should be clear that $q(\tau^+)$ is an element from the set of states that may be reached after the occurrence of any event that is *enabled* at state $x_h(\tau)$, since not *all* events may possibly happen at *any* state. Upon the occurrence of an event, some continuous state variables $x_r \in x$ may also *reset* their values, or more generally, may discontinuously change their values at the time instants where events happen.

These systems will be studied in further chapters of this part of the book. This chapter focuses attention to the modeling and analysis of DES only, so that from now on, events become the basic observables.

4   Modeling of DES

In modeling DES, we distinguish different levels of abstraction, depending on the temporal aspects that we consider in the model. These levels are often known as the *logic*, or *untimed* level, and the *timed* level.

4.1   Logic level

In the first level of abstraction we are concerned only about the sequences of events that can be generated by the system. If E is the set of all possible events that the system may generate, issue, etc., defined as

$$E = \{e_\alpha, e_\beta, \dots\} \tag{12}$$

then at this level, a *string s* (or *trace*, or *sequence of events*) of the system represents the ordered (and finite) set of events that have happened in the system so far.

$$s = e_1 e_2 e_3 \dots e_{n-1} e_n \quad e_i \in E \quad \forall\, i = 1, \dots, n \tag{13}$$

As we will see in the next Section, the even set E is often referred to as the *system alphabet*, and the set of all possible strings s that may be generated by a given system G is called the *system language* $\mathcal{L}(G)$.

At this level we have no information about time. For example, from a given system trace s like the one in Eq. 13 we cannot conclude how long the system took between the occurrence of the second and the third event.

Consider again the queueing example, with $E = \{arrival, departure\}$ as set of events. A specification stating that a piece should not wait in the queue for more than t time units cannot be modelled or controlled at this level. If time information is relevant for the control problem specifications, then it is necessary to introduce more information in the model.

4.2   Timed level

At the next abstraction level, beside the event sequence, we also specify the time instant at which events take place. Given an event set as in Eq. 12, and time $t \in \mathbb{R}^+$, the alphabet will now be a subset of $E \times \mathbb{R}^+$, and we will talk about the *timed language* of the system.

$$s = (e_1, \tau_1)(e_2, \tau_2) \ldots (e_n, \tau_n) \tag{14a}$$

$$e_i \in E, \; \tau_i \in \mathbb{R}^+, \quad \forall i \in 1, \ldots, n \tag{14b}$$

$$\tau_i < \tau_{i+1}, \quad \forall i \in 1, \ldots, n-1 \tag{14c}$$

A string s is now composed of pairs event–time, with the only condition that $\tau_i < \tau_{i+1}$ in order to guarantee the monotonic progress of time. If we allow more than one event to happen simultaneously, we could either allow $\tau_i = \tau_{i+1}$, or redefine Eqs. 14a–14c as

$$s = (E_1, \tau_1)(E_2, \tau_2) \ldots (E_n, \tau_n) \tag{15a}$$

$$E \supseteq E_i \in 2^E, \; \tau_i \in \mathbb{R}^+ \quad \forall i \in 1, \ldots, n \tag{15b}$$

$$\tau_i < \tau_{i+1}, \quad \forall i \in 1, \ldots, n-1 \tag{15c}$$

In this framework at a given time instant τ_i, instead of considering the occurrence of a single event e_i we allow the simultaneous occurrence of (in principle) *any* subset of E.

In this new framework we still do not allow the occurrence of two or more instances of the same event at the same time instant. It is possible to extend the framework to deal with this new situation.

$$s = (B_1, \tau_1)(B_2, \tau_2) \ldots (B_n, \tau_n) \tag{16a}$$

$$B_i : E \to \mathbb{N} \tag{16b}$$

$$\tau_i \in \mathbb{R}^+ \quad \forall i \in 1, \ldots, n \tag{16c}$$

$$\tau_i < \tau_{i+1}, \quad \forall i \in 1, \ldots, n-1 \tag{16d}$$

That is, instead of allowing the occurrence of events $e_i \in E$ or subsets of events $E_i \subseteq E$, we allow now the occurrence of *multisets*, or *bags* of events, where the same event may appear more than once. A bag B may be defined as a function from E to the set of natural numbers $\mathbb{N}$, where $B(e_i)$ denotes the number of times that e_i appears in B (known as the *cardinality* of e_i in B).

Of course, assumptions of Eqs. 16a–16d, where we allow the occurrence of multisets of events at a given time instant, is the most general framework we can think of. How general we make our model will greatly determine the complexity of the problems that can be posed in the study of the system.

5 Languages and automata

In this Section some formalisms to model and analyze DES at the logic (untimed) level are introduced. *Languages* and *automata*, and their relationship will be treated. Establishing an analogy with CS it could be said that languages are to DES what signals are to CS. Similarly, automata are to DES what state space models are to CS.

5.1 Languages

We use notation similar to that in Section 4.1:

- E: event set (alphabet)
- string: $s = e_1 e_2 e_3 \ldots e_n$ $e_i \in E$, $\forall i$
- ε : empty string
- E^+: set of all non–empty finite strings
- $E^* = E^+ \cup \{\varepsilon\}$

A language L is defined as an arbitrary subset of E^*. A string $t \in E^*$ is said to be a *prefix* of a string $s \in E^*$ iff there exists a string $u \in E^*$ such that $s = tu$. Denote by $\text{pref}(s)$ the set of all strings that are prefix of s. A string $u \in E^*$ is said to be a *suffix* of a string $s \in E^*$ iff there exists a string $t \in E^*$ such that $s = tu$. Denote by $\text{suff}(s)$ the set of all strings that are suffix of s. Note that for any $s \in E^*$

$$s = s\varepsilon \implies s \in \text{pref}(s),\ \varepsilon \in \text{suff}(s)$$
$$s = \varepsilon s \implies \varepsilon \in \text{pref}(s),\ s \in \text{suff}(s)$$

5.2 Some operations on languages

Concatenation: The concatenation of two languages $L_1, L_2 \subseteq E^*$ is defined as

$$L_1 L_2 = \{s \in E^* \mid \exists\, s_1 \in L_1, s_2 \in L_2 : s = s_1 s_2\} \tag{17}$$

The interpretation of this operation is the successive execution of two different kind of tasks. For example, if all the strings in L_1 represent the behavior of a subsystem S_1, and similarly L_2 the behavior of subsystem S_2, the language $L_1 L_2$ represents the complete execution of one task of system S_1 immediately followed by an execution of S_2.

Kleene closure: The *Kleene closure* of a language L is defined as

$$L^* = \{\varepsilon\} \cup L \cup LL \cup LLL \cup \cdots \tag{18}$$

The interpretation of this operation is the successive execution of tasks of the same kind. For example, if language L models the execution of some batch production system, language L^* represents the same batch production restarted over and over again, as many times as desired (including zero times, since $\varepsilon \in L^*$).

Prefix closure: The *prefix closure* of a language L is defined as

$$\overline{L} = \{s \in E^* \mid \exists\, t \in E^* : st \in L\} \tag{19}$$

A language L is said to be *prefix closed* iff $L = \overline{L}$. Note that the inclusion $L \subset \overline{L}$ always holds, because for any $s \in L$, we have that $s \in \text{pref}(s)$, thus $s \in \overline{L}$. It is easy to prove that for any language $L \subseteq E^*$, $\overline{\overline{L}} = \overline{L}$.

Example 5. Given $E = \{a, b\}$, let us consider the languages L_1 and L_2 defined as

$$L_1 = \{a, b, bbb\} \qquad L_2 = \{\varepsilon, a, aa, aab, b\}$$

If we obtain the prefix closure of L_1 and L_2 we will easily realize that L_2 is indeed prefix closed, but L_1 is not.

$$\overline{L_1} = \{\varepsilon, a, b, bb, bbb\} \supsetneq L_1 \qquad \overline{L_2} = \{\varepsilon, a, aa, aab, b\} = L_2$$

String projection: Given $E_o \subseteq E$ the *projection* of a string $s \in E^*$ over E_o^* as

$$P_{E_o} : E^* \to E_o^* \quad \begin{cases} P_{E_o}(\varepsilon) = \varepsilon \\ P_{E_o}(se) = \begin{cases} P_{E_o}(s)e & \text{if } e \in E_o \\ P_{E_o}(s) & \text{else} \end{cases} \end{cases} \tag{20}$$

The projection $P_{E_o}(s)$ removes from s all the events that do not belong to E_o.

String inverse projection: Given $E_o \subseteq E$ the *inverse projection* $P_{E_o}^{-1}$ of a string $s \in E_o^*$ over E^*, as a function $P_{E_o}^{-1} : E_o^* \to 2^{E^*}$ defined as

$$P_{E_o}^{-1}(s) = \{t \in E^* : P_{E_o}(t) = s\} \tag{21}$$

The inverse projection of a string s is the set of all strings t whose projection results in s.

Language projection, and inverse projection: In a similar fashion, it is possible to generalize the projection and inverse projection operation on strings to deal with languages. Defining $L \subseteq E^*$, $K \subseteq E_o^*$, $E_o \subseteq E$, the projection and inverse projection of languages are defined as

$$P_{E_o}(L) = \{s \in E_o^* \mid \exists t \in L : P_{E_o}(t) = s\} \tag{22a}$$
$$P_{E_o}^{-1}(K) = \{s \in E^* \mid \exists t \in K : P_{E_o}(s) = t\} \tag{22b}$$

In general

$$P_{E_o}\left(P_{E_o}^{-1}(K)\right) = K, \quad P_{E_o}^{-1}\left(P_{E_o}(L)\right) \supseteq L$$

Synchronous product: Given $E = E_1 \cup E_2$, $L_1 \subseteq E_1^*$, $L_2 \subseteq E_2^*$, and the projection operations $P_{E_1} : E^* \to E_1^*$, and $P_{E_2} : E^* \to E_2^*$, the *synchronous product* between L_1 and L_2 is defined as

$$L_1 \parallel L_2 = \{s \in E^* : P_{E_1}(s) \in L_1 \wedge P_{E_2}(s) \in L_2\} \tag{23}$$

The interpretation of $L_1 \parallel L_2$ is the following: assume that L_1 represents the behavior of subsystem S_1 whereas L_2 represents the behavior of subsystem S_2. Then, $L_1 \parallel L_2$ contains the strings that represents the simultaneous execution of subsystems S_1 and S_2. If these two subsystems share common events (i.e., $E_1 \cap E_2 \neq \emptyset$), they both are supposed to synchronize on these events, that is, these events have to be performed simultaneously by both subsystems.

5.3 Automata

An *automaton* is a representation of a DES. By representing a DES as an automaton, we are implicitly defining also the system language, as mentioned in 4.1. We will only deal with finite automata, i.e. automata whose state space is a finite set. An automaton may be defined as a tuple $G = (Q, E, \delta, D, q_0, Q_m)$ whose meaning is the following:

$$
\begin{aligned}
Q : \quad & \text{state set} \\
E : \quad & \text{event set} \\
\delta : Q \times E \to Q : \quad & \textit{transition} \text{ function} \\
D : Q \to 2^E : \quad & \textit{active event} \text{ function} \\
Q \ni q_0 : \quad & \text{initial state} \\
Q \supseteq Q_m : \quad & \text{set of \textit{marked} states}
\end{aligned}
$$

Q is the set of all possible states of the DES, i.e., the state space. If Q is a finite set, we will talk about *finite* automaton.

δ is the transition function, or *next state* function: if the current system state is $q \in Q$ and the event $e \in E$ happens, $\delta(q, e)$ will be the new system state after the occurrence of e. In general δ is a partially defined function, in the sense that $\delta(q, e)$ does not need to be defined for *every* $q \in Q$, and *every* $e \in E$. For example, considering again the model of a queue with $E = \{arrival, departure\}$, and assuming that the state q_{empty} represents the state where there exists no pieces waiting in the queue, it is clear that $\delta(q_{empty}, departure)$ has to be necessarily undefined. This is related to the D function.

D is the active event function, that is, the set of events that are allowed to happen for a given state. Hence, $D(q)$ represents the set of events $e \in D(q) \subseteq E$ that make function $\delta(q, e)$ to be defined.

q_0 is the initial state, i.e. the state at which the system is supposed to be in at the beginning of its operation.

The marked state set Q_m is a particular subset of Q. The elements in Q_m are typically states that represent the completion of a specific task, or a state that is considered *acceptable* from some point of view. For example, in a batch production plant model, a marked state may represent a state where the last batch was successfully completed, and machines are in a safe state to turn them off. This is why this set is often known as the *final state* set (this does not mean, of course, that the system stops when it reaches one of these states).

A finite automaton is often graphically represented as a directed graph. This is illustrated in the following example.

Example 6. Let us consider the graph shown in Fig. 7

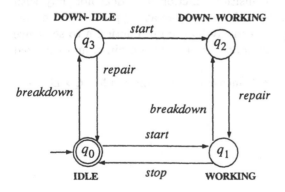

Fig. 7. Directed graph representation for automaton of Example 6

The directed graph representation of an automata, like in Fig. 7, is also known as a *state transition* diagram (bold capital labels are not actually part of the representation; they have been added to states just for the sake of clarity).

Fig. 7 represents the finite automaton whose state set is $Q = \{q_0, q_1, q_2, q_3\}$, initial state q_0, event set $E = \{start, stop, breakdown, repair\}$, and marked state set $Q_m = \{q_0\}$. Nodes in this graph represent states, and labeled and directed arcs (*arrows*) represent transitions between pairs of states.

Final states are represented by double circle nodes. The initial state is pointed by an arrow that is not labeled and is not connected to any other state.

The complete transition function for this example is

$$\delta(q_0, start) = q_1 \qquad \delta(q_1, stop) = q_0 \qquad \delta(q_1, breakdown) = q_2$$
$$\delta(q_2, repair) = q_1 \qquad \delta(q_0, breakdown) = q_3 \qquad \delta(q_3, start) = q_2$$
$$\delta(q_3, repair) = q_0$$

This function is sometimes represented by means of a table. In this example:

δ	q_0	q_1	q_2	q_3
start	q_1	—	—	q_2
stop	—	q_0	—	—
breakdown	q_3	q_2	—	—
repair	—	—	q_1	q_0

The symbol — means that the function is not defined for that pair (state,event).

The active event set function D is also represented in the graph. For a given state q the set $D(q)$ can be seen as the labels of the arrows that point outwards from q. For instance, in Fig. 7 $D(q_1) = \{stop, breakdown\}$.

It is possible to extend the transition function δ to deal not only with events, but also with event strings. We will denote as $\delta^*(q, s)$ the state that is reached from q by applying successively transition function δ to sequence of states successively reached from q and the events contained in s (in fixed order).

Assuming $e \in E, s \in E^*$, the *extended transition* function $\delta^* : Q \times E^* \to Q$ is defined as

$$\delta^*(q, \varepsilon) = q$$
$$\delta^*(q, es) = \begin{cases} \delta^*(\delta(q, e), s) & \text{if } e \in D(q) \\ \text{undefined} & \text{else} \end{cases} \tag{24}$$

We will write $\delta^*(q, s)!$ to denote that the state $\delta^*(q, s)$ is defined, i.e., that all successive applications from the state q of transition function δ to events in string s are well defined.

5.4   Some operations on automata

This subsection introduces some operations (among the many unary and binary operations that can be defined) on automata, that will be useful to understand the concepts introduced in the last Section of this chapter.

Accessible part: $Ac(G)$. The *accessible part* of an automaton G is an automaton obtained by removing from G those states that cannot be reached from its initial state q_0 after a finite number of state transitions. Formally, given an automaton $G = (Q, E, \delta, D, q_0, Q_m)$, the $Ac(G)$ operation is defined as

$$Ac(G) = (Q_{ac}, E, \delta_{ac}, D_{ac}, q_0, Q_{ac,m})$$
$$Q_{ac} = \{q \in Q : \exists s \in E^* : \delta^*(q_0, s) = q\}$$
$$\delta_{ac} = \delta \,|_{Q_{ac} \times E \to Q_{ac}}$$
$$D_{ac} = D\,|_{Q_{ac} \to 2^E}$$
$$Q_{ac,m} = Q_m \cap Q_{ac}$$

Example 7. Let us consider the automaton of Fig. 8. By simple visual inspection it is easy to realize that state q_3 is not accessible from initial state q_0 (there exists no arrow pointing to q_3). Similarly, it is easy to see that any other state is accessible, since there exists a string $s \in E^*$ that makes it reachable from q_0.

The accessible part of this automaton would be obtained by removing state q_3, and its associated transitions.

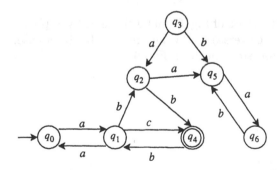

Fig. 8. Example of automaton to illustrate *Ac*, *CoAc*, and *Trim* operations

Coaccessible part: $CoAc(G)$. The coaccessible part of an automaton G is an automaton obtained from G by removing those states from which final states cannot be reached, after a finite number of state enabled transitions. Formally, given an automaton $G = (Q, E, \delta, D, q_0, Q_m)$, the $CoAc(G)$ operation is defined as

$$CoAc(G) = (Q_{coac}, E, \delta_{coac}, D_{coac}, q_{0,coac}, Q_m)$$
$$Q_{coac} = \{q \in Q : \exists s \in E^* : \delta^*(q, s) \in Q_m\}$$
$$\delta_{coac} = \delta|_{Q_{coac} \times E \to Q_{coac}}$$
$$D_{coac} = D|_{Q_{coac} \to 2^E}$$
$$q_{0,coac} = \begin{cases} q_0 & \text{if } q_0 \in Q_{coac} \\ \text{undefined} & \text{else} \end{cases}$$

Example 8. Let us consider again the automaton of Fig. 8. Just like in Example 7, we can easily realize that states q_5 and q_6 are not coaccessible. The only final state (q_4) cannot be reached from $\{q_5, q_6\}$. If the system falls into any of these two states, it will be unable to leave it and reach a final state. This situation is known as *livelock*, in contrast to a *deadlock* state: if the system gets into a deadlock state, it will not be able to progress any more (in the case of DES this means that no more events will ever happen). On the other hand, in a livelock situation the system may still continue changing state by generating events, but it will never reach a final state.

Trim operation: *Trim(G)*. The trim operation transforms automaton G into another automaton (a part of G) that is both accessible and coaccessible. The automaton *Trim(G)* can be expressed in terms of accessible and coaccessible operations, which may permute. Formally, given an automaton $G = (Q, E, \delta, D, q_0, Q_m)$, the *Trim(G)* operation is defined as

$$Trim(G) = CoAc(Ac(G)) = Ac(CoAc(G))$$

Example 9. The trim part of automaton of Fig. 8 can be obtained by applying the *Ac* and *CoAc* operators (or vice versa) to this automaton. In either case, this operation results in the automaton shown in Fig. 9.

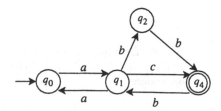

Fig. 9. Trim part of automaton of Fig. 8

Parallel composition: $G_1 \parallel G_2$. Given two different automata $G_1 = (Q_1, E_1, \delta_1, D_1, q_{0,1}, Q_{m,1})$, $G_2 = (Q_2, E_2, \delta_2, D_2, q_{0,2}, Q_{m,2})$, the *parallel composition* of G_1 and G_2, denoted as $G_1 \parallel G_2$, is the automaton defined as follows:

$$G_1 \parallel G_2 := (Q_1 \times Q_2, E_1 \cup E_2, \delta, D, (q_{0,1}, q_{0,2}), Q_{m,1} \times Q_{m,2})$$

where

$$\delta\left((q_1, q_2), e\right) = \begin{cases} (\delta_1(q_1, e), \delta_2(q_2, e)) & \text{if } e \in D_1(q_1) \cap D_2(q_2) \\ (\delta_1(q_1, e), q_2) & \text{if } e \in D_1(q_1)\backslash E_2 \\ (q_1, \delta_2(q_2, e)) & \text{if } e \in D_2(q_2)\backslash E_1 \\ \text{undefined} & \text{otherwise} \end{cases}$$

and

$$D(q_1, q_2) = (D_1(q_1) \cap D_2(q_2)) \cup (D_1(q_1)\backslash E_2) \cup (D_2(q_2)\backslash E_1)$$

In words, the parallel composition of automata G_1 and G_2 results in an automaton that simultaneously models the behavior of both systems. Private events, that is, events in $E_1\backslash E_2$ and $E_2\backslash E_1$ for G_1 and G_2 respectively, may be executed by their corresponding system at any moment they are enabled by the current state value at their corresponding component (subautomaton) (see lines 2 and 3 of the definition of function δ). On the other hand, events in $E_1 \cap E_2$ are shared by both systems, and are allowed to happen only when they are enabled in both systems (see first line of the definition of δ).

The parallel composition operation is one of the most powerful tools for the systematic obtaining of large models for complex systems, starting from small subsystems models whose correctness is easy to verify just at a glance.

It is possible to prove that

$$\mathcal{L}(G_1 \parallel G_2) = \mathcal{L}(G_1) \parallel \mathcal{L}(G_2)$$

5.5 Languages vs. automata

Both languages and automata can be seen as a formal definition of a DES. In both cases all the possible sequence of events that can be generated by the system are well defined.

Any automaton G implicitly defines two languages, the *language generated* by the automaton, denoted as $\mathcal{L}(G)$, and the *language marked* by the automaton, denoted as $\mathcal{L}_m(G)$. Formally, given an automaton $G = (Q, E, \delta, D, q_0, Q_m)$ these languages are defined as

$$\mathcal{L}(G) = \{s \in E^* \ : \ \delta^*(q_0, s)!\} \tag{25}$$

$$\mathcal{L}_m(G) = \{s \in E^* \ : \ \delta^*(q_0, s) \in Q_m\} \tag{26}$$

From these definitions it is clear that $\mathcal{L}_m(G) \subseteq \mathcal{L}(G)$.

Finite automata are a particularly interesting formalism for representing DES. This interest stems from the fact that most problems that can be posed for a DES can be solved by means of a computer algorithm in a polynomial time if the DES is represented by means of a finite automaton. Thus, an immediate question is how large the set of languages that can be represented by finite automata is.

Unfortunately, a finite automata cannot represent *every* language. The set of languages that can be represented by an automaton is known as the set of *regular languages*. This set is quite restrictive compared to the set of *all* possible languages, but is general enough to be important in practice to represent the behavior of many DES.

6   Introduction to supervisory control

This section gives a brief and necessarily incomplete introduction to supervisory control. Supervisory control must guarantee that the behavior of the overall plant is always acceptable. Here acceptable may mean that no deadlocks (or livelocks) occur, and/or that certain variables always remain within a specified safe set, and/or that eventually all the tasks are completed.

The plant manager describes certain specifications of the closed loop plant. The open loop behavior $\mathcal{L}(G)$ may contain strings that are not acceptable in the sense that they violate some of the specifications. The supervisor will then influence the plant behavior by sometimes blocking certain events, which would be allowed according to the plant model G. By blocking some controllable transitions the supervisor eliminates all those trajectories that lead to a violation of a specification. Supervisory control is usually implemented by blocking as few controllable events as possible. This allows the maximal freedom for the lower level controllers to optimize their actions.

Since the supervisory controller must act autonomously on the overall plant, it is important to observe that the specifications are hard constraints on the plant behavior. The supervisor must ensure that all future trajectories satisfy the specifications. This is different from a stochastic approach, where one would only require that the plant satisfies the specifications most of the time or that it minimizes some average cost. Moreover, as it has been already noted, the supervisor must have a global model of the plant when predicting future trajectories. This explains why the supervisor usually works with a composition of many interacting models. Computational tractability usually dictates the use of abstract discrete event models for most of the components.

6.1   General framework

We will consider a DES $G = (Q, E, \delta, D, q_0, Q_m)$ as defined in Section 5, where the event set E may be partitioned into two disjoint subsets

$$E = E_c \cup E_{uc}$$

E_c is the set of *controllable* events. Any event that may be disabled by the controller belongs to this set. On the other hand, E_{uc} is the set of *uncontrollable* events. Events in E_{uc} cannot be prevented from happening by the controller by means of any control action over the system.

Similarly, event set E accepts also another partition into two different disjoint subsets

$$E = E_o \cup E_{uo}$$

E_o represents the set of *observable* events. Any event in E_o can be *seen*, or *observed*, by the controller upon its occurrence. On the other hand, E_{uo} is the set of *unobservable* events, that is, those events whose occurrence cannot be noticed by the controller. As it can be intuitively guessed, the existence of uncontrollable and/or unobservable events in the system introduces extra complexity to the supervisory control problem.

Fig. 10 shows the general feedback loop structure for supervisory control.

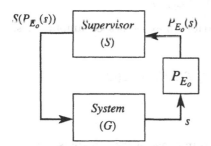

Fig. 10. General supervisory control loop

Since, in general, unobservable events may exist, it is clear that the system output that is fed back to the supervisor will not be the whole event string s generated by the system G. If we define the projection operation $P_{E_o} : E^* \to E_o^*$, the information available for the supervisor will be $P_{E_o}(s)$. The control action of the supervisor will be denoted as $S(P_{E_o}(s))$, where S may be seen as a function $S : E_o^* \to 2^E$. As mentioned at the beginning of this section, the goal of this control action is to disable some events. We will assume that $E_{uc} \subseteq S(P_{E_o}(s)) \subseteq E$ is the set of events *allowed* by the supervisor after the observed system behavior $P_{E_o}(s)$. In this framework it is assumed that the control action is immediately updated (but not necessarily changed) just after the occurrence of any observable event.

Thus, when an event string s has been generated by the system G, then the set of allowed events is

$$S(P_{E_o}(s)) \cap D(\delta^*(q_0, s))$$

that is, events that are allowed by the supervisor and that are also active in the system.

One of the most important issues related to partial observability are system failures, which are rarely observable by the controller. One important research area nowadays in supervisory control is the problem of *diagnosis*, that is, how to infer that some unobservable event has (possibly or certainly) happened just by looking at the observable behavior of the system. Issues related to partial observability will not be treated in this introduction. Thus, from this point on it is assumed that all events are observable.

We will denote as $\mathcal{L}(S/G)$ the language generated by the closed loop system. Assuming $s \in E^*$, $e \in E$, the language $\mathcal{L}(S/G)$ may be defined as

$$\varepsilon \in \mathcal{L}(S/G)$$
$$s \in \mathcal{L}(S/G), se \in \mathcal{L}(G), e \in S(s) \iff se \in \mathcal{L}(S/G)$$

The marked language of the closed loop system consists of the strings that are allowed by the supervisor and that are marked in the open loop system:

$$\mathcal{L}_m(S/G) = \mathcal{L}(S/G) \cap \mathcal{L}_m(G)$$

In general, the following inclusion relations may be stated among these languages

$$\emptyset \subseteq \mathcal{L}_m(S/G) \subseteq \overline{\mathcal{L}_m(S/G)} \subseteq \mathcal{L}(S/G) \subseteq \mathcal{L}(G)$$

6.2   Controllability

One of the first questions that can be stated in supervisory control is, given a DES $G = (Q, E, \delta, D, q_0, Q_m)$, and a desired closed loop language K, is how to decide whether there exists a supervisor S such that $\mathcal{L}(S/G) = \overline{K}$.

The following theorem provides necessary and sufficient conditions for the existence of such a supervisor.

Theorem 1 (Controllability Theorem).
Given a DES $G = (Q, E, \delta, D, q_0, Q_m)$, $E_{uc} \subseteq E$ and $\emptyset \subsetneq K \subseteq \mathcal{L}(G)$ there exist supervisor S such that $\mathcal{L}(S/G) = \overline{K}$ if and only if

$$\overline{K} E_{uc} \cap \mathcal{L}(G) \subseteq \overline{K}$$

This Condition on K is known as the *controllability condition*, and it can be read in the following terms: if a string $s \in K$ admits a prefix $t \in$ pref(s) that can be extended with an uncontrollable event $e \in E_{uc}$, and the resulting string te belongs to the system language $\mathcal{L}(G)$, then string te should also belong to $\overline{K}$. In other words: if the supervisor cannot prevent a string from happening, that string should be legal (that is, the string satisfies the specification K). If this condition holds, the supervisor S that makes $\mathcal{L}(S/G) = \overline{K}$ is defined as

$$S(s) = [E_{uc} \cap D\left(\delta^*(q_0, s)\right)] \cup \{e \in E_c \; : \; se \in \overline{K}\}$$

After observing any event string s, the supervisor enables all uncontrollable events that are also active in the system (left hand side of union operator) and those controllable events whose occurrence keeps the system inside the desired behavior (right hand side of union operator).

In general, one may find that the desired language K is not controllable. A natural choice then is to take a subset or superset of K, depending on the problem specification. If K represents the largest allowed language, then one could try to take the largest controllable sublanguage of K. On the other hand, if K represents the smallest[1] required language that should be included in the closed loop system language, and the latter should be the smallest possible, one could then try to take the smallest controllable superlanguage of K.

The *supremal controllable sublanguage* of K is denoted as $K^{\uparrow C}$. The *infimal prefix–closed controllable superlanguage* of K is denoted as $K^{\downarrow C}$.

The prefix–closure condition on $K^{\downarrow C}$ is a technical condition allowing a constructive definition of $K^{\downarrow C}$: it is the intersection of all the prefix–closed controllable superlanguages of K. The set of controllable languages is *not* closed under intersection, but the set of prefix–closed controllable languages is.

For any $\emptyset \subsetneq K \subseteq \mathcal{L}(G)$, the following inclusion relations hold

$$\emptyset \subseteq K^{\uparrow C} \subseteq K \subseteq \overline{K} \subseteq K^{\downarrow C} \subseteq \mathcal{L}(G)$$

Depending on whether K is or not prefix–closed and/or regular, there exist different methods to compute $K^{\downarrow C}$ and $K^{\uparrow C}$.

6.3 Blocking

Another important issue related to supervisory control is *blocking*. A DES is said to be in a blocking state if, from that state, it cannot reach a marked state. Formally, given a DES $G = (Q, E, \delta, D, q_0, Q_m)$, it is said to be *nonblocking* iff

$$\overline{\mathcal{L}_m(G)} = \mathcal{L}(G)$$

The inclusion $\overline{\mathcal{L}_m(G)} \subseteq \mathcal{L}(G)$ always holds, because any prefix of a string $s \in \mathcal{L}_m(G)$ has to belong necessarily to the language generated by the system $\mathcal{L}(G)$. Thus, the actual nonblocking condition is $\overline{\mathcal{L}_m(G)} \supseteq \mathcal{L}(G)$, whose interpretation is the following: any string s generated by the system, $s \in \mathcal{L}(G)$,

---

[1] When we say *largest* and *smallest* we are referring to set inclusion among languages

can be extended by another string $t \in E^*$, such that the string st leads the system to a marked state.

The nonblocking condition can also be checked on the automaton modelling the system. System G is nonblocking iff

$$Ac(G) \subseteq CoAc(G)$$

All accessible states of G should also be coaccessible, i.e., the system may eventually reach a marked state from any accessible state.

The following theorem provides necessary and sufficient conditions for the existence of a nonblocking supervisor.

Theorem 2 (Nonblocking Controllability Theorem). *Given a DES* $G = (Q, E, \delta, D, q_0, Q_m)$, $E_{uc} \subseteq E$, *and* $\emptyset \subsetneq K \subseteq \mathcal{L}_m(G)$, *there exists a supervisor* S *such that*

$$\mathcal{L}_m(S/G) = K, \quad \text{and} \quad \mathcal{L}(S/G) = \overline{K}$$

iff

1) $\overline{K} E_{uc} \cap \mathcal{L}(G) \subseteq \overline{K}$ *(controllability condition)*
2) $K = \overline{K} \cap \mathcal{L}_m(G)$ *($\mathcal{L}_m(G)$-closure)*

(note that, in this case, language K does not represent the desired generated language, but the desired *marked* language).

Condition 1 corresponds to the controllability condition of Theorem 1 for language $\overline{K}$ (recall that $\overline{\overline{K}} = \overline{K}$). Condition 2 is known as the $\mathcal{L}_m(G)$-*closure* condition. Since $K \subseteq \overline{K}$ and $K \subseteq \mathcal{L}_m(G)$ (by the assumptions of Theorem 2), inclusion $K \subseteq \overline{K} \cap \mathcal{L}_m(G)$ always holds. Inclusion $K \supseteq \overline{K} \cap \mathcal{L}_m(G)$ is what does not hold in general, and it can be read in the following terms: if a string $s \in K$ (which is a *marked string* by assumption, since $K \subseteq \mathcal{L}_m(G)$) admits a prefix $t \in \text{pref}(s)$ which is also a marked string (i.e., $t \in \mathcal{L}_m(G)$), then string t should also belong to K.

6.4 Supervisory control specifications

In the typical specification of most supervisory control problems, one defines one or more *required* and/or *allowed* languages, either marked or not, for instance L_{req}, L_{all}, $L_{m,req}$, and $L_{m,all}$, and specifies as requirements one or more of the following four inclusion relations

$$L_{req} \subseteq \mathcal{L}(S/G) \subseteq L_{all}, \qquad L_{m,req} \subseteq \mathcal{L}_m(S/G) \subseteq L_{m,all}$$

Of course, controllability and nonblockingness are usually part of the problem specification as well.

When we set as specification the largest allowed language, we usually require the supervisor to be *maximally permissive*, in the sense that is should allow the system to behave as free as possible, while keeping it inside the allowed behavior. Formally, if L_{all} is the largest allowed language, a supervisor S such that $\mathcal{L}(S/G) \subseteq L_{all}$ is said to be maximally permissive if for any other supervisor S_{other}

$$\mathcal{L}(S_{other}/G) \subseteq L_{all} \quad \Longrightarrow \quad \mathcal{L}(S_{other}/G) \subseteq \mathcal{L}(S/G)$$

Summary

In this Chapter we have introduced the fundamentals of DES. We have seen why DES are indeed a particular kind of dynamic systems, and why ordinary differential equations and difference equations are not suitable tools to model these systems. How DES interact with continuous systems, greatly determines what we can call a *hybrid system* in the strict sense. A couple of formalisms to describe DES at the logic (*untimed*) level of abstraction have been introduced. These are *languages* and *automata*. Some of the most important operations between languages and automata have been introduced, and the existing relationship between these two formalisms has also been presented. Finally, the last Section presented a brief introduction to *supervisory control* where two of the most important problems, *controllability* and *blocking*, were considered. Two of the most relevant theorems that provide necessary and sufficient conditions for these problems have also been quoted.

Recommended references

For a much more complete introduction to DES, the reader is strongly encouraged to see [1]. This book presents a very comprehensive, self contained, and excellent introduction to the fundamentals of DES. It deals not only with the concepts quoted in this Chapter, but also with some other modeling formalisms like Petri nets and other abstraction levels in the modeling of DES like *stochastic models*.

For an introduction to hybrid systems, the reader is suggested to see the following Chapters of this Part, and references therein, in this book. Also, in the *Lecture Notes in Computer Science* and *Lecture Notes in Control and Information Sciences* series books edited by Springer, some nice books on this topic can be found. For example, see [6] and [7]. Some journals edit also special issues on this topic. For example, for a nice discussion on hybrid systems, the reader is suggested to see the guest editors' presentation of [5]. In [5] can also be found good papers that represent different approaches in the study of hybrid systems.

The study of languages and automata theory is also perfectly covered by [1], specially from the control systems point of view. For a good introduction of this topic from the computer science point of view and the study of *formal languages*, see [3].

The initial work on supervisory control is [2], although in [1] the reader can also find an excellent introduction to this topic. For an introduction to the problem of supervisory control under partial observability and failure diagnosis, see [4,8] and references therein.

References

1. Cassandras, C.G., Lafortune, S., *Introduction to Discrete Event Systems*, Kluwer Academic Publishers, Boston 1999.
2. Ramadge, P.J., Wonham, W.M., "The Control of Discrete Event Systems", *Proceedings of the IEEE*, **77-1**, 81–98, 1989.
3. Hopcroft, J.E., Ullman, J.D. *Introduction to Automata Theory, Languages, and Computation*, Addison–Wesley, Reading, MA 1998.
4. Sampath, M., Sengupta, R., Lafortune, S., Sinnamohideen, K., Teneketzis, D. "Failure Diagnosis Using Discrete Event Models", *IEEE Transactions on Control Systems Technology*, **4**, 105–124, 1996.
5. Antsaklis, P.J., Nerode, A., (eds.) *IEEE Transactions on Automatic Control: Special Issue on Hybrid Control Systems*, **43–4**, April 1998.
6. van der Schaft, A., Schumacher, H., *An Introduction to Hybrid Dynamical Systems*, Lecture Notes in Control and Information Sciences, **251**, Springer, 2000.
7. Vaandrager, F.W., van Schuppen, J.H., (eds.) *Hybrid Systems: Computation and Control*, Second International Workshop HSCC'99, Lecture Notes in Computer Science, **1569**, Springer, 1999.
8. Cassandras, C.G., Lafortune, S., "Discrete event systems: The state of the art and some recent trends", in E. Datta (ed.), *Applied and computational control, signals and circuits*, **83–147**, Birkhäuser, 1998.

Petri Nets Models of Timed Discrete Event Plants

René K. Boel

SYSTeMS Group, Electrical Engineering Department, Universiteit Gent, Belgium

1 Introduction

The goal of this chapter is to introduce Petri nets as a state-based modelling formalism for discrete event systems. As explained in the preceding chapter, the state of a discrete event system belongs to a countable set of possible values. The state remains constant, except at the asynchronously occurring instants along the time axis, when an event takes place. A mathematical model of such a plant must specify the order in which events can occur, and in the case of timed models, it must also specify the time instants when the events can occur. Translating this statement in the language of state-based models, this means the following: for each current value of the state the model must specify which events are allowed to occur (and, in the timed models, when they can occur); given that an event occurs, the model must also specify what the next state is going to be.

The preceding chapter has introduced automata as state-based modelling formalism for describing the set of sequences of events that can be generated by a DES plant. In each state the model specifies a set of transitions which are enabled, and with each transition which is enabled in the current state, there corresponds a mapping of the current state into the next state, given that the transition occurs. Events are associated to the occurrence of transitions. The problem with automata is that the state space is a completely unstructured set without any ordering relation between its elements. The set of enabled transitions, and the "next state" function, are necessarily defined by enumeration, for each value of the current state. It is in general impossible to express the set of enabled transitions via algebraically expressed constraints. Neither is there in general an algebraic expression for calculating the next state.

Enumerating sets becomes computationally infeasible for realistic plant models consisting of many components. Each component expresses certain constraints on the ordering of the events, via the set of enabled transitions and via the "next state" function associated to these enabled transitions. Feasible trajectories of the overall plant must satisfy the constraints imposed by each component separately.

In this chapter Petri nets are used as model of automata which can be represented in a compact graphical form, and where the state space is nicely structured as a vector with integer components. Sets of enabled transitions and "next state" functions are represented using linear functions of the state, with integer valued coefficients. Section 2 presents some examples illustrating how Petri nets can be used to model typical discrete event systems, such as flexible manufacturing systems and communication protocols, and how the structure of the state vectors can be used for analysis of plant properties and for control design. These examples also illustrate the limitations to the modelling power of Petri nets, due to the fact that the model only specifies sets of enabled transitions, but never expresses that a transition is forced to occur.

Some elements of the set of feasible trajectories, satisfying all the model constraints, may have properties that are not acceptable for the proper operation of the plant. Specifications expressing safety conditions, avoiding uneconomical operating conditions, etc. can often be expressed via linear inequalities that must be satisfied by all reachable states. In order to allow autonomous plant operation these specifications must be enforced strictly by a high level, supervisory controller (as introduced in control theory by Ramadge and Wonham in [2]). In section 3 the combination of graphical and linear algebraic analysis tools for Petri nets will be used for the synthesis of maximally permissive supervisory controllers.

As explained in the preceding chapter discrete event models, such as automata and Petri nets, only enforce the order of occurrence of events. Practical applications often require the specification not only of the ordering of events, but also of the real time when events can take place. This is the case for example when a manufacturing system is handling perishable items, which have to leave the plant before a certain due date. For this purpose section 4 adds constraints to the Petri net model which specify that an enabled transition must be executed in a certain time interval, measured from the time when the transition became enabled. Timed Petri net models represent much more general discrete event systems, since it is possible to express events that are forced to occur. Section 4 presents the modelling formalism for timed Petri nets, and discusses some examples of the use of timed Petri nets for the modelling of communication and transportation networks. Section 4 also illustrates how the use of timed models influences the problem of control synthesis. Deadlock prevention and task scheduling are used as case studies.

2   Petri net models

2.1   Petri net semantics

A marked Petri net (P, T, F_+, F_-, m_0) is a graph, with the following elements:

- $\#P$ *places* $p \in P$ (denoted by circles in the graphical representation)
- $\#T$ *transitions* $t \in T$ (denoted by bars in the graphical representation)
- a set F_+ of directed arcs pointing from some transitions to some of the places (F_+ can be interpreted as a relation, or as a subset of $T \times P$). In

the graphical representation these arcs are lines starting at t and ending with an arrow at p.

- a set F_- of directed arcs pointing from some places to some of the transitions (F_- can be interpreted as a relation, or a subset of $P \times T$). In the graphical representation these arcs are lines starting at p and ending with an arrow at t.

- the *initial marking* $m_0 \in \mathbb{N}^{\#P}$, a column vector of $\#P$ nonnegative integers, each element counting the number of *tokens* that initially are present in the places $p \in P$. The $m_0(p)$ tokens in p are represented graphically by $m_0(p)$ dots in the circle p.

In order to simplify the notation we introduce the following sets: ${}^{\bullet}t = \{p \in P \mid (p,t) \in F_-\}$ and $t^{\bullet} = \{p \in P \mid (t,p) \in F_+\}$ is the set of input (resp. output) places of transition t. Input and output transitions of a place p are defined as ${}^{\bullet}p = \{t \in T \mid (t,p) \in F_+\}$ and $p^{\bullet} = \{t \in T \mid (p,t) \in F_-\}$.

Example 1: production unit in FMS: Figure 1 is an example of a Petri net with 4 places and 2 transitions. It represents a production unit consisting of 3 machines, an input buffer and an output buffer. It is one component of a larger plant. The input (left) and output (right) places are the buffers, where workpieces are stored before and after the operation carried out by this production unit. In the example there is initially one workpiece waiting in the input buffer, and there are 2 finished products waiting in the output buffer. The two other places represent the status of the 3 equivalent machines in this production unit; the tokens in p_{free} count the number of idle machines (initially 2), while the token in p_{busy} correspond to 1 busy machine. The transitions represent resp. the start of an operation (machine starts working on a workpiece in the input buffer) and the completion of an operation (machine completes work on a workpiece, puts it in the output buffer and returns to the idle condition).

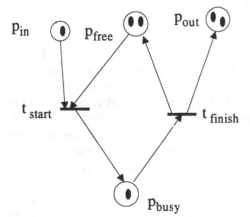

Fig. 1. Petri net model of a production unit with input and output buffers

A Petri net models the occurrence of an event in a discrete event plant by the execution of a corresponding transition $t \in T$. In example 1 the execution of transition t_{start} corresponds to the start event, while the execution of transition t_{finish} corresponds to the completion event. The marking m of the places determines the state of the plant. It will become clear that the marking does indeed determine the set of all possible future state trajectories that can be reached. This justifies the use of the term "state". In example 1 the marking $m = [m_{in}, m_{free}, m_{busy}, m_{out}] = [1, 2, 1, 2]$ indicates that there is one workpiece waiting in the input buffer, while two items are in the output buffer; 2 machines are idle while 1 machine is busy.

The dynamics of the sequence of events generated by a marked Petri net is determined as follows. Given a current value m of the state, a transition is enabled if each of its input places contains at least one token:

$$t \text{ is enabled} \iff \forall p \in {}^\bullet t : m(p) \geq 1$$

The event corresponding to t can occur provided that the constraints $m(p) \geq 1$ are simultaneously satisfied for all input places of t. Let $T_{enabled}(m)$ be the set of enabled transitions given the current value $(m(p), p \in P)$ of the state. The next event that will occur corresponds to a transition $t_{next} \in T_{enabled}(m)$, selected arbitrarily among the transitions in $T_{enabled}(m)$. The next marking, following the execution of t_{next}, is obtained by removing one token from all the input places of t_{next} and by adding one token to each of the output place of t_{next}. The affine function $f : \mathbb{N}^{\#P} \to \mathbb{N}^{\#P}$ generating the next state $m_{next} = f(m, t_{next})$ is defined as follows:

First update the marking m according to

$$\forall p \in {}^\bullet t_{next} : m_{inter}(p) = m(p) - 1$$
$$\forall p \notin {}^\bullet t_{next} : m_{inter}(p) = m(p)$$

Then update the marking $m_{inter}(p)$ according to:

$$\forall p \in t_{next}^\bullet : m_{next}(p) = m_{inter}(p) + 1$$
$$\forall p \notin t_{next}^\bullet : m_{next}(p) = m_{inter}(p)$$

Example 1 continued: For the marking indicated in Figure 1 both transitions are enabled. If t_{start} is selected as next event to occur, then the next state is $m_{next} = [0, 1, 2, 2]$; if t_{finish} is selected as the next event then $m_{next} = [1, 3, 0, 3]$. In the first case, the only transition that is enabled under the state m_{next} is t_{finish}; in the second case the only enabled transition is t_{start}. It is easy to verify that this expresses a condition on the language generated by the automaton, represented by the Petri net in example 1: the number of times that event t_{start} is executed is equal to $m(p_{idle}) - m_0(p_{idle})$ + the number of times that t_{finish} is executed.

It is easy to enumerate the set of all states that can be reached in example 1, since transition t_{start} can be executed at most once. Indeed p_{input} contains only one token, and there is no input transition to p_{input} that could add more tokens. The reader should verify that the set of allowed traces of this discrete event model is the prefix-closure of (see section 5.1 of the preceding chapter):

$$\{(t_{start}, t_{finish}, t_{finish}, t_{finish}), (t_{finish}, t_{start}, t_{finish}, t_{finish})\}$$

The set of reachable states is

$$\{[1, 2, 1, 2], [0, 1, 2, 2], [1, 3, 0, 3], [0, 2, 1, 3], [0, 3, 0, 4]\}.$$

It is easy to observe some general rules that hold along any state trajectory generated by the model of example 1 : $m(p_{in})$ is monotonely decreasing, while $m(p_{out})$ is monotonely increasing. Moreover it can be verified for all reachable states that $m(p_{idle}) + m(p_{busy}) = m_0(p_{idle}) + m_0(p_{busy}) = 3 =$ the number of machines in the production unit. The proof of this invariance property is very easy: the dynamics of the Petri net allows only two transitions; if t_{start} occurs then $m(p_{idle})$ is reduced by 1 while $m(p_{busy})$ is increased by 1 keeping the sum constant, if t_{finish} occurs the opposite state change takes place, again keeping the sum constant. Note that not every state that satisfies this equality can be reached. Another invariant property that can be demonstrated in the same way is that $m(p_{output}) - m_0(p_{output}) = m_0(p_{busy}) - m(p_{busy}) +$ the number of times that t_{start} occurred. This is also equal to the number of times that t_{finish} occurred.

Transition t_{start} expresses a synchronization requirement. The corresponding "start" event can only occur when two conditions are satisfied simultaneously: there must be a workpiece waiting in the input buffer, and there must be an idle machine. Such synchronization requirements can be used to build large models using models of smaller components. A transition t_{syn} that appears in more than one component model can only be executed (the corresponding event can only occur) when the transition t_{syn} is enabled in all the components where it appears. Suppose that starting work on a workpiece that is waiting in the input buffer requires the availability of a crane, used also for other operations, in order to move the workpiece from the input buffer to the machine. Then one has to add to the model of the overall plant a second Petri net component describing the location and the availability of the crane. In this second Petri net the transition t_{start} will also appear, and it will have an input place that is marked by a token only when the crane is not carrying a load and is in the right position for picking up the workpiece from the input buffer.

A place p with several output transitions represents choice in the Petri net. If $m(p) \geq 1$ then one of the transitions $t \in {}^{\bullet}p$ can be selected as next event to occur (provided all the other synchronizing enabling conditions of this t are also satisfied). In the transport model choice represents the fact that the plant supervisor can decide to send the crane to different locations, depending on the need of the production schedule. In a very simple model the token (= crane) could move either to the place on the left or to the place on the right from its present place. Choice can also be used to model plant breakdowns and repairs, as illustrated in Figure 2, which provides a more detailed model of the production unit of Figure 1. Tokens in the place p_{bd} model a machine that is broken down. When the machine is repaired this can happen either by transition t_{rep} with loss of the workpiece that was in the machine at the time of the breakdown, or by transition $t_{rp,2}$ which corresponds to a repair where the work on the workpiece can be continued in p_{busy}.

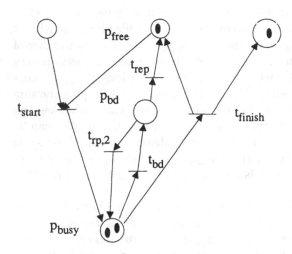

Fig. 2. Petri net model of a production unit, with machine breakdown and repair

Comparing Figure 1 and Figure 2, one sees that, depending on the purpose of the model, it may be possible to use different Petri net models with different levels of detail. Some of the invariant properties which have been proven for the simpler model may remain true, while others require refinement. The invariant property expressing that the total number of machines remains constant now becomes: $m(p_{idle}) + m(p_{busy}) + m(p_{bd}) = m_0(p_{idle}) + m_0(p_{busy}) + m_0(p_{bd})$. The fact that p_{bd} is a choice place implies that sometimes workpieces may get lost, and therefore the relationship between the number of times that t_{start} occurs and the number of times that t_{finish} occurs no longer holds; it is replaced by an upper bound: the number of times that event t_{start} is executed is at least $m(p_{output}) - m_0(p_{output})$ + the number of times that t_{finsh} is executed.

In any practical application the input buffer in Figure 1 is the output buffer of some other production unit (or receives external arrivals). A Petri net model of a plant consists of several production units, as represented in Figure 1 or 2, connected via common places, as illustrated on top of Figure 3. This Figure illustrates how a large plant can be represented via the interaction of several smaller Petri net components. Often when the user is analyzing the behaviour of the overall plant, global properties such as the arrival and departure times of workpieces in a production unit, can be studied using a more abstract model of each production unit. Each production unit can be represented by one single transition, with intermediate buffers, as shown on the bottom of Figure 3. This abstraction represents the progress of the workpieces in a plant without failures of production units. Since this model is untimed it is clear that to an outside observer it will not matter whether there are 3 slow machines in the production unit, or whether there is one single machine that is 3 times as fast. However the abstraction would not hold in case that the machines can break down, and workpieces may sometimes be lost. Then the abstracted models at the bottom should also include this choice.

Models of large plants are obtained via the composition of several Petri net components. Using common places as illustrated above is one way of describing the interaction between components. Common, synchronizing transitions form another method for describing the interaction between different components. In that case there are transitions t that appear in different components. The common transition t can be executed only when it is enabled in each component. Basically each component in the model adds further constraints to the allowed sequences of events that can occur. Some components represent one "geographically" coherent part of the plant (and modelling is often quite simple because the Petri net looks very similar to the physical plant lay-out of that part). Other components are purely logical, expressing certain constraints on the order of the events, imposed for example by the supervisor. Components of this type can represent scheduling decisions, rules for allocating scarce resources to competing tasks, etc.

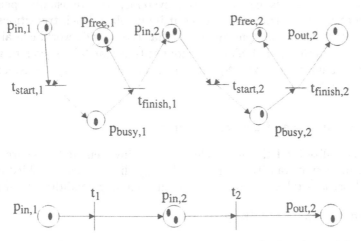

Fig. 3. Petri net model (detailed and abstracted) of a plant with two production units

Consider a flexible manufacturing plant, as an example of compositional modelling. For each type of machines there is one Petri net component modelling the availability of the machines (there are as many tokens as there are machines of that type); another Petri net models the evolution of the workpieces of a given type, starting with the arrival of raw materials up to the finished product (each place in this Petri net corresponds to a certain intermediate condition of the workpiece); this component represents the recipe used for the production of that particular type of part. Yet another Petri net component models the transportation system moving the workpieces from one location to another. Consider as a typical event the start t_{start} of a certain operation on a machine of a given type. This transition t_{start} appears in 3 Petri nets. It can be executed only when it is enabled in each of these components. This requires that there is at least one machine of that type available, that a workpiece of that type is waiting in some buffer, and that the transportation system is ready to move a workpiece of the appropriate type from its buffering position to the input position of that machine. Another Petri net component could model the decision to start the first operation for production of a given finished product, as a response to external requests (or orders) for different finished products. This is a component representing the logic of a supervisory controller.

It should be remarked that Petri nets were introduced in computer science in order to model the evolution of programs being executed on a processor. Most of the literature on Petri nets deals with closed models, where all places (resp. all transitions) have at least one input transition (resp. place) and at least one output transition (resp. place). Petri nets are also often used for logical models in fault detection in large plants (e.g. nuclear power stations). There too models are usually closed. In control engineering applications the Petri nets that one encounters as components of models of manufacturing systems, transportation systems, communication systems, etc. are usually open in the sense that there are transitions without input place (such transitions are always enabled - they represent an action that the outside world can always force to occur in the model), places without output transitions (counting how many times some event happened), etc. In this paper we mainly consider such open Petri nets.

2.2   Algebraic analysis of general Petri nets

It is easy to generalize the Petri net model by assigning weights to the arcs. If an arc from place p to transition t has weight $w_{p,t}$ then t is only enabled if p contains at least $w_{p,t}$ tokens. In other words the enabling condition is now:

$$t \text{ is enabled} \iff \forall p \in {}^{\bullet}t : m(p) \geq w_{p,t}$$

On executing transition t the model dynamics now remove $w_{p,t}$ tokens from each of its input places $p \in {}^{\bullet}t$. If an arc from transition t to place p has weight $w_{t,p}$ then executing transition t adds $w_{t,p}$ tokens to p. This extension of the simple Petri net model allows more compact models of discrete event plants in some cases.

Assume from now on that the Petri net under consideration does not have any self-loops; that is: there are no places that are at the same input and output place to the same transition: $\forall t \in T : {}^\bullet t \cap t^\bullet = \emptyset$. Under this assumption it is possible to find an algebraic description of a Petri net which is equivalent to the graphical description. The evolution of the Petri net is governed by the enabling conditions as defined above, and the "next state" function $f(m,t) = m_{next}$ can be expressed, without use of $m_{intermediate}$, as follows:

$$\forall p \in {}^\bullet t_{next} : m_{next}(p) = m(p) - w_{p,t}$$
$$\forall p \in t^\bullet_{next} : m_{next}(p) = m(p) + w_{t,p}$$
$$\forall p \notin {}^\bullet t_{next} \cup t^\bullet_{next} : m_{next}(p) = m(p)$$

Because there are no self-loops the Petri net is uniquely specified by an *incidence matrix* $E \in \mathbb{N}^{\sharp P \times \sharp T}$ with elements $w_{t,p} - w_{p,t}$ on the t-th column and the p-th row. The p-th row expresses the arcs interconnecting place p to transitions, while the t-th column expresses all the arcs connecting t to places. Because there are no self-loops either $w_{p,t}$ or $w_{t,p}$ or both are 0. This is the property that implies that E uniquely specifies all the arcs of the Petri net graph, and their weight. In particular it is possible to uniquely decompose E into $E = E_+ - E_-$ where E_+ contains all the elements of the form $w_{t,p}$ while E_- contains all the elements of the form $w_{p,t}$. A Petri net will from now on be denoted by $(P, T, (w_{t,p}, w_{p,t}), m_0)$ with the obvious replacement of F_+, F_- by $w_{t,p}, w_{p,t}$.

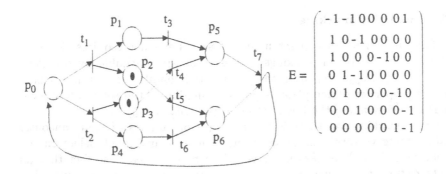

$$E = \begin{pmatrix} -1 & -1 & 0 & 0 & 0 & 0 & 1 \\ 1 & 0 & -1 & 0 & 0 & 0 & 0 \\ 1 & 0 & 0 & 0 & -1 & 0 & 0 \\ 0 & 1 & -1 & 0 & 0 & 0 & 0 \\ 0 & 1 & 0 & 0 & 0 & -1 & 0 \\ 0 & 0 & 1 & 0 & 0 & 0 & -1 \\ 0 & 0 & 0 & 0 & 0 & 1 & -1 \end{pmatrix}$$

Fig. 4. Incidence matrix of a simple Petri net

Let us further extend the model of Petri nets by allowing several transitions to be executed simultaneously. A *firing vector* σ is a positive integer valued vector of dimension $\sharp T$. Its t-th component counts how many times the transition t is fired when σ is executed. The firing vector σ can be executed simultaneously if all the enabling conditions of all its transitions are satisfied simultaneously, i.e. if $\forall p \in P : m(p) \geq \sum_t w_{p,t}.\sigma(t)$. This constraint imposes the requirement that all the places p contain enough tokens to allow each transition t to remove $\sigma(t).w_{p,t}$ tokens from p. This extension allows for more compact models, since it is not necessary to define a new transition corresponding to the case where several events are executed simultaneously. Note that this generalization does not change the set of reachable states. If a state is reachable by execution of a family of simultaneously enabled transitions, then these transitions are also individually enabled and can be executed one by one in any order. This leads to the same marking.

The enabling condition for the firing vector σ can also be expressed in vector form (where $\geq$ denoting componentwise inequality):

$$m \geq E_-.\sigma \tag{1}$$

When all the transitions in σ are executed simultaneously then the state jumps from m to m_{next} given by

$$m_{next} = m + E.\sigma \tag{2}$$

or elaborated per row: $m_{next}(p) = m(p) + \sum_t (w_{t,p} - w_{p,t}).\sigma(t)$.

These equation allow the use of linear algebraic tools for analyzing Petri nets. One should however be very careful **not** to interpret equation (2) as a difference equation. The "next state" equation in (2) is only defined when the enabling conditions (1) are satisfied, and these constraints are not included if (2) is treated as a difference equation.

2.3   Reachable markings

Verifying properties of a plant modelled as a Petri net requires the evaluation of the set of all markings (states) that are reachable by the Petri net $(P, T, (w_{t,p}, w_{p,t}), m_0)$ from the given initial condition m_0. Equivalently, proving properties of the plant requires evaluation of the set of all allowed sequences of events of the Petri net $(P, T, (w_{t,p}, w_{p,t}), m_0)$.

Given marking m_0 there is a set $\Sigma(m_0)$ of families of simultaneously enabled firing vectors (consisting of one or more transitions); select one element $\sigma_1 \in \Sigma(m_0)$ as next firing vector to be executed. The fact that all the transitions t in σ_1 can be executed simultaneously $\sigma(t)$ times, given that the state of the model is m_0, is denoted by $m_0[\sigma_1 \rightarrow$. When σ_1 occurs the value of the state jumps from m_0 to $m_1 = m_0 + E.\sigma_1$. This state change is denoted as $m_0[\sigma_1 \rightarrow m_1$. Once the marking m_1 is reached, a new set $\Sigma(m_1)$ of simultaneously enabled firing vectors is obtained; select $\sigma_2 \in \Sigma(m_1)$: then

$m_1[\sigma_2 \to m_2$ where $m_2 = m_1 + E.\sigma_2 = m_0 + E.(\sigma_1 + \sigma_2)$. One can summarize this in the notation $m_0[\sigma_1 \to m_1[\sigma_2 \to m_2$.

The set of all states that can be reached in one step from an initial condition $m_0 \in M_0$ for some subset M_0 of markings, is denoted by

$$\mathcal{R}_1(P, T, w_{t,p}, w_{p,t}, M_0) = \{m \mid \exists m_0 \in M_0, \exists \sigma \in \Sigma(m_0) \ s.t. m_0[\sigma \to m]\}$$

The set of markings reachable in n steps is defined inductively:

$$\mathcal{R}_{k+1}(P, T, (w_{t,p}, w_{p,t}), M_0) = \mathcal{R}(P, T, (w_{t,p}, w_{p,t}), \mathcal{R}_k(P, T, (w_{t,p}, w_{p,t}), M_0))$$

If m is in $\mathcal{R}_n(P, T, (w_{t,p}, w_{p,t}), M_0)$ then there exists a sequence of firing vectors $\sigma_k, k = 1, 2, \ldots, n$ such that

$$m_0[\sigma_1 \to m_1[\sigma_2 \to m_2[\sigma_3 \to \ldots m_{n-1}[\sigma_n \to m$$

Using the convention that $\emptyset \in \Sigma(m)$ for any state m guarantees that the sets of reachable markings $\mathcal{R}_n(P, T, w_{t,p}, w_{p,t}, M_0)$ are monotonely increasing in n. Hence the set of markings that can eventually be reached by a given marked Petri net, after an arbitrary number of steps, is well defined as

$$\mathcal{R}_\infty(P, T, (w_{t,p}, w_{p,t}), M_0) = \cup_{n=1}^\infty \mathcal{R}_n(P, T, (w_{t,p}, w_{p,t}), M_0)$$

When no confusion is possible we will write $\mathcal{R}_\infty(P, T, (w_{t,p}, w_{p,t}), m) = \mathcal{R}_\infty$. It is easy to see from the above expressions that, if a marking m is reachable via the consecutive execution of an arbitrary number of simultaneously enabled firing vectors, then there must exist a vector $x = \sum_i \sigma_i \in \mathbb{N}^{\#T}$ such that $m = m_0 + E.x$. The set $\mathcal{R}_\infty$ of reachable states for the Petri net $(P, T, (w_{t,p}, w_{p,t}), m_0)$ is a subset of the set

$$\tilde{\mathcal{R}} = \{m \mid \exists x \in \mathbb{N}^{\#T} : m = m_0 + E.x\} \supset \mathcal{R}_\infty$$

In general $\mathcal{R}_\infty$ is a strict subset of $\tilde{\mathcal{R}} = \{m_0 + E.x, x \geq 0\}$ since there may be values $x \geq 0$ for which there does not exist a sequence $\{\sigma_1, \sigma_2, \ldots, \sigma_n\}$ of successively enabled transitions generating $x = \sum_{i=1}^n \sigma_i$. Consider e.g. $m_1 = m_0 + E.x$ for $m_0^T = (0, 0, 1, 1, 0, 0, 0)$, and $x^T = (1, 1, 0, 2, 2, 0, 2)$ in the example of Figure 4. Then the marking $m_1 = (0, 1, 0, 0, 1, 0, 0) \in \tilde{\mathcal{R}}$ is not reachable because the structure of the Petri net clearly shows that either p_1 and p_2 are marked together, or p_3 and p_4 are marked together, after t_7 has been executed for the first time. Hence m_1 is not reachable.

Define the support $sp(x)$ of a firing vector x as the set of transitions t such that $x(t) > 0$. When the firing vector corresponding to x is executed, then all the transitions t in $sp(x)$ and only those transitions are executed.

The set $\tilde{\mathcal{R}}$ is very easy to work with when verifying properties of a plant. Clearly it would be very useful if one could reduce the set $\tilde{\mathcal{R}}$ so that it contains fewer unreachable markings. Some unreachable markings can be eliminated from $\tilde{\mathcal{R}}$ by imposing constraints, such as invariants properties, proven via whatever technique is available.

A place invariant is a function $V(m)$ that remains constant under any enabled firing vector:

$$\{\, m \in \mathcal{R}_\infty \Longrightarrow V(m) = V(m_0)\} \Longleftrightarrow$$
$$\{\, m \in \mathcal{R}_\infty,\, \sigma \in \Sigma(m) \Longrightarrow V(m + E.\sigma) = V(m)\}$$

Proving that a function $V(m)$ is a place invariant is easy; if $\forall m \in \tilde{\mathcal{R}}, \forall \sigma s.t. m \geq E.\sigma : V(m + E.\sigma) = V(m)$ then $V(m)$ is certainly a place invariant. However this requirement may be too strict, since it does not have to be satisfied for an unreachable marking m. Moreover the hard part is in guessing what functions are possible place invariants (compare this situation to proving properties of a classical differential equation via invariants).

Linear place invariants are functions of the form $V(m) = m^T.g$ for some $\sharp P$-vector g. It is possible to generate all linear place invariants by considering all vectors g in the null space of E^T. Indeed $\forall x \geq 0 : m^T.g = m_0^T.g + x^T.E^T.g$ if $E^T.g = 0$. Unfortunately linear place invariants defined in this way do not exclude any unreachable markings from $\tilde{\mathcal{R}}$ since the derivation above does not take into account that the equality $x.E^T.g = 0$ only must hold for those vectors x such that $m_0 \geq E_-.x$. And this is exactly the constraint that is not expressed by $\tilde{\mathcal{R}}$ either.

It is possible to get other interesting place invariants by looking for traps and siphons in the Petri net. A trap $Q \subset P$ is a subset of places such that every transition that has an input place in Q also has a output place in Q :

$$t \in \cup_{p \in Q}\, p^\bullet \Longrightarrow t^\bullet \cap Q \neq \emptyset$$

The firing of any transition that can remove a token from a place in Q must add a token to some place in Q. Hence a trap that contains at least one token under the initial marking m_0 will always contain one token $(\sum_{p \in Q} m_0(p) > 0 \Rightarrow \forall m \in \mathcal{R}_\infty : \sum_{p \in Q} m_0(p) > 0)$.

Similarly one can define a siphon S as a subset of places such that $t \in \cup_{p \in Q}\, {}^\bullet p \Longrightarrow \exists p \in Q \ni t \in p^\bullet$. A siphon that initially contains no token will remain empty forever. In order to find all traps one has to find all the solutions of the inequality $x.E \geq 0, x \geq 0, x \in \mathbb{N}^{\sharp P}$. If x is one of these solutions, then the support of x (that is the set of places corresponding to a strictly positive value of $x(p)$) forms a trap. A siphon is obtained by reversing the inequality to $x.E \leq 0$. The support of a solution x of $x.E = 0$ is simultaneously a trap and a siphon, and is at the same time a (linear) place invariant.

In order to define these concepts in a more graphical way, we introduce paths and cycles. A path is an ordered sequences $\{p_0, t_1, p_1, \ldots, t_n\}$ or $\{t_0, p_1, t_1, \ldots\}$ such that for any pair (t_n, p_{n+1}) (resp. (p_n, t_n)) there is an arc in the graph defined by F_+ (resp. F_-) connecting these two consecutive elements of the path. A cycle is a path where no place or transition occurs twice in the sequence, and where the last element is connected to the first element (i.e. $(t_n, p_0) \in F_+$ or $(p_n, t_0) \in F_-$). Traps and siphons can also be recognized by considering the graph of the Petri net. A trap only contains cycles or incoming paths. A siphon only has cycles or outgoing paths.

The following theorem states conditions under which existence of a non-negative integer solution to the equation $m - m_0 = E.x$ implies that m is indeed reachable:

Theorem 1: Let (P, T, F, m_0) be a marked Petri net, such that every cycle is a trap, and let $0 \leq x \in \mathbb{N}^{\#T}$ be a solution to the equation $m - m_0 = E.x$ Define the subnet that contains all transitions in the support $sp(x)$ of x, and all their input and output places. Restrict m_0 to this subnet, and assume that all siphons in the subnet have at least one token under m_0. Then m is reachable from m_0, i.e. $m \in \mathcal{R}_\infty$.

This result can be further improved if the Petri net is completely acyclic, i.e. if the graph of (P, T, F) contains no cycle at all. In that case the network can be partitioned in layers, with tokens always moving from the top layer down to a lower layer. Assume that the acyclic Petri net does not have any entrance or exit transitions (i.e. no tokens ever enter or leave the net) then the union of the last k layers always is a trap; the union of the first n layers is a siphon. Theorem 1 could be applied. However the following stronger result of Ichikawa and Hiraishi (see [11]) holds:

Theorem 2: If the Petri net (P, T, F) is acyclic, then for any initial marking m_0

$$\tilde{R}(P, T, F, m_0) = \mathcal{R}_\infty(P, T, F, m_0)$$

Further results on how to describe the difference between $\tilde{R}$ and $\mathcal{R}_\infty$ can be found in the work of Silva et al. [15] and of Desel and Esparza [7]. If one can prove that a certain marking m is not in $\tilde{R}$, then m is definitely not reachable. If the solutions to the equations $m - m_0 = E.x$ were taken in the space of rational (or real) variables, then the following theorem of Farkas (or equivalently the Fredholm alternative as stated in linear functional analysis) could be used to prove that there is no solution to the equation $m - m_0 = E.x$:

Farkas' lemma: Let A be a matrix and b a vector of appropriate dimensions, with rational elements; then either $A.x = b$, $x \geq 0$, $x \in \mathbb{N}^{\#P}$ has a solution or $y.A = 0$, $y.b < 0$, $y \in \mathbb{N}^{\#T}$ has a solution but it is not possible that both of these equations have a solution simultaneously.

The following theorem proves that some markings are not reachable:

Theorem 3: For a marked Petri net (P, T, F, m_0), assume that $y.E = 0, y.(m - m_0) < 0$ has a solution. The m is not reachable.

Vectors y which solve the equation $y.E = 0$ are called transition invariants, because they corresponds to sequences of transitions which, when executed one after the other or simultaneously, keep the linear invariant $y.m$ constant.

2.4   Limitations of Petri net modelling

Petri nets can model in a compact, and often easily understandable, way many of the constraints which determine the allowed sequences of events in a discrete event system. However Petri nets cannot model those cases where the marking of a place disables certain transitions. Of course one could extend the Petri net formalism by defining disabling arcs between some places p and some transitions t. Such a disabling arc would add the constraint to the model that the transition t cannot be executed whenever the place p is marked, irrespective of the enabling conditions. It can be shown that such a Petri net formalism with disabling arcs can describe the same sets of allowed sequences as can be described by Turing machines. Petri nets with disabling arcs have the same modelling power as Turing machines.

Since it is known that many interesting questions about reachability are undecidable for Turing machines, it is for most purposes not appropriate to introduce these disabling arcs into the model. Of course if one can prove that the set $\mathcal{R}_\infty(P, T, (w_{t,p}, w_{p,t}), m_0)$ of reachable states is finite, then any property of reachable states can be proven by enumeration. In this case one can model disabling connections by adding to the model many enabling arcs from any state that is not disabling. But such a Petri net will generally not be easy to interpret, since many extra places have to be introduced, and one loses all the advantages of a compact state representation.

In order to appreciate the limitations on the modelling that this imposes, consider the following simple example of a production unit with two machines and two types of tasks to be executed. The simplest mode of operating the plant is that machine i carries out tasks of type i, for $i = 1, 2$ as indicated in Figure 5a, consisting of 2 independent Petri net components; each component has one place where jobs are waiting for an idle machine and one place where a token indicates that the machine carrying out tasks of that type is idle. If there are many tasks of type 1 arriving, and few tasks of type 2, then the machine working on jobs of type 1 will be overloaded while the machine operating on type 2 tasks may be idle most of the time.

More flexible plant operation can be achieved if both machines can carry out tasks of both types. This is shown in Figure 5b. There is now a common place holding the tokens which correspond to idle machines. An idle machine can carry out tasks of either of the 2 types, whenever a job arrives. The model does not specify any rule for allocating machines to type 1 or to type 2 tasks in those cases where there is competition for idle machines. This allocation decision is left to the scheduler, which is not modelled in this case.

One possible rule for allocating machines is the alternating priority rule. Each machine will alternately work on a job of type 1 and then on a job of type 2, then again on type 1, etc. This is modelled as a Petri net in Figure 5c. This mode of operation is very inefficient in case the arrival streams of jobs are very irregular, if one stream of jobs has a much higher intensity than the other.

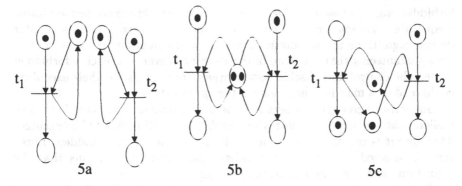

Fig. 5. Different operation modes of two machines, representable via Petri nets

There are many much more efficient rules for allocating machines to jobs, such as fixed priority for one stream of tasks, or priority for the longest waiting line, or priority to type 1 unless the waiting line of type 2 exceeds a given threshold, etc. In each of these cases the supervisory rule cannot be modelled as a Petri net since it requires that a transition (the start of a low priority job) is disabled in certain states. This cannot be implemented using a Petri net.

The set of reachable states is known to be bounded if there is an upper bound to the length of each waiting line (the buffers are finite, and arrivals to a full buffer are not allowed to enter). In that case it is of course possible to construct a Petri net, with a rather complicated structure, that will enable transitions only when the corresponding task has priority (the reader is urged to try this as an exercise). This requires very many places to be added to the Petri net model. The Petri net model then looses its advantage of compact modelling. The relation between the Petri net graph and the physical plant layout is completely lost. Moreover analysis tools based on linear algebraic equations will have too many variables to be useful. This discussion shows that it is dangerous to try to represent all components of a large plant by the same Petri net modelling formalism. It is useful to use automata, or other modelling formalisms, for those components where the set of places and transitions would become too large for easy interpretation and analysis. Good modelling tools allow different modelling formalisms to be used for different components.

3   Supervisory control synthesis

3.1   Maximally permissive control laws

This section discusses the implementation of supervisory controllers, as defined in the preceding chapter, for plants modelled as Petri nets. The constraints on the plant behaviour are modelled via the specification of sets of forbidden states. These forbidden state specifications are expressed via linear inequalities that must not be satisfied by the marking vector. This chapter derives algorithms for automatic synthesis of maximally permissive control laws guaranteeing that the state of the Petri net never can reach a forbidden value. The control laws disable certain transitions, whenever their execution could lead to a marking that violates the specifications.

Supervisory control only sets out the constraints for the lower level controllers that may actually select which enabled transition should be executed. Therefore it is important that the smallest possible set of forbidden transitions is selected by the control synthesis algorithm. This means that the algorithm must generate a maximally permissive control law.

Formally we define the control problem as follows. For any discrete event plant modelled as a Petri net, the set $\mathcal{R}_\infty(P, T, (w_{t,p}, w_{p,t}), M_0)$ of reachable markings, and the set of all feasible sequences of events, are well defined, as shown in the previous section. All these feasible trajectories must satisfy certain requirements, such as safety conditions, or conditions that are imposed by economic considerations of the plant operation. Assume that these specifications are expressed by a set M of forbidden states. The plant will behave properly if no state in M can ever be reached:

$$\mathcal{R}_\infty(P, T, (w_{t,p}, w_{p,t}), M_0) \cap M = \emptyset$$

If this intersection is not empty, then further constraints must be imposed on the behaviour of the plant. These further constraints are generated by a supervisor. A supervisor is a component of the plant model that will disable certain transitions, based on observations of the current state. Observations of the state are often implemented by observation of the sequence of events that has taken place up to the current point in time. In a Petri net model this supervisor can be realized by adding extra input places for some transitions in the Petri net. These supervisory places, with output towards transition t, must not contain a token whenever the supervisor must disable transition t. The transitions that have been executed in the past must put the tokens into these supervisory places.

This section develops algorithms designing such *supervisory controllers* given a Petri net plant model $(P, T, w_{t,p}, w_{p,t}, m_0)$, and given a set M of *"forbidden states"*. Based on these observations, and on the internal state of the supervisor, the largest set of enabled transitions must be synthesized by the supervisor. In that case the supervisor realizes a maximally permissive feedback control law.

Remark: Compare this formulation of the Petri net control problem to a classical control problem. A supervisory controller influences all the components of the plant. In each component a local controller can apply control values, provided these are allowed by the supervisor. The supervisory controller selects control values u such that a forbidden set of states $B^c = X - B$ outside some set B is never reached; in other words the controlled system must have B as an invariant set. Assume the system is represented by I interacting components $\dot{x}_{i,t} = f_i(x_t,,u), x_0$, where $x_t = (x_{1,t}, x_{2,t}, \ldots, x_{I,t})$. The control values must lie in some prespecified set U: $u \in U$. The solutions corresponding to an input function $u(t)$ are denoted by $x(t, u(.), x_0)$. A good control law restricts the control variables $u(t) \in g(x_t) \subset U$ to the subset $g(x_t)$ chosen so that the boundedness property is true: $\forall t : u(t) \in g(x_t), x_0 \in B \Rightarrow \forall t : x(t, u(.), x_0) \in B$. Usually a control law $g(x_t)$ selects just one value, but one can equally well design stabilizing controllers by specifying a set of allowed control values. If some other decision maker (e.g. a local controller acting under the supervision of the stabilizing controller) selects at any time t an arbitrary control value in the set $g(x_t)$, then the system will remain stable (think of a scalar linear plant that will be stabilized for any controller guaranteeing that $\forall t : u(t).x_t < 0$. The local decision maker can use the remaining freedom to select a value among the allowed control values $g(x_t)$ in order to minimize some local cost criterion. The supervisor basically only sets out the margins within which each of the local controllers has to operate, in order to achieve acceptable plant behaviour.

In this chapter we consider controlled Petri nets. The set T of transitions is partitioned in a controllable set T_c and an uncontrollable set $T_u = T - T_c$. Uncontrollable transitions in T_u can always be executed when they are enabled by the current marking of the Petri net. In practical applications there are always transitions that cannot be influenced by the supervisor, such as component failures, or external arrivals of requests for service like connection requests in a communication system. Controllable transitions in T_c on the other hand can only be executed when they are simultaneously enabled by the marking of the Petri net and when the external supervisor also allows the execution of the transition. In a graphical representation of a Petri net the controlled transitions are recognized by the fact that they have a rectangular control input box (see fig. 6).

The set of forbidden states is specified by a disjunction of linear predicates:

$$M = \cup_{i=1}^I M_i \text{ where } M_i = \{m \in \mathbb{N}^{\#P} \mid m^T.f_i > b_i\}$$

for some vectors f_i and scalars b_i To simplify the analysis we assume that all the elements of f_i and b_i are positive. For more general forbidden sets the reader is referred to [12] The state is acceptable only if none of the inequalities are satisfied, i.e. if all the specifications are satisfied simultaneously.

The problem of synthesizing a maximally permissive control law is easiest in the special case where all transitions can be disabled by the external supervisor: $T = T_c$. This is treated in the next subsection. Control can then be achieved by adding control places to the net, which are connected to existing transitions in such a way that the specifications become place invariants for the extended net. This has the advantage that the controlled net is again a Petri net, and similar analysis and control synthesis techniques can be applied to the controlled net in order to enforce further specifications.

Control laws for partially controllable models are basically "model predictive" in the sense that the Petri net model is used to calculate at each state what is the set of all markings that can be reached through the consecutive execution of uncontrollable transitions. This uncontrollably reachable set depends on the current state. The set of initial states such that the uncontrollably reachable set originating in that state has a non-empty intersection with the forbidden set M is treated as the new forbidden set.

Consider the forbidden set specification $M = \{m \in \mathbb{N}^{\#P} \mid m^T.f > b\}$. In order to determine whether this specification is satisfied one only needs to consider the marking of the places in the set $support(f) = \{p \in P \mid f(p) > 0\}$. Let $up(p)$ be the set of all paths ending in the place p such that all the transitions in the path $up(p)$ are uncontrollable:

$$up(p) = \{(t_0, p_1, t_1, \ldots t_{n-1}, p) \mid \forall \ell : (t_{\ell-1}, p_\ell) \in F_+,$$
$$(p_\ell, t_{\ell-1}) \in F_- \text{ and } t_\ell \in T_u\}$$

Assume that a firing vector σ including at least one controllable transition has been executed, leading to a state m. Prior to the next execution of a controllable transition the forbidden set M can only be reached by the execution of some uncontrollable transitions in an uncontrollable path in $up(p)$. The *influencing net* is defined as the subnet $PN_{inf} = (P_{inf}, T_{inf}, F_{+,inf}, F_{-,inf})$ of the Petri net (P, T, F_+, F_-) containing all the places in $up(p)$ for some $p \in support(f)$. These are all the places that have an uncontrollable path towards a place in $support(f)$, including the places in $support(f)$ themselves. The influencing net also includes all the uncontrollable transitions connected to at least one of these places, and all the arcs connecting the places and transitions selected for inclusion in the influencing net. The initial marking of PN_{inf} is obtained by restricting m_0 to P_{inf}. In Figure 6 the dashed line surrounds the influencing subnet corresponding to the forbidden set specification which depends only on the state of the place p.

All the transition in PN_{inf} are uncontrollable. In between consecutive executions of controllable transitions tokens can only reach a place in $support(f)$ via the execution of a sequence of uncontrollable transitions, enabled by tokens which are present in places in PN_{inf}. Tokens in places outside PN_{inf} can always be prevented from reaching $support(f)$ by disabling a controllable transition.

The evolution of a partially controlled Petri net can be interpreted as a dynamic game, with "nature" selecting firing sequences of transitions that are either uncontrollable or that are allowed by the current control value (= set of allowed controllable transitions). Of course these transitions can only fire if they enabled by the marking of the Petri net. Each time a controllable transition (or a set of simultaneously enabled transitions including a controllable transition) is executed a sensor observes the new state $m_{obs,n}$. The second player, opposing nature and corresponding to the supervisor, then selects a new subset $u_{allow}(m_{obs,n}) \subset T_c$, of controllable transitions. Nature can then

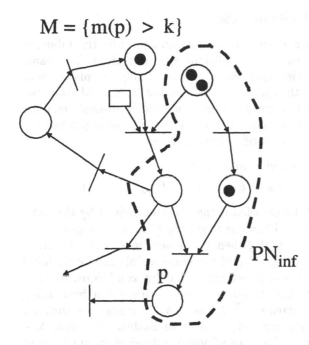

$M = \{m(p) > k\}$

PN_{inf}

p

Fig. 6. Controlled Petri net with forbidden set, and its influencing net

again select any sequence of state enabled transitions in $T_u \cup u_{allow}$ until the next execution of a controllable transition. Nature can select from this set a sequence of transitions, trying to reach a forbidden state in M. As soon as this sequence has been executed by the opponent, including at least one final controllable transition, a new state $m_{obs,n+1}$ is is observed by the sensors. The supervisor will again select a set $u_{allow}(m_{obs,n+1})$, and the opponent will again try to reach a forbidden marking $m_{obs,n+2} \in M$. The set of markings reachable by the opponent, up to to the next execution of a controlled transition in $u_{allow}(m_{obs,n+1})$, is given by $\mathcal{R}_\infty(P, T_u \cup u_{allow}(m_{obs,n+1}), F_+, F_-, m_{obs,n+1})$. From the definition of M it follows immediately that it suffices to consider the reachable set of the influencing net

$$\mathcal{R}_{inf}(m_{obs,n+1}) = \mathcal{R}_\infty(P_{inf}, T_{inf}, F_+, F_-, m_{obs,n+1} \mid P_{inf}).$$

The goal of the supervisor is indeed to ensure that

$$\mathcal{R}_{inf}(m_{obs,n+1}) \cap \{M \mid P_{inf}\} = \emptyset$$

The supervisor must block all controllable transitions that might lead from a marking in $\mathcal{R}_{inf}(m_{obs,n+1})$ to M. At the same time the supervisor must limit the evolution of the Petri net as little as possible. Hence a maximally permissive control law blocks only a subset of the controllable transitions that have output places in P_{inf}. This subset should moreover be as small as possible, in order to allow lower level controllers as much freedom as possible in selecting transitions for execution.

3.2   Completely controllable models

When all the transitions in a Petri net are controllable, then the influencing net reduces trivially to $support(f)$ without any transitions. The game is played without opponent. In other words the maximally permissive control law must only prevent that a forbidden marking is reached from the observed current state by the execution of one single simultaneously enabled set of (controllable) transitions. Since the specifications are expressed via linear inequalities it is natural to verify whether the set

$$\mathcal{R}_1 \cap M = \{ x \in \mathbb{N}^{\# T} \mid x \geq 0 \, m = m_0 + E.x,$$
$$m_0 \geq E_-.x, \ (m_0 + E.x)^T.f_i > b_i \}$$

is empty. This will certainly be satisfied if the superset defined by the same conditions, but omitting the enabling condition $m_0 \geq E_-.x$, is empty.

If the set $\mathcal{R}_1 \cap M$ is not empty, then the supervisory controller must add extra constraints to the firing rules of the system. This can be achieved by adding extra input places, and hence extra enabling conditions, to some transitions. These extra constraints must prevent a transition from firing whenever this firing would increase $m^T.f_i$ above b_i. Transitions t that are input transitions to a place $p \in support(f_i)$ must be disabled as soon as $b_i - (m+E.\sigma(t))^T.f_i < 0$. Here $x = \sigma(t)$ is the $\#T$ vector with all elements 0 except for a 1 in the place corresponding to t. Whenever $b_i - (m + E.\sigma(t))^T.f_i \geq 0$ a maximally permissive control law must allow execution of t.

In order to obtain this property one must enlarge the Petri net model with an extra place $p_{c,i}$ which initially contains $b_i - m_0^T.f_i$ tokens. In the extended net, with the extra place $p_{c,i}$, the place invariant $m^T.f_i + m(p_{c,i} = b_i$ must hold. Since $m(p_{c,i}) \geq 0$ this is equivalent to the specification.

The firing of a transition t belonging to ${}^\bullet p$ for $p \in support(f_i)$ must remove $f_i(p)$ tokens from $p_{c,i}$ while the firing of a transition t that belongs to $p^\bullet$ for $p \in support(f_i)$ must add $f_i(p)$ tokens to $p_{c,i}$. Adding the place $p_{c,i}$ to the Petri net means that one row is added to the incidence matrix E. This extra row must be of the form $E_{\# P+1} = -f_i^T.E$. This insures that the last row $E_{\# P+1}$ of the extended incidence matrix "adds $f_i^T.E_{.,j}$ tokens" to the control place $p_{c,i}$ whenever the execution of a transition t_j removes $f_i^T.E_{.,j}$ tokens from $support(f_i)$ (where $E_{.,j}$ denotes the j-th column of the incidence matrix E). Similarly the structure of the last row $E_{\# P+1}$ of the extended incidence matrix "removes $f_i^T.E_{.,j}$ tokens" from the control place $p_{c,i}$ whenever the execution of a transition t_j adds $f_i^T.E_{.,j}$ tokens to $support(f_i)$. Hence the added control place $p_{c,i}$ ensures that the place invariant $m^T.f_i + m(p_{c,i}) = b_i$ is enforced.

In order to enforce I linear inequality constraints $f_i^T.m \leq b_i, \forall i = 1, \ldots, I$ it suffices to add I rows to the incidence matrix E, one row per inequality. Each of these rows is constructed as described in the preceding paragraph. Since the original Petri net has no self-loops, and since the construction of the control law has not introduced any self-loops, it is easy to draw the controlled

Petri net that enforces the I specifications simultaneously. Figure 7 shows a very simple example of a control law enforcing the specification that two places together never contain more than one token. The original Petri net is drawn in full lines, the additional control places and control arcs are drawn in dashed lines.

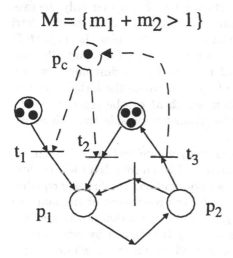

$$M = \{m_1 + m_2 > 1\}$$

Fig. 7. Control place enforcing specification $m_1 + m_2 \leq 1$

The control law obtained by adding these control places is maximally permissive since it will never disable a transition unless the firing of that transition would lead to a forbidden marking. Of course some of these added rows of the incidence matrix may be unnecessary since they correspond to place invariants that are already enforced by the original Petri net (or by other control places). It is computationally hard to check for redundant rows in an incidence matrix. However since the extra control place corresponding to these redundant rows will always contain at least one token, they will never disable a transition, and these redundant rows only make the controller more complicated than necessary, but they do not reduce the set of reachable markings.

3.3 Partially controlled Petri nets

It has been shown in section 3.1 that for a partially controlled Petri net the supervisory control decision must ensure that a larger set M^* of forbidden states is avoided. Every marking m_0 in M^* must be such that the execution of any state-enabled sequence of uncontrollable transitions cannot lead to a marking M. The evolution of a partially controlled Petri net can be interpreted as the alternating execution of a collection of enabled transitions - including at least one state-enabled and control-enabled controllable transition - followed by the execution of a sequence of sequentially state-enabled uncontrollable transitions. The marking m_0 reached by the execution of an enabled collection of transitions including at least one controllable transition is forbidden if there exists an allowable (sequentially state-enabled) sequence of uncontrollable transitions whose execution leads from m_0 to a forbidden marking in M. The set M^* contains all the markings for which the model based prediction of the reachable markings includes a forbidden marking.

Moreover the game-theoretic interpretation has shown that only the execution of uncontrollable transitions within the influencing net - the sub-Petri net PN_{inf} containing all uncontrollable paths leading to a place in $support(f)$ - must be considered. Designing a supervisory controller for a partially controllable Petri net with forbidden set M is equivalent to designing a supervisory controller for a fully controllable Petri net, where the forbidden set is the extended set M^* of all markings from which M can be reached via the execution of a sequence of uncontrollable transitions inside the influencing net.

In the literature very general unions and/or intersections of linear inequalities have been used as specification of forbidden sets. Here we consider only the simplest case of one forbidden set specified via the linear inequality $M = \{f^T.m > b\}$. Let $m_{inf} = m \mid P_{inf}$ be the restriction of the marking to the influencing net PN_{inf}, and let f_{inf} be the restriction of the vector f to places in PN_{inf}. This restricted vector f_{inf} is obtained by removing all zeros from f. The extended set M^* contains all markings m_0 whose restriction $m_{0,inf}$ is such that there exists a sequence of uncontrollable transitions in PN_{inf} leading to a marking such that $f^T.m = f_{inf}^T.m_{inf} > b$. The results of section 3.2 will be applicable if M^* can be represented via unions of sets specified by linear inequalities. In this subsection we consider a few special cases where this is possible.

If the influencing net PN_{inf} is acyclic, then theorem 2 can be applied. Let E_{inf} be the incidence matrix of the influencing net PN_{inf}. Then there exists a marking $m \in M$ that is reachable from m_0 if and only if there exists a vector $x \in \mathbb{N}^{\sharp T_{inf}}$ such that $m_{inf} = m_{0,inf} + E_{inf}.x$. A marking m_{inf} corresponding to M^* will be such that there exists a vector $x \in \mathbb{N}^{\sharp T_{inf}}$ such that

$$f_{inf}^T.m_{0,inf} + f_{inf}^T.E_{inf}.x > b$$

In order to determine whether a marking belongs to M^* or not one has to solve a linear integer programming problem, with $\sharp T_{inf}$ variables.

One special case that has been solved in the literature is the case where the influencing net is a state machine. Then the influencing net PN_{inf} has no synchronizing transitions: each transition $t \in T_{inf}$ has at most one input place and at most one output place. Tokens move independently of each other inside the influencing net when this is a state machine. The number of tokens inside PN_{inf} remains constant. This allows for a graphical analysis of the allowed markings (see [3] and [9]) Consider all the tokens inside PN_{inf} under $m_{0,inf}$. Solving the linear integer programme then corresponds to finding the worst possible distribution of these tokens among the places in $support(f)$. This solution can be obtained explicitly: it is possible to determine a vector $f^* \in \mathbb{N}^{\#P_{inf}}$ and a nonnegative number b^* such that

$$M^* = \{m_0 \mid f_{inf}^{*T}.m_{0,inf} > b^*\}$$

For details on these calculations see [11]. This paper also treats more general influencing nets, where one only requires that each transition in the influencing net has at most one input place.

Another special case of interest is the class of marked graphs. Assume that the influencing net PN_{inf} does not contain any choice places - each place has at most one input transitions and one output transition - then each token in P_{inf} can follow only one (uncontrollable) path towards $support(f)$. Using this property it is possible to express the forbidden set M^* by expressing the maximum number of tokens along any uncontrollable path leading to $support(f)$. this case has been treated initially by Holloway and Krogh [8]. The extended forbidden set M^* is then, under additional conditions on paths connecting places in $support(f)$, the union of sets determined by linear inequalities. A marking is acceptable only if it does not violate any of the linear inequalities.

If the forbidden set can be described by the union of forbidden sets then it is easy to determine the maximally permissive control law. It suffices to determine the maximally permissive control law u_i corresponding to each of the inequalities $f_i^{*T}.m_{inf} > b_i^*$. The maximally permissive control law then disables any controllable transition that is disabled by at least one of the supervisors that have been designed. In other words $u_{all} = \cap u_i$.

For general influencing nets PN_{inf} the extended forbidden set M^* can be described as the intersection of unions of sets determined by linear inequalities. In general it is not possible then to decompose the control design problem. Conditions under which supervisory control design problems can be decomposed into solving simpler problems (relating to simpler specifications, and smaller influencing nets) are treated in [13]

4   Timed Petri net models

4.1   Timed specification of plant models

The untimed Petri net models introduced in the preceding sections allow the representation of the allowed sequences of events in a plant. The components of the model express precedence constraints on the ordering of the events that can take place. The models specify in each state - that is after a given sequence of events has been observed - which events can be executed, and which state is reached next after the execution of a particular event. In many practical applications the model should also describe constraints on the time when a certain enabled transition is allowed to occur. In order to achieve this goal the present section extends the Petri net model of section 2 by defining how the enabling conditions of certain transitions depend on the state of several stopwatches. These stopwatches all run in synchronization with a global, real-time clock, but they can be started and stopped when certain events happen.

A timed Petri net model includes for each enabling of transition t a clock with state $c_{n,t}$, which is started whenever the transition t becomes enabled. This means that the state of the clock is initialized to $c_{n,t} = 0$ as soon as all the input places in $\bullet t$ simultaneously contain a token for the n-th. From that point in time on the clock is incremented synchronously with a real time clock. The clock is stopped when the transition t is executed, or when the execution of some other transition removes tokens from a place in $\bullet t$ thereby disabling the transition t before it is executed.

Consider as an example the problem of finding an optimal schedule for the operation of flexible plants such as a steel plant [4], or an electroplating plant [16]. Each batch that must be produced requires the execution of a sequence of different operations, requiring a certain amount of time, specified by a recipe. Each work position in the plant can execute a certain subset of all the operations that are included in the many different recipes that can be executed by the plant. Scarce resources are needed in the work positions for carrying out a particular operation, and transportation resources are needed in order to transport workpieces from one work position to the next work position The precedence constraints among these different operations, the resource availability constraints, and the upper and lower bounds on the duration of each operation, together constitute the recipe for a particular type of product. Availability of resources in a work position, availability of a transportation resource, and the precedence constraints encoded by the recipes can be represented by an untimed Petri net.

However constraints on the time duration of each operation require time dependent enabling rules. Moreover the recipe often includes other strong timing constraints. In the steel factory the operations on a batch of steel - from the start of conversion up to the continuous casting - must be completed within a certain time, in order to avoid that the batch cools down too much. Moreover the continuous casting machine at the output must operate uninterruptedly, imposing as a further specification a maximal time distance between the arrival of successive batches at the continuous casting machine.

The untimed Petri net models, which have been defined so far in this chapter, can be used to obtain supervisory controllers, which limit the allowed sequences of events. These supervisory controllers impose constraints on the plant operation, ensuring that certain specifications related to safety of the plant or avoidance of deadlock, are always met. If the Petri net model is extended with timing information, then it is also possible to consider optimization problems such as finding an optimal schedule. The optimal schedule minimizes the time span required for the execution of a collection of batches, in such a way that all plant specifications are met. An optimal schedule must maximize the throughput of a plant by selecting which event to execute and when to execute it, taking into account the constraints imposed by the supervisory controller.

It is important to realize that the inclusion of time information in the model may in fact allow for more liberal supervisory controllers. This is illustrated clearly in the deadlock avoidance problem. A good candidate schedule must be such that it can never lead to a deadlock. In a deadlock situation the plant has reached a state where no transition at all is enabled. This problem can be treated by the methods of the preceding section if one defines the set of all states where no transition is enabled as the forbidden set M.

In an untimed model scheduling decisions can only determine the order in which events (like "start task" or "move work piece from a to b") are executed. Any enabled transition however can be delayed for an arbitrarily long time. The scheduler cannot assume that an enabled event will take place before an arbitrarily large number of other enabled transitions is executed. Deadlock avoidance algorithms [6] divide the Petri net in "conflict resolution" sections. Deadlock can be avoided by imposing the constraint that tokens are only allowed to enter the next "conflict resolution" section after the critical resources, necessary to execute tasks in that section, have become available.

If a timed Petri net model is used, then it is possible that the model guarantees that a resource will become available before some fixed time in the future. Consider the state at time τ where work pieces are ready to enter a "conflict resolution" section Using the untimed model the operations in this "conflict resolution" section can start only after a critical resource is released by some other operation. The timed model on the other hand can specify that resource R will be needed by the operations after at least Δ time units, and the state of the timed model can also guarantee that resource R will be released before time $\tau + \Delta$. Then the deadlock avoidance algorithm for the timed model may allow the start event of the operations of one single "conflict resolution" section to be executed. In an untimed model this start event of the operations would have to be blocked because the model would not give any information on lower or upper bounds on the time when the resource R would become free or would be needed, but in the timed model this information is available, and can be used to increase the throughput.

Clearly the set of allowed trajectories (sequences of events or sequences of markings) is smaller for the timed discrete event model, compared to the untimed discrete event model. A solution to the supervisory control problem for an untimed Petri net will enforce all the specifications for the corresponding timed Petri net model. However the untimed solution may not be maximally permissive. Indeed the set of reachable states in the untimed model may be smaller than for the untimed model because certain events that are feasible in the untimed model may not be feasible in the timed. Certain executable paths in the untimed model may not be executable because they include an event that is pre-empted by some other simultaneously enabled events.

In the next subsection the model of a timed Petri net will be introduced more formally. This model will be illustrated for the case of a steel plant. The model will be constructed via interacting components, illustrating how modelling of large plants becomes tractable. Finally we will briefly discuss the difficulties in designing supervisory control laws for timed Petri nets.

4.2   Timed Petri net model

A *timed Petri net* (TPN) is a six-tuple (P, T, F, L, U, M_0) in which (P, T, F) is a Petri net. The functions L and $U : T \to \mathcal{R} \cup \{\infty\}$ ($\forall t \in T : 0 < L(t) \leq U(t)$) associate to each transition $t \in T$ a time interval $[L(t), U(t)]$. If t becomes state-enabled at time $\theta_e \in \mathcal{R}$ – being state-enabled is defined as in the untimed case by the condition that all places in $^\bullet t$ contain a token given the state M_θ – then the transition *must* fire at some time $\theta_f \in [\theta_e + L(t), \theta_e + U(t)] \subset \mathcal{R} \cup \{\infty\}$, provided t did not become disabled because the firing of another transition removed a token from a place in $^\bullet t$. Notice that t is forced to fire if it is still state-enabled at $\theta_e + U(t)$. Untimed Petri nets are a special case of TPNs with $\forall t \in \mathcal{T} : U(t) = \infty$.

The *state* of a timed Petri net at time $\theta \in \mathcal{R}$ is a map $M_\theta : P \to \mathcal{L}_{(-\infty, \theta]}$ where $\mathcal{L}_{(-\infty, \theta]}$ denotes the set of all bags with elements in $(-\infty, \theta]$. A bag [14] is a generalization of a set, allowing the same element to occur more than once. If $p \in P$ contains $m_\theta(p) \in Z_+$ tokens at time θ, the bag $M_\theta(p) := \{\tilde{\theta}_1, \ldots, \tilde{\theta}_{m_\theta(p)}\}$ enumerates the arrival times of these $m_\theta(p)$ tokens. All these arrival times are necessarily smaller than θ. The set of all possible states, denoted by $\mathcal{M}$, satisfies some other constraints imposed by the dynamics of the model, to be specified below.

The state information at time θ indicates that transition t became state-enabled at time $\theta_e(t) = \max_{p \in ^\bullet t} \min M_\theta(p)$. By convention the minimum of an empty bag is ∞. Transition t must fire at some time $\theta_f(t)$ in the interval $[\max_{p \in ^\bullet t} \min M_{\theta_f(t)}(p) + L(t), \max_{p \in ^\bullet t} \min M_{\theta_f(t)}(p) + U(t)]$. Execution of t at $\theta_f(t)$ changes the distribution of tokens as in the untimed case. The state changes as follows: from each place $p_1 \in ^\bullet t$ a token with value $\min M_{\theta_f(t)}(p_1)$, is removed from the bag $M_{\theta_f(t)}(p_1)$, and to each place $p_2 \in t^\bullet$ a token with value $\theta_f(t)$ is added to the bag $M_{\theta_f(t)}(p_2)$. The state of a Timed Petri net (TPN) changes only when a transition is executed. Any state $M_\theta \in \mathcal{M}$ must be such that $\forall t \in T : \max_{p \in ^\bullet t} \min M_\theta(p) + U(t) \geq \theta$ since otherwise the firing interval of some transition t would have been pre-empted by the firing of another transition, using up some token necessary for the firing of t. When a bag $B = t_1, \ldots, t_k$ of transitions is simultaneously enabled then all transitions in the bag can be executed simultaneously. Notice that this can happen at time $\theta \in \cap_{t_i \in B}[\theta_e(t) + L(t), \theta_e(t) + U(t)]$ provided $\forall p \in P : m(p) \geq \sum_{i=1}^{k} I_{p \bullet t_i}$. When interpreting a TPN as a timed discrete event system, the firing of a transition at time θ_f in the TPN corresponds to the occurrence of the corresponding event at time θ_f in the plant. A sequence of consecutive events $t_{i_1}, t_{i_2}, \ldots$ with their firing times $\theta_1 \leq \theta_2 \leq \ldots$ — an element of the language of the timed automaton — is allowable if for each k the transition t_{i_k} can be executed at θ_k, leading to a state M_{θ_k} such that $t_{i_{k+1}}$ is executable at θ_{k+1}. In particular, it is necessary that execution of $t_{i_{k+1}}$ is not pre-empted by the forced execution of another transition prior to $\theta_{i_{k+1}}$.

The language of a TPN model does not automatically impose a lower bound on the interval between two consecutive firings of the same transition. If the physical model has constraints requiring such a minimum delay $L(t)$ between two consecutive executions of t then one has to add to the model an extra loop around t, i.e. add a place p_0 to the net such that $^\bullet p_0 = p_0^\bullet = \{t\}$. In general the model of a TPN does not exclude Zeno behaviour, where infinitely many events happen in a finite interval of time. In order to make sure that the set of allowed trajectories does not include such Zeno behaviour, it is sufficient that in every cycle of the Petri net (P, T, F) there is at least one transition t with a non-zero lower bound $L(t) > 0$.

A *controlled timed Petri net (CtlTPN)* is a six-tuple (P, T, F, L, U, T^c) in which (P, T, F, L, U) is a timed Petri net. The control can add further constraints to the plant model by shrinking the interval where some controllable transitions must fire. Let $I_{\mathcal{R}}$ denote the set of closed intervals on $(0, \infty]$. A control value C is a function $C : T \to I_{\mathcal{R}}$ such that $C(t) \subset [L(t), U(t)]$, reduces the interval of possible firing times for transition t from $[L(t), U(t)]$ to $[C_l(t), C_u(t)]$. This means that the firing of a state-enabled transition cannot be delayed forever, unless the transition is remote. A transition t with $C_u(t) = U(t) = \infty$ is called a remote transition.

The physical limitations of the controller and the plant specify the set of allowed values $C_l(t)$ and $C_u(t)$. Obviously the constraints $L(t) \leq C_l(t) \leq C_u(t) \leq U(t)$ must always be satisfied. For an uncontrollable transition t the upper and lower bounds cannot be changed: $C_l(t) = L(t)$ and $C_u(t) = U(t)$. For other transitions, called delayable transitions, only the lower bound can be changed: $C_u(t) = U(t), C_l(t) \leq U(t)$. A delayable, remote transition is called a fully controllable transition. The set of all allowable control values is denoted by $\mathcal{C}$.

4.3   Timed Petri net model of a steel plant

In order to illustrate the modelling power of timed Petri nets we consider as a case study the problem of developing a model of a steel plant. The model is intended for automatic generation of optimal schedules of tasks in the steel plant, where all the schedules satisfy a number of specifications. The operations of the steel plant must be scheduled so that it produces a prespecified sequence of batches of steel of different qualities (each batch characterised by a recipe that determines the sequence of operations to be carried out on the batch, and by a lower and an upper bound on the casting time). The optimal schedule must minimize the completion time of the last batch in the sequence, while satisfying all model constraints specified below.

The plant layout is shown on top of Figure 8. The plant is so large that the complete model must be split in different components. The main components correspond to the different stages of the operation on each batch. Each of these stages of operation is carried out in a different geographical location, and hence these 3 components also correspond to geographically separated parts of the plant.

The treatment of batches can be subdivided in 3 consecutive phases (corresponding to 3 different areas in the physical plant, and also corresponding to the 3 different components of the model):

- conversion and initial stage of metallurgical treatment (the model assumes there are two convertors, and there is one common resource that can be used for one of the convertors at a time only; there is also a cleaning operation of the convertors that must be executed from time to time; the model also includes the first metallurgical treatments, that are common to all qualities of steel that are being produced)

- metallurgical treatments specific to the quality of steel (recipe) to be produced (recipes differ by the path followed in this part of the plant, and by the time requirements of the operations; some qualities of steel only require operations in one or two positions, while other qualities require use of the special position on the right of the top track);

- continuous casting of the batches (modelling a buffer, the loading of the turret, and the casting itself).

Each of these phases naturally corresponds to one component in the modular (or compositional) mathematical model. On the bottom of Figure 8 these components are represented as timed Petri nets. The exact formal semantics of a timed Petri net are set out in the preceding subsection. However it will be shown below that the constraints that are quality dependent require that the model be extended further to include "coloured" tokens, where the tokens have a certain value (more general than the binary value present/absent in the usual Petri net formalism). These furtehr constraints allow for a compact model. Of course one could also use a different timed Petri net for each quality.

It should be observed that each component represents local constraints on the evolution of the overall plant. Of course there are also global cosntraints, which depend on varaibles in several components. The life time of a batch, specified to be below a certain maximum to avoid excessive cooling, depends on the start time of the "start conversion" transition in component 1 and the completion of the "casting" operation $th6$ in component 3. When designing a feasible schedule the requests for an overhead crane come from component 1, 2 and 3. Whether these requests are feasible depends therefore on what happens in all 3 components.

The three components are interconnected in an acyclic way, making scheduling easy as illustrated in [4]. The output of component 1 is the input of component 2, and the output of component 2 is the input of component 3. These three components represent the foreground events of the plant model. Convertor and casting machine are the most expensive components, which are likely to be the bottlenecks. The performance measure - the execution time of the start of the casting of the last batch - can be expressed in terms of event times in module 3 only. The life time of a batch, and the uninterrupted operation of the casting machine depend mainly on events in these 3 components. All the major scheduling decisions relate to the events in these 3 modules. Supervisory control must impose certain constraints enforcing global specifications, such as the maximal life time of each batch, or the availability of resources used in different components such as overhead cranes, on the operation of the individual components.

Besides this acyclic graph of foreground components, the plant model also consists of several other components representing the transport of empty steel ladles, the movement of the two cranes (running on the same rail and hence subject to "no collision" constraints) which transport the steel ladles from one position to the next position, and the availability of a car on each of the 4 rails connecting neighbouring positions for metallurgical treatment. These background components have inputs from the continuous casting component and the metallurgical treatment component, and send output to the convertor and metallurgical treatment components. Hence they create cycles in the model which will complicate the synthesis procedure of the schedule.

Each component of the model is a timed Petri net. The lower and upper bounds corresponding to a transition are determined by the physical limitations on the time needed to execute a certain event, and by the constraints imposed by the recipes. In the representation of component 3, $th5$ corresponds to the time needed to turn the turret of the casting machine. This is the delay between the completion of the casting of the $n - 1$-th batch, and the start of the casting of the n-th batch. This is always fixed, and lower and upper bound are therefore the same. The lower and upper bounds for transition $th6$ on the other hand correspond to the casting time of a batch. This is a variable that depends on the quality of steel being produced in the n-th batch, and is moreover a controllable variable. This shows that a further extension of the timed Petri net model is needed. Tokens can have an extra property, their "colour" which corresponds in this example to the quality of steel being produced by the batch represented by the token. This extension is needed in components 2 and 3 where the recipe determines the sequence of operations and the duration of the metallurgical operations, and of the casting.

Whether the extensions, of including time and colour, on top of the basic Petri net model are useful depends on whether one can develop analysis and control synthesis tools for these extended models. Some results on verification

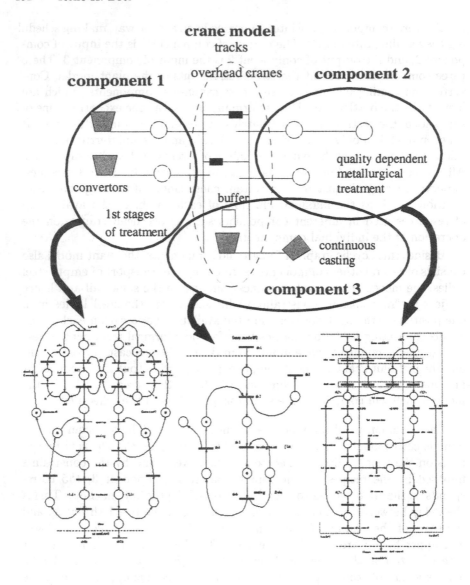

Fig. 8. Plant Layout, and timed Petri net models of main components

of properties for such coloured timed Petri nets are available in the literature
[17]. The work on defining good algorithms, combining the graphical repre-
setnation fo a Petri net, with algebraic analysis tools, is the topic for some
ongoing research. The main contribution of this subsection is to illustrate
how such models can be used for representing large plants in a compact, easy
to understand, model.

4.4 Control of timed Petri nets

For untimed Petri nets control actions can only block a controllable transition. For timed Petri nets the control actions can be much more detailed. The transitions can be delayed for a certain time, or may be forced to happen at some point in time.

Again one can define forbidden states or forbidden markings inorder to achieve safe of proper behaviour of the plant. However control cannot arbitrarily delay execution of events, and hence the forbidden set must again be transformed on a forbidden set of states. Even if the specification is in terms of a set of forbidden markings M, the forbidden set M^* of states that must be avoided, in order to guarantee that uncontrollable execution of events can never lead to a forbidden marking, will depend on the full state description (including timing information). It is therefore clear that control synthesis for timed Petri nets is quite complicated. Some open problems on constructing an "influencing net" have been posed in [5]. Some simple case where the forbidden set M^* can be constructed have been obtained in [10]

5 Acknowledgments

The author is grateful to B. Bordbar, F. Montoya, and G. Stremersch, and to the colleagues in the ESPRIT project "VHS - Verification of Hybrid Systems", for the numerous discussions, and for all their interesting comments and suggestions.

References

1. Cassandras C.G., Lafortune S., *Introduction to Discrete Event Systems*, Kluwer Academic Publishers, Boston 1999.
2. Ramadge P.J., Wonham W.M., "The Control of Discrete Event Systems", *Proceedings of the IEEE*, **77-1**, 81–98, 1989.
3. Boel R., Ben-Naoum L., and Van Breusegem V., On forbidden state problems for a class of controlled Petri nets, *IEEE Transactions on Automatic Control*, 40(10):1717–1731, 1995.
4. Boel R., Montoya F., "Modular Synthesis of efficient schedules in a timed discrete event plant", in *Proceedings of the 40th IEEE Conference on Decision and Control*, Sydney, 2000.
5. Boel R., Stremersch G., "Forbidden State Control Synthsesis for timed Petri ent Models", in M. Vidyasagar, V. Blondel, E. Sontag and J. Willems, eds., *Open Problems in Mathematical Systems and Control Theory*, pp. 61–66, Springer Verlag London, 1998.
6. Banaszak Z., Krogh B., Deadlock avoidance in flexible manufacturing systems with concurrently competing process flows. *IEEE Transactions on Robotics and Automation* 38(7):724–734, 1990.
7. Desel J., Esparza, J., *Free choice Petri Nets*, Cambridge Tracts in Theoretical Computer Science, Cambridge University Press, 1995.

8. Holloway L., Krogh B., Synthesis of feedback control logic for a class of controlled Petri nets, *IEEE Trabsactions on Automatic Control*, 35(5):514–523, 1990.

9. Li Y., Wonham W.M., Control of vector discrete-event systems, part 2: controller synthesis, *IEEE Transactions on Automatic Control*, 39(3):512–531, 1994.

10. Stremersch G., Boel R., On the influencing net and forbidden state control of timed Petri nets with forced transitions. In *Proceedings of the 37th IEEE Conference on Decision and Control*, pp. 3287–3292, 1998.

11. Stremersch G., Boel R., Structuring acyclic Petri nets for reachability analysis and control. *Discrete Event Dynamical Systems: Theory and Applications*, 2000, in print.

12. Stremersch G., On the union of legal down–sets in supervisory control of Petri nets. In *Proceedings of the IEEE Conference on Decision and Control*, Sydney, 2000.

13. Stremersch G., Boel R., Decomposition of the supervisory control problem for Petri nets under preservation of maximal permissiveness, *IEEE Transactions on Automatic Control*, 2000, to appear.

14. Peterson J., *Petri Net Theory and the Modelling of Systems* Prentice–Hall, 1981.

15. Silva M., Teruel E., Colom J., "Linear Algebraic and Linear Programming Techniques for the Analysis of Place/Transition Net Systems", in *Proceedings of the 1998 Conference on Applications and Theory of Petri Nets*, Lecture Notes in Computer Sciences, Springer Verlag, 1999, pp. 309–375.

16. Spacek P., Manier M.-A., El Moudni A., "Control of an Electroplating Line in the Max and Min Algebras", *Systems Science*, vol. 30, no. 7, July 1999, pp. 759–778.

17. van der Aalst W., Interval Timed Coloured Petri Nets and their Analysis, In:*Proceedings of the 14th International Conferernce on the Applications and Theory of Petri Nets*, Springer Verlag, 1993.

Modelling and Analysis of Hybrid Dynamical Systems*

Arjan J. van der Schaft[1] and Johannes M. Schumacher[2]

[1] Faculty of Mathematical Sciences, University of Twente, Enschede,
The Netherlands
[2] Department of Econometrics and Operations Research, Tilburg University,
Tilburg, The Netherlands

Abstract. In this paper we discuss features of hybrid dynamical systems based on the hybrid automaton model. Next we describe an alternative way of representing hybrid systems by event-flow formulas. Finally, a special class of hybrid dynamical systems, called complementarity systems, is discussed, and conditions for existence and uniqueness of solutions are reviewed.

1 Introduction

Generally speaking, hybrid systems are mixtures of real-time (continuous) dynamics and discrete events. These continuous and discrete dynamics not only coexist, but *interact* and changes occur both in response to discrete, instantaneous, events and in response to dynamics as described by differential or difference equations of time. A main difficulty in the discussion of hybrid systems is that they encompass in some sense every possible dynamical system we can think of.

Various scientific communities with their own approaches and motivations are involved in the research on hybrid systems. At least the following three communities can be distinguished.

First there is the *computer science* community that looks at a hybrid system primarily as a discrete (computer) program interacting with an analog environment. In this context also the terminology *embedded systems* is being used. A main aim is to extend standard program analysis techniques to systems which entail some kind of continuous dynamics. The emphasis is often on the discrete event dynamics, with the continuous dynamics of a relatively simple form. A key issue is that of *verification*.

A second community involved in the study of hybrid systems is the *modeling and simulation* community. Physical systems can often operate in different *modes*, and the transition from one mode to another sometimes can be idealized as an instantaneous, discrete, transition. Examples include electrical circuits with switching devices such as (ideal) diodes and transistors, and mechanical systems subject to inequality constraints as encountered e.g. in

---

* This paper is based on material contained in [30,31].

robotics. Since the time scale of the transition from one mode to another is often much faster than the time scale of the dynamics of the individual modes, it may be advantageous to model the transitions as being instantaneous. This time instant is called an *event time*. Basic issues then concern the well-posedness of the resulting hybrid system, e.g. the existence and uniqueness of solutions, and the ability to efficiently *simulate* the multi-modal physical system.

A third community contributing to the area of hybrid systems is the *systems and control* community. In general, most cost in current control system development is spent on ad-hoc systems integration, and validation techniques that rely on exhaustively testing of complex control systems. One can think of hierarchical systems with a discrete decision layer and a continuous implementation layer (e.g. supervisory control or multi-agent control). Thus there is a clear need for systematic hierarchical design methodologies based on hybrid systems. Additional motivation for the study of hybrid systems is provided from different angles. Classical *switching control* schemes and *relay control* immediately lead to hybrid systems. For *nonlinear control* systems it is known that in some important cases there does not exist a *continuous* stabilizing state feedback, but that nevertheless the system can be stabilized by a switching control. Finally, *discrete event systems theory* can be seen as a special case of hybrid systems theory. In many areas of control, e.g. in power converters and in motion control, control strategies are inherently hybrid in nature.

In view of the wide range of hybrid system associations it is clear that any presentation of the subject is bound to be biased. The present survey paper, based on [30], emphasizes the dynamical systems aspects of hybrid systems.

From a general system-theoretic point of view one can look at hybrid systems as systems having two different types of ports along which they interact with their environment. First type of ports are the *communication ports*. The *discrete* variables associated with these ports are symbolic in nature, and represent "data-flow". The strings of symbols at these communication ports in general are not directly related with real (physical) time; there is only a sequential ordering.

Second type of ports are the *physical ports* (with "physical" interpreted in the broad sense; perhaps "analog" would be a more appropiate terminology). The variables at these ports are usually continuous variables, and related to physical measurement. Also the flow of these variables is directly related to physical time. In principle the signals at the physical ports may be discrete time signals (or sampled-data signals), but in most cases they will be ultimately *continuous time* signals.

Thus a hybrid system can be regarded as a combination of *discrete* (or symbolic) dynamics and *continuous* dynamics. The main problem in the definition and representation of a hybrid system is precisely to specify the *interaction* between this symbolic and continuous dynamics.

A key issue in the formulation of hybrid systems is the often required *modularity* of the hybrid system description. Indeed, because we are inherently dealing with the modeling of *complex* systems, it is very important to model a complex hybrid system as the *interconnection* of simpler (hybrid) subsystems. This implies that the hybrid models that we are going to discuss are preferably of a form that admits easy interconnection and composition. Besides this notion of *compositionality* other important (related) notions are those of "reusability" and "hierarchy". Another terminology that is used in this context is that of *"object-oriented modeling"*.

2 Definitions of hybrid systems

We start with a reasonably generally accepted "working definition" of hybrid systems, which already has proved its usefulness. This definition, called the *hybrid automaton* model, provides the framework and terminology to discuss a range of typical features of hybrid systems. At the end of this section we discuss alternative ways of modeling hybrid systems.

2.1 Continuous and symbolic dynamics

In order to motivate the hybrid automaton definition, we recall the "paradigms" of continuous and symbolic dynamics; namely, state space models described by differential equations for continuous dynamics, and finite automata for symbolic dynamics. Indeed, the definition of a hybrid automaton *combines* these two paradigms.

Definition 1 (Continuous-time state-space models).
A continuous-time state-space system is described by a set of *state* variables x taking values in $\mathbb{R}^n$ (or, more generally, in an n-dimensional state space manifold X), and a set of *external* variables w taking values in $\mathbb{R}^q$, related by a mixed set of differential and algebraic equations of the form

$$F(x, \dot{x}, w) = 0 \tag{1}$$

Here $\dot{x}$ denotes the time-derivative of x. Solutions of (1) are all (sufficiently smooth) time functions $x(t)$ and $w(t)$ satisfying

$$F(x(t), \dot{x}(t), w(t)) = 0$$

for (almost) all times $t \in \mathbb{R}$ (the continuous-time axis).

Of course, the above definition encompasses the more common definition of a continuous-time input-state-output system

$$\begin{aligned}
\dot{x} &= f(x, u) \\
y &= h(x, u)
\end{aligned} \tag{2}$$

where we have split the vector of external variables w into a sub-vector u taking values in $\mathbb{R}^m$ and a sub-vector y taking values in $\mathbb{R}^p$ (with $m + p = q$), called respectively the vector of *input* and *output* variables. The only algebraic equations in (2) are those relating the output variables y to x and u, while generally in (1) there are additional algebraic constraints on the state space variables x.

A main advantage of general continuous-time state space systems (1) over continuous-time input-state-output systems (2) is the fact that the first class is closed under interconnection, while the second class in general is not. In fact, modeling approaches that are based on modularity (viewing the system as the interconnection of smaller sub-systems) almost invariably lead to a mixed set of differential and algebraic equations. Of course, in a number of cases it may be relatively easy to eliminate the algebraic equations in the state space variables, in which case (if we can also easily split w into u and y) we can convert (1) into (2).

We note that Definition 1 does not yet completely specify the continuous-time system, since (on purpose) we have been rather vague about the precise *solution concept* of the differential-algebraic equations (1). For example, a reasonable (but not the only possible!) choice is to require $w(t)$ to be piece-wise continuous (allowing for discontinuities in the "inputs") and $x(t)$ to be continuous and piecewise differentiable, with (1) being satisfied for almost all t (except for the points of discontinuity of $w(t)$ and non-differentiability of $x(t)$.)

Next we give the standard definition of a finite automaton (or *finite state machine*, or *labeled transition system*).

Definition 2 (Finite automaton). A finite automaton is described by a triple (L, A, E). Here L is a finite set called the *state space*, A is a finite set called the *alphabet* whose elements are called *symbols*. E is the *transition rule*: it is a subset of $L \times A \times L$ and its elements are called *edges* (or transitions, or events).

A sequence $(l_0, a_0, l_1, a_1, \ldots, l_{n-1}, a_{n-1}, l_n)$ with $(l_i, a_i, l_{i+1}) \in E$ for $i = 1, 2, \ldots, n - 1$ is called a trajectory or *path*.

The usual way of depicting an automaton is by a *graph* with vertices given by the elements of L, and edges given by the elements of E. Then A can be seen as a set of *labels* labeling the edges. Sometimes they are called *synchronization* labels, since interconnection with other automata takes place via these (shared) symbols. One can also specialize Definition 2 to *input-output* automata by associating with every edge *two* symbols, namely an *input* symbol i and an *output* symbol o, and by requiring that for every input symbol there is only one edge originating from the given state with this input symbol. (Sometimes, such automata are called *deterministic* input-output automata.) Deterministic input-output automata can be represented by equations of the following form:

$$l^\sharp = \nu(l, i)$$
$$o = \eta(l, i) \tag{3}$$

where $l^\sharp$ denotes the new value of the discrete state *after* the event takes place, resulting from the old discrete state value l and the input i.

Often the definition of a finite automaton also entails the explicit specification of a subset $I \subset L$ of *initial states* and a subset $F \subset L$ of *final states*. A path $(l_0, a_0, l_1, a_1, \ldots, l_{n-1}, a_{n-1}, l_n)$ is then called a *successful path* if in addition $l_0 \in I$ and $l_n \in F$.

In contrast with the continuous time systems defined in Definition 1 the solution concept (or *semantics*) of a finite automaton (with or without initial and final states) is completely specified; the behavior of the finite automaton are all (successful) paths. In theoretical computer science parlance the definition of a finite automaton is said to entail an "operational semantics", completely specifying the formal language generated by the finite automaton.

Note that the definition of a finite automaton is conceptually not very different from the definition of a continuous-time state space system. Indeed we may relate the state space L with the state space X, the symbol alphabet A with the space W (where the external variables take their values), and the transition rule E with the set of differential-algebraic equations given by (1). Furthermore the paths of the finite automaton correspond to the *solutions* of the set of differential-algebraic equations. The analogy between continuous-time input-state-output systems (2) and input-output automata (3) is obvious, with the differentiation operator $\frac{d}{dt}$ replaced by the "next state" operator $\sharp$.

A (minor) difference is that in finite automata one usually considers (as in Definition 2) paths of *finite* length, while for continuous-time state space systems the emphasis is on solutions over the whole time-axis $\mathbb{R}$. This could be remedied by adding to the finite automaton a *source state* and a *sink state* and a blank label, and by considering solutions defined over the whole time-axis $\mathbb{Z}$ which "start" at minus infinity in the source state and "end" at plus infinity in the sink state, while producing the blank symbol when remaining in the source or sink state. Also the set I of initial states and the set F of final states in some definitions of a finite automaton do not have a direct analogon in the definition of a continuous-time state space system.

2.2 Hybrid automaton

Combining Definitions 1 and 2 leads to the following type of definition of a hybrid system.

Definition 3 (Hybrid automaton, [1]). A *hybrid automaton* is described by a septuple $(L, X, E, A, W, Inv, Act)$ where

- L is a finite set, called the set of *discrete states* or *locations*. They are the *vertices* of a graph.
- X is the continuous state space of the hybrid automaton in which the continuous state space variables x take their values. For our purposes $X \subset \mathbb{R}^n$ or X is an n-dimensional manifold.
- E is a set of edges called transitions (or *events*). Every edge is defined by a four-tuple $(l, Guard_{ll'}, Jump_{ll'}, l')$, where $l, l' \in L$, $Guard_{ll'}$ is a subset of X and $Jump_{ll'}$ is a relation defined by a subset of $X \times X$. The transition from the discrete state l to l' is *enabled* when the continuous state x is in $Guard_{ll'}$, while during the transition the continuous state x jumps to a value x' given by the relation $(x, x') \in Jump_{ll'}$.
- A is a finite set of symbols labeling the edges. Each edge is associated with a symbol by a labeling function λ from E to A.
- $W = \mathbb{R}^q$ is the continuous communication space in which the continuous external variables w take their values.
- Inv is a mapping from the locations L to the set of subsets of X, that is $Inv(l) \subset X$ for all $l \in L$. Whenever the system is at location l, then the continuous state x must satisfy $x \in Inv(l)$. The subset $Inv(l)$ for $l \in L$ is called the *location invariant* of location l.
- Act is a mapping that assigns to each location $l \in L$ a set of differential-algebraic equations F_l, relating the continuous state variables x with their time-derivatives $\dot{x}$ and the continuous external variables w:

$$F_l(x, \dot{x}, w) = 0 \qquad (4)$$

The solutions of these differential-algebraic equations are called the *activities* of the location.

Clearly, the above definition is very much based on Definition 2, with the discrete state space L now being called the space of locations. (Note that the set of edges E in Definition 3 also defines a subset of $L \times A \times L$.) In fact, Definition 3 extends Definition 2 by associating with every vertex (location) a continuous dynamics (whose solutions are the activities), and by associating with every transition $l \to l'$ also a possible jump in the continuous state.

Note that the *state* of a hybrid automaton consists of a discrete part $l \in L$ and a continuous part in X. Furthermore, the *external variables* consist of a discrete part taking their values a in A and a continuous part w taking their values in $\mathbb{R}^q$. Also, the dynamics consists of discrete transitions (from one location to another), together with a continuous part evolving in the location invariant.

It should be remarked that the above definition of a hybrid automaton has the same ambiguity as the definition of a continuous-time state-space system, since it still has to be complemented by a precise specification of the solutions (activities) of the differential-algebraic equations associated with every location. In fact, in the original definitions of a hybrid automaton (see e.g. [1])

the activities of every location are explicitly given, instead of implicitly generating them as the solutions to the differential-algebraic equations. On the other hand, for somebody acquainted with differential equations it is rather restricted to immediately specify continuous dynamics by time functions from $\mathbb{R}^+$ to X. Indeed, continuous time dynamics is almost always described by sets of differential-algebraic equations, and only in exceptional cases (such as linear dynamical systems) one can obtain explicit solutions.

The description of a hybrid automaton is summarized in Figure 1.

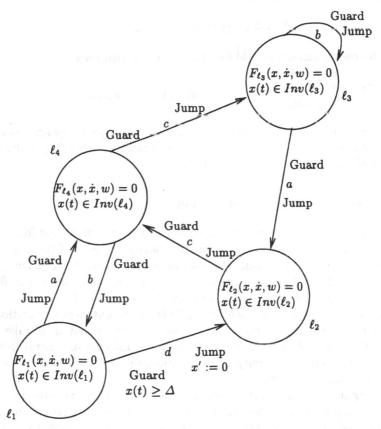

Fig. 1. Hybrid automaton.

A reasonable definition of the *trajectories* (or solutions, or in computer science terminology, the *runs* or *executions*) of a hybrid automaton is as follows. A continuous trajectory (l, δ, x, w) associated with a location l consists of a non-negative time δ (the *duration* of the continuous trajectory), a piecewise continuous function $w : [0, \delta] \to W$, and a continuous and piecewise differentiable function $x : [0, \delta] \to X$ such that

- $x(t) \in Inv(l)$ for all $t \in (0, \delta)$,
- $F_l(x(t), \dot{x}(t), w(t)) = 0$ for all $t \in (0, \delta)$ except for points of discontinuity of w.

A *trajectory* of the hybrid automaton is an (infinite) sequence of continuous trajectories

$$(l_0, \delta_0, x_0, w_0) \overset{a_0}{\to} (l_1, \delta_1, x_1, w_1) \overset{a_1}{\to} (l_2, \delta_2, x_2, w_2) \overset{a_2}{\to} \ldots$$

such that at the *event times*

$$t_0 = \delta_0, \quad t_1 = \delta_0 + \delta_1, \quad t_2 = \delta_0 + \delta_1 + \delta_2, \ldots$$

the following inclusions hold for the discrete transitions

$$x_j(t_j) \in Guard_{l_j l_{j+1}} \qquad \text{for all } j = 0, 1, 2, \ldots$$
$$(x_j(t_j), x_{j+1}(t_j)) \in Jump_{l_j l_{j+1}}$$

Furthermore, to the j-th arrow $\to$ in the above sequence (with j starting at 0) one associates a symbol (label) a_j, representing the value of the discrete "signal" at the j-th discrete transition.

2.3   Features of hybrid dynamics

Note that the trajectories of a hybrid automaton exhibit the following features. Starting at a given location the continuous part of the state evolves according to the continuous dynamics associated with this location, provided it remains in the location invariant. Then, at some time instant in $\mathbb{R}$, called an *event time*, an event occurs and the discrete part of the state (the location) *switches* to another location. This is an *instantaneous* transition which is *guarded*, that is, a necessary condition for this transition to take place is that the guard of this transition is satisfied. Moreover in general this transition will also involve a *jump* in the continuous part of the state. Then, after the instantaneous transition has taken place, the continuous part of the state, starting from this new continuous state, will in principle evolve according to the continuous dynamics of the new location. Thus there are two phenomena associated with every event, a *switch* and a *jump*, describing the instantaneous transition of, respectively, the discrete and the continuous part of the state at such an event time.

A basic issue in the specification of a hybrid system is the *specification of the events and event times*. First, the events may be *externally induced* via the labels (symbols) $a \in A$, this leads to *controlled* switchings and jumps. Secondly, the events may be *internally induced*, this leads to what is called *autonomous* switchings and jumps. The occurrence of internally induced events is determined by the guards and the location invariants. Whenever the location invariants are going to be violated then the hybrid automation *has* to switch to a new location, with possible resetting of the continuous state. At such an event time the guards will determine to *which* locations the transition is possible. (There may be more than one; furthermore, it may be possible to switch to the *same* location.)

If the location invariants are not going to be violated then *still* discrete transitions may take place if the corresponding guards are being satisfied. That is, if at a certain time instant the guard of a discrete transition is satisfied then this may cause a possible event time. This may lead to a large class of trajectories of the hybrid automaton, and a tighter specification of the behavior of the hybrid automaton will critically depend on a more restrictive definition of the guards.

Many questions naturally come up in connection with the analysis of the trajectories of a hybrid automaton:

- It could happen that, after some time t, the system ends up in a state $(l, x(t))$ from which there is no continuation, that is, there is no possible continuous trajectory from $x(t)$ and no possible transition to another location. In the computer science literature this is usually called "*deadlock*". Clearly, this is an undesirable phenomenon, because it means that the system is "stuck".

- The set of durations δ_i may get smaller and smaller for increasing i even to such an extent that $\sum_{i=0}^{i=\infty} \delta_i$ is finite, say τ. This means that τ is an accumulation point of event times. In the computer science literature this is called *Zeno* behavior (referring to Zeno's paradox of Achilles and the turtle). This may not be a totally undesirable behavior of the hybrid system. In fact, as long as the continuous and discrete parts of the state will converge to a unique value at the accumulation point τ, then we can re-initialize the hybrid system at τ at these limits, and let the system run as before starting from the initial time τ. The bouncing ball is a situation like this.

- In principle the durations of the continuous trajectories are allowed to be zero, thereby covering the occurrence of *multiple events*. In this case the underlying time-axis of the hybrid trajectory has a structure which is more complicated than $\mathbb{R}$ containing a set of event times: a certain time instant $t \in \mathbb{R}$ may correspond to a *sequence* of sequentially ordered transitions, all happening at this same time instant t (called a multiple event time).

- It may happen that the hybrid system gets stuck at such a multiple event time by switching indefinitely between different locations (and not proceeding in time). This is sometimes called *livelock*. Such a situation occurs if a location invariant is (going to be) violated and a guarded transition takes place to a new location in such a way that the new continuous state does not satisfy (or is imminent to violate) the location invariants of the new location, while the guard for the transition to the old location is satisfied. In some cases this problem can be resolved by creating a *new* location, with a continuous dynamics that "averages" the continuous dynamics of the locations between which the infinite switching occurs. The classical Filippov's notions of solutions of discontinuous vector fields can be interpreted in this way.

- In general the set of trajectories (runs) of the hybrid automaton may be very large, especially if the guards are not very strict. For certain purposes (e.g. verification) this may not be a problem, but in other cases one may wish the hybrid system to be *deterministic* or *well-posed*, in the sense of having *unique* solutions for given discrete and continuous "inputs" (assuming that we have split the vector w of continuous external variables into a vector of continuous inputs and outputs, and that every label a actually consists of an input label and an output label). Especially for *simulation* purposes this may be very desirable, and in fact one would in certain cases dismiss a hybrid model as inappropriate if it does not have this well-posedness property. On the other hand, non-deterministic discrete-event systems are very common in computer science. A simple example of a hybrid system not having unique solutions will be given in Section 3. (A "physical" example of the same type exhibiting non-uniqueness of solutions is the classical Painlevé example of a stick sliding (subject to Coulomb friction) with one end on a table, see e.g. [8]).

- The solution concept of the continuous-time dynamics associated to a location may itself be problematic, especially because of the possible presence of algebraic constraints. In particular, in some situations one may want to associate *jump behavior* with these continuous-time dynamics. Within the hybrid framework this can be incorporated as internally induced events, where the system switches to the same location but with a reset of the continuous state.

Remark 1. Of course, the definition of a hybrid automaton can still be generalized in a number of directions. A particularly interesting extension is to consider *stochastic* hybrid systems, such as described by *piecewise-deterministic Markov processes*, see e.g. [10]. In this notion the event times are determined by the system reaching certain boundaries in the continuous state space (similar to the notion of location invariants), and/or by an underlying *probability distribution*. Furthermore, also the resulting discrete transitions together with their jump relations are assumed to be governed by a probability distribution.

2.4   Generalized hybrid automaton

In Definition 3 of a hybrid automaton there is still an apparent asymmetry between the continuous and the symbolic (discrete) part of the dynamics. Furthermore, the location invariants and the guards play a very much related role in the specification of the discrete transitions. The following generalization of Definition 3 takes the location invariants, the set of edges E and the labeling function λ together, and symmetrizes the definition of a hybrid automaton. (The input-output version of this definition is due to [23].)

Definition 4 (Generalized hybrid automaton). A *generalized hybrid automaton* is described by a sixtuple (L, X, A, W, R, Act) where L, X, A, W and Act are as in Definition 3, and R is a subset of $(L \times X) \times (A \times W) \times (L \times X)$. A continuous trajectory (l, a, δ, x, w) associated with a location l and a discrete external symbol a consists of a non-negative time δ (the *duration* of the continuous trajectory), a piecewise continuous function $w : [0, \delta] \to W$, and a continuous and piecewise differentiable function $x : [0, \delta] \to X$ such that

- $(l, x(t), a, w(t), l, x(t)) \in R$ for all $t \in (0, \delta)$,
- $F_l(x(t), \dot{x}(t), w(t)) = 0$ for almost all $t \in (0, \delta)$ (except for points of discontinuity of w).

A *trajectory* of the generalized hybrid automaton is an (infinite) sequence

$$(l_0, a_0, \delta_0, x_0, w_0) \to (l_1, a_1, \delta_1, x_1, w_1) \to (l_2, a_2, \delta_2, x_2, w_2) \to \cdots$$

such that at the event times

$$t_0 = \delta_0, t_1 = \delta_0 + \delta_1, t_2 = \delta_0 + \delta_1 + \delta_2, \ldots$$

the following inclusions hold:

$$(l_j, x_j(t_j), a_j, w_j(t_j), l_{j+1}, x_{j+1}(t_j)) \in R, \text{ for all } j = 0, 1, 2, \ldots$$

The subset R encompasses the notions of the location invariants, guards, and jumps of Definition 3 in the following way. To each location l we associate the location invariant

$$Inv(l) = \{(x, a, w) \in X \times A \times W \mid (l, x, a, w, l, x) \in R\}.$$

(Abusing notation, x and w denote here elements of X, respectively W, instead of *variables* taking their values in these spaces.) Furthermore, given two locations l, l' we obtain the following guard for the transition from l to l':

$$Guard_{ll'} = \{(x, a, w) \in X \times A \times W \mid \exists x' \in X, (l, x, a, w, l', x') \in R\}$$

with the interpretation that the transition from l to l' can take place if and only if $(x, a, w) \in Guard_{ll'}$. Finally, the associated jump relation is given as

$$Jump_{ll'}(x, a, w) = \{x' \in X \mid (l, x, a, w, l', x') \in R\}.$$

Note that the resulting location invariants, guards as well as jump relations are in principle of a more general type than in Definition 3, since they all may depend on the continuous and discrete external variables. On the other hand, it can be readily seen that any set E of edges and location invariants as in Definition 3 can be recovered from a suitably chosen set R as in Definition 4. Therefore, Definition 4 indeed does generalize Definition 3.

A further generalization of the definition of a hybrid automaton would be to allow the continuous state spaces associated with every location to be different. (This is now to some extent captured in the location invariants or the subset R.)

2.5   Hybrid time axes

A conceptual problem in Definitions 3 and 4 is the formalization of the notion of the time evolution corresponding to a hybrid trajectory, and in particular the inbedding of the event times in the continuous time axis $\mathbb{R}$. Indeed, literally it is stated that at the same "physical" time t_j the system is at different locations and that the continuous variables x take different values at this time instant t_j (namely, their values "just before" and "just after" the event has taken place). Obviously this is a mathematical inadequacy which needs to be repaired. This problem is further aggrevated by the fact that the durations δ_j of the continuous trajectories are allowed to be equal to zero, causing *multiple events*, in which case the system is at more than two different locations and x takes more than two values at the same time instant.

Furthermore, the notion of hybrid trajectory as employed in Definition 4, as well as in Definition 3, does not cover hybrid trajectories which have the property that their event times have an accumulation point but still the trajectory does progress in time *after* this accumulation point. The following formalization of the intuitive notion of a hybrid time evolution takes into account these considerations. Let $\mathbb{R}$ be the continuous-time axis. The *hybrid time axis* corresponding to a hybrid trajectory will be specified by a set $\mathcal{E}$ of *time events*. A time event in $\mathcal{E}$ consists of an *event time* $t \in \mathbb{R}$ together with a *multiplicity* $m(t)$, which is an element of $\mathbb{N} \cup \infty$, where $\mathbb{N}$ is the set of natural numbers $1, 2, 3, \ldots$. The time event will be denoted by the *sequence*

$$(t^{0\natural}, t^{1\natural}, t^{2\natural}, \ldots, t^{m(t)\natural}),$$

specifying the sequentially ordered "discrete transition times" at the *same* continuous time instant $t \in \mathbb{R}$ (the event time). For simplicity of notation we will sometimes write $t^{\natural}, t^{\natural\natural}, t^{\natural\natural\natural}, \ldots$ for $t^{1\natural}, t^{2\natural}, t^{3\natural}, \ldots$.

A time event with multiplicity equal to 1 is just given by a pair

$$(t^{0\natural}, t^{\natural})$$

with the interpretation of denoting the time instants "just before" and "just after" the event has taken place. If the multiplicity of the time event is larger than 1 (a *multiple time event*) then there are some "intermediate time-instants" ("all at the same event time t") ordering the sequence of discrete transitions taking place at t.

2.6   Event-flow formulas

In some sense the definition of a generalized hybrid automaton as given in Definition 4 drifts away from the explicit specification of location invariants, guards and discrete transitions, in the original definition of a hybrid automaton (Definition 3), by summarizing them into one abstract set R. A next step is to generate this set R by means of *equations*, and also to include in this description of R the differential(-algebraic) equations representing the activities. This leads to a methodology of what we shall call *event-flow formulas* (EFFs), and which seems to be an attractive way of modeling hybrid systems with a substantial continuous time dynamics.

In general, one should distinguish between the *syntax* of a description format (the rules that determine what is to be considered as a well-formed description) and its *semantics*, which in the case of dynamical systems can be interpreted as the notion of solution. In principle different semantics can be attached to the same syntax.

We now first describe the syntax of EFFs. We start with a finite index set V whose elements are called *variables*. The set V is the disjoint union of four subsets denoted by X, P, W, and S. The variables in X are called *continuous state variables*, those in P are called *discrete state variables*, those in W are *continuous communication variables*, and those in S are *discrete communication variables*. To each of the variables there is an associated *range space*. For the continuous variables we let this space be the real line; it would not be difficult to generalize to the case of differentiable manifolds. The range spaces of the discrete variables are finite sets which are denoted by L_i ($i = 1, \ldots, k$) in the case of state variables and by A_i ($i = 1, \ldots, r$) in the case of communication variables. The sets L_i are the *locations*, whereas the sets A_i are usually called *alphabets*.

For each continuous state variable $x \in X$ we introduce a new variable denoted by $\dot{x}$ which also has the real line as its range space; as the notation suggests, the symbol will be used in the semantics to express differentiation with respect to time. The set of new variables that is obtained in this way will be denoted by $\dot{X}$. Likewise, for each continuous state variable $x \in X$ and each discrete state variable $p \in P$ we introduce new variables $x^\sharp$ and $p^\sharp$ which will be used to express update operations. Both new variables have the same range space as the variables from which they are derived. The new sets of variables that are thus obtained will be denoted by $X^\sharp$ and $P^\sharp$. We write $V' := V \cup \dot{X} \cup X^\sharp \cup P^\sharp$.

Let V_0 be a subset of V'. A *valuation* of V_0 is a mapping that assigns to each element of V_0 an element of its associated range space. If the elements of V_0 are given a fixed order, then valuations of V_0 can be written as vectors whose length is the number of elements of V_0.

A *clause* over V_0 is a mapping that assigns to each valuation of V_0 the value TRUE or FALSE. In applications, a clause is typically given by an arithmetic or logical expression. As a trivial example, if $V_0 = \{x, x^\sharp\}$ (taken in this order), then a clause over V_0 is for instance given by the expression $x^\sharp = x + 1$, which returns TRUE for the valuation $(0, 1)$ and FALSE for the valuation $(0, 0)$. The semantics to be developed below is based in particular on clauses over variables in $X \cup P \cup W \cup \dot{X}$ (*flow clauses*) and clauses over variables in $X \cup P \cup S \cup X^\sharp \cup P^\sharp$ (*event clauses*). If ϕ is a clause over V_0 then we also say that V_0 is the *span* of ϕ, and we write span(ϕ)$= V_0$.

Finally we can express the notion of an event-flow formula.

Definition 5. An *event-flow formula*, or EFF, is a Boolean formula whose terms are clauses.

Next we come to the *semantics* of event-flow formulas. EFFs are intended to represent the set of possible evolutions of systems that are partly described by differential equations and partly by update operations. Now, updates take place at *event times* that may be different for different evolutions. Moreover, not all variables in a hybrid system need to be updated at the same time; it may happen that an event is local to some subsystem. As a consequence, we need a concept of time that is considerably more complicated than the usual model based on the real line. In [31] we have taken a fairly radical point of view to equip each variable with its own (hybrid) time axis. Then the joint evolutions of all variables that occur in a given EFF are considered, and we define in what sense an EFF can be satisfied by such a joint evolution. In the process we obtain an overall time axis which however is not in general a totally ordered set; this reflects the idea of *partial synchronization*. We refer to [31] for a detailed treatment. The resulting semantics is more complicated than the one given before in [30], because we have chosen to work with different time axes for each variable rather than with a uniform time axis. The benefit is that we can now define composition in a simpler way than in [30].

Definition 6. The *parallel composition* of two EFFs ϕ_1 and ϕ_2 is defined by $\phi_1 \parallel \phi_2 := \phi_1 \wedge \phi_2$.

In this way one may in fact unambiguously define the parallel composition of an arbitrary number of EFFs.

Finally we note that the framework presented in [31] might easily be extended to allow the variables to be defined only on *part* of the physical time axis, in the spirit of Benveniste's "presences" [6].

2.7 Discussion of representations

Many formalisms for the description of hybrid systems have been proposed in the literature; see for instance [2,7,1,23,24,6,19]. Here we only discuss the hybrid automaton model and EFFs as treated before.

Definitions 3 and 4 of a (generalized) hybrid automaton do provide workable representations of hybrid systems for various aims. First of all, they offer a clear picture of hybrid dynamics, which is very useful for exposition and theoretical analysis. A favorable feature of the hybrid automaton model is that the *semantics* of the model is quite explicit, as we have seen above. Furthermore, for a certain type of hybrid systems and for certain applications, the hybrid automaton representation can be quite effective.

Nevertheless, a drawback of the hybrid automaton representation is its tendency to become rather complicated. This is foremost due to the fact that in the hybrid automaton model it is necessary to specify all the locations and all the transitions from one location to another, together with all their guards and jumps (or to completely specify the subset R of Definition 4). If the number of locations grows, this usually becomes an enormous and error-prone task. Other related types of (graphical) representations of hybrid systems that have been proposed in the literature, such as differential (or dynamically colored) *Petri nets*, may be more efficient than the hybrid automaton model in certain cases but have similar features.

For hybrid systems arising in a *"physical domain"* it seems natural to use representation formalisms such as event-flow formulas, which are closer to first principles physical modeling. First principles modeling of dynamical systems almost invariably leads to sets of *equations*, differential or algebraic. Furthermore, the hybrid nature of such systems is usually in first instance described by "if-then" or "either-or"statements, in the sense that in one location of the hybrid system a particular subset of the total set of differential and algebraic equations has to be satisfied, while in another location a different subset of equations should hold. Thus, while in the (generalized) hybrid automaton model the dynamics associated with every location are in principle completely independent, in most "physical" examples the *set of equations* describing the various activities or *modes* (continuous dynamics associated to the locations) will remain almost the same, replacing one or more equations by some others.

Seen from this perspective, the (generalized) hybrid automaton model (and other similar descriptions of hybrid systems) may be quite far from the kind of model one obtains from physical first principles modeling, and the translation of the modeling information provided by equations, inequalities and logical statements into a complete specification of all the locations of the hybrid automaton together with all the possible discrete transitions and the complete continuous-time dynamics of every location may be a very tedious operation for the user. This becomes especially clear in an object-oriented modeling approach, where the interconnection (or composition) of hybrid automata may easily lead to a rapid growth in number of locations, and a rather elaborate (re-)specification of the resulting hybrid automaton model obtained by interconnection. Thus from the user's point of view an interesting alternative for efficiently specifying "physical" hybrid systems is to look for possibilities of specifying such systems primarily *by means of equations*, as in the framework of event-flow formulas. The setting of event-flow formulas is close to that of simulation languages such as Modelica™ [11], Some of the modeling constructs in Modelica relating to hybrid systems do in fact have the form of event-flow formulas. Synchronous languages like LUSTRE [13] and SIGNAL [5] are also related, be it more distantly since these languages operate in discrete-time; see [6] for an approach to general hybrid systems inspired by the SIGNAL language.

The formalism of event-flow formulas results in rather *implicit* representations of a hybrid system, as opposed to the almost completely *explicit* representations provided by the (generalized) hybrid automaton model. An EFF does not directly provide a recipe for generating solutions; it is rather a testing device that determines whether a proposed solution is valid or not. In fact an EFF is just a list of all the laws satisfied by a given system. In the terminology of [36], EFFs are kernel representations rather than image representations. Descriptions of this form are user-friendly in the sense that they facilitate specification, but they do pose a challenge to the developers of simulation software.

Within the framework of event-flow formulas one still strives for *complete* specifications of the hybrid system under consideration. In some examples of multi-modal physical systems, as e.g. arising in robotics and power-converters, the initial description of the hybrid system obtained from first principles modeling is *incomplete*, especially with regard to the specification of the discrete dynamics. In fact, one would like to *automatically* generate a complete event-flow formula description based on this initial, incomplete, description, together with some additional information. In Section 3 we will briefly describe such a framework for a special class of hybrid systems, called *complementarity* hybrid systems.

2.8   Existence and uniqueness of solutions

Hybrid systems provide a rather wide modeling context, so that there are no easily verifiable necessary and sufficient conditions for well-posedness of general hybrid dynamical systems. It is already of interest to give sufficient conditions for well-posedness of particular classes of hybrid systems (such as complementarity systems as described in Section 3). The advantage of considering special classes is that one can hope for conditions that are relatively easy to verify. In a number of special cases, such as mechanical systems or electrical network models, there are moreover natural candidates for such sufficient conditions.

Uniqueness of solutions will always be understood in the sense of what is sometimes called *right uniqueness*, that is, uniqueness of solutions defined on an interval $[t_0, t_1)$ given an initial state at t_0. It can easily happen in general hybrid systems, and even in complementarity systems, that uniqueness holds in one direction of time but not in the other; this is one of the points in which discontinuous dynamical systems differ from smooth systems. To allow for the possibility of an initial jump, one may let the initial condition be given at t_0^-.

We have to distinguish between *local* and *global* existence and uniqueness. Local existence and uniqueness, for solutions starting at t_0, holds if there exists an $\varepsilon > 0$ such that on $[t_0, t_0 + \varepsilon)$ there is a unique solution starting at the given initial condition. For global existence and uniqueness, we require that for given initial condition there is a unique solution on $[t_0, \infty)$. If local uniqueness holds for all initial conditions and existence holds globally, then uniqueness must also hold globally since there is no point at which solutions can split. However local existence does not imply global existence. This phenomenon is already well-known in the theory of smooth dynamical systems; for instance the differential equation $\dot{x}(t) = x^2(t)$ with $x(0) = x_0$ has the unique solution $x(t) = x_0(1 - x_0 t)^{-1}$ which for positive x_0 is defined only on the interval $[0, x_0^{-1})$. Some growth conditions have to be imposed to prevent this "escape to infinity". In hybrid systems, there are additional reasons why global existence may fail; in particular we may have an accumulation of mode switches (Zeno behavior). Although the occurrence of an accumulation of mode switches would seem to be exceptional, no general conditions are known at present which exclude this phenomenon. We use the term *well-posedness* to refer to local existence and uniqueness of solutions for all feasible initial conditions (i. e. initial conditions for which none of the inequality constraints are violated).

As already noted in [28] it is not difficult to find examples of hybrid systems that exhibit nonuniqueness of smooth continuations. For a simple example of this phenomenon within a switching control framework, consider the plant

$$\begin{aligned}
\dot{x}_1 &= x_2, \qquad y = x_2 \\
\dot{x}_2 &= -x_1 - u
\end{aligned} \qquad\qquad (5)$$

in closed-loop with a switching control scheme of relay type

$$\begin{aligned}
u(t) &= -1, &&\text{if } y(t) > 0 \\
-1 &\leq u(t) \leq 1, &&\text{if } y(t) = 0 \\
u(t) &= 1, &&\text{if } y(t) < 0.
\end{aligned} \qquad\qquad (6)$$

(this could be interpreted as a mass-spring system subject to a "reversed" – and therefore non-physical – Coulomb friction.) It will shown in the next section that such a variable-structure system can be modelled as a *complementarity system*. Note that from any initial (continuous) state $x(0) = (x_1(0), x_2(0)) = (c, 0)$, with $|c| \leq 1$, there are three possible smooth continuations for $t \geq 0$ that are allowed by the equations and inequalities above:

(i) $x_1(t) = x_1(0), \quad x_2(t) = 0, \quad u(t) = -x_1(0),$
    $-1 \leq u(t) \leq 1, \quad y(t) = x_2(t) = 0.$
(ii) $x_1(t) = -1 + (x_1(0) + 1)\cos t, \quad x_2(t) = -(x_1(0) + 1)\sin t,$
    $u(t) = 1, \quad y(t) = x_2(t) < 0.$
(iii) $x_1(t) = 1 + (x_1(0) - 1)\cos t, \quad x_2(t) = -(x_1(0) - 1)\sin t,$
    $u(t) = -1, \quad y(t) = x_1(t) > 0.$

So the above closed-loop system is not well-posed as a dynamical system. If the sign of the feedback coupling is reversed, however, there is only one smooth continuation from each initial state. This shows that well-posedness is a non-trivial issue to decide upon in a hybrid system, and in particular is a meaningful performance characteristic for hybrid systems arising from switching control schemes.

3 Complementarity systems

In several examples of hybrid physical systems the modes are determined by
pairs of so-called "complementary variables". Two scalar variables are said
to be *complementary* if they are both subject to an inequality constraint,
and if at all times at most one of the inequalities can be strict. The most
obvious example is that of the ideal diode. In this case the complementary
variables are the voltage across the diode and the current through it. When
the voltage drop across the diode is negative the current must be zero, and the
diode is said to be in *nonconducting mode*; when the current is positive the
voltage must be zero, and the diode is in *conducting mode*. There are many
more examples of hybrid systems in which mode switching is determined by
complementarity conditions. We call such systems *complementarity systems*.
As we shall see, complementarity conditions arise naturally in a number of
applications; moreover, in several applications one may rewrite a given system
of equations and inequalities in complementarity form by a judicious choice
of variables.

As a matter of convention, we shall always normalize complementary variables in such a way that both variables in the pair are constrained to be non-
negative; note that this deviates from standard sign conventions for diodes. So
a pair of variables (u, y) is said to be subject to a complementarity condition
if the following holds:

$$u \geq 0, \quad y \geq 0, \quad \begin{vmatrix} y = 0 \\ u = 0 \end{vmatrix} \tag{7}$$

where | denotes disjunction. Often we will be working with several pairs
of complementary variables. For such situations it is useful to have a vector
notation available. We shall say that two vectors of variables (of equal length)
are *complementary* if for all i the pair of variables (u_i, y_i) is subject to a
complementarity condition. In the mathematical programming literature, the
notation

$$0 \leq y \perp u \geq 0 \tag{8}$$

is often used to indicate that two vectors are complementary. Note that the
inequalities are taken in a componentwise sense, and that the usual interpretation of the "perp" symbol (namely $\sum_i y_i u_i = 0$) does indeed, in conjunction
with the inequality constraints, lead to the condition $\{y_i = 0\} \vee \{u_i = 0\}$ for
all i. Alternatively, one might say that the 'perp' is also taken componentwise.

Therefore, complementarity systems are systems whose flow conditions
can be written in the form

$$f(\dot{x}, x, y, u) = 0 \tag{9a}$$
$$0 \leq y \perp u \geq 0. \tag{9b}$$

In this formulation, the variables y_i and u_i play completely symmetric roles. Often it is possible to choose the denotations y_i and u_i in such a way that the conditions actually appear in the convenient "semi-explicit" form

$$\dot{x} = f(x, u) \tag{10a}$$
$$y = h(x, u) \tag{10b}$$
$$0 \leq y \perp u \geq 0. \tag{10c}$$

The flow conditions (9a) or (10a) still have to be supplemented by appropriate event conditions which describe what happens when there is a switch between modes. In some applications it will be enough to work with the default event conditions that require continuity across events; in other applications one needs more elaborate conditions.

In the mathematical programming literature, the so-called *linear complementarity problem* (LCP) has received much attention; see the book [9] for an extensive survey. The LCP takes as data a real k-vector q and a real $k \times k$ matrix M, and asks whether it is possible to find k-vectors u and y such that

$$y = q + Mu, \quad 0 \leq y \perp u \geq 0. \tag{11}$$

The main result on the linear complementarity problem that will be used below is the following: the LCP above has a unique solution u for all q if and only if M is a P-matrix, cf.[27], [9, Thm. 3.3.7]. (A matrix M is a P-matrix if all its *principal minors* are positive. Given a matrix M of size $k \times k$ and two nonempty subsets I and J of $\{1, \ldots, k\}$ of equal cardinality, the (I, J)-*minor* of M is the determinant of the square submatrix $M_{IJ} := (m_{ij})_{i \in I, j \in J}$. The principal minors are those with $I = J$.)

3.1 Examples

Example 1 (Circuits with ideal diodes).

A large amount of electrical network modeling is carried out on the basis of ideal lumped elements: resistors, inductors, capacitors, diodes, and so on. There is not necessarily a one-to-one relation between the elements in a model and the parts of the actual circuit; for instance, a resistor may under some circumstances be better modeled by a parallel connection of an ideal resistor and an ideal capacitor than by an ideal resistor alone. The standard ideal elements should rather be looked at as a construction kit from which one can quickly build a variety of models.

To write the equations of a network with (say) k ideal diodes in complementarity form, first extract the diodes so that the network appears as a k-port. For each port, we have a choice between denoting voltage by u_i and current by y_i or vice versa (with the appropriate sign conventions). Usually it is possible to make these choices in such a way that the dynamics of the k-port can be written as

$$\dot{x} = f(x, u), \quad y = h(x, u).$$

For linear networks, one can actually show that it is *always* possible to write the dynamics in this form. To achieve this, it may be necessary to let u_i denote voltage at some ports and current at some other ports; in that case one sometimes speaks of a "hybrid" representation, where of course the term is used in a different sense than in "hybrid systems". Replacing the ports by diodes, we obtain a representation in the semi-explicit complementarity form (10a).

Example 2 (Mechanical systems with unilateral constraints).
Mechanical systems with geometric inequality contraints, as often occurring in robotics, are given by equations of the following form (see [28]), in which $\frac{\partial H}{\partial p}$ and $\frac{\partial H}{\partial q}$ denote column vectors of partial derivatives, and the time arguments of q, p, y, and u have been omitted for brevity:

$$
\begin{aligned}
\dot{q} &= \frac{\partial H}{\partial p}(q,p) & q &\in \mathbb{R}^n, \quad p \in \mathbb{R}^n \\
\dot{p} &= -\frac{\partial H}{\partial q}(q,p) + \frac{\partial C^T}{\partial q}(q)u, & u &\in \mathbb{R}^k \\
y &= C(q), & y &\in \mathbb{R}^k \\
y &\geq 0, \quad u \geq 0, \quad y^T u = 0.
\end{aligned}
\tag{12}
$$

Here, $C(q) \geq 0$ is the column vector of geometric inequality constraints, and $u \geq 0$ is the vector of Lagrange multipliers producing the constraint force vector $(\partial C/\partial q)^T(q)u$. (The expression $(\partial C^T/\partial q)$ denotes an $n \times k$ matrix whose i-th column is given by $\partial C_i/\partial q$.) The complementarity conditions in this case express that the i-th component of u_i can be only non-zero if the i-th constraint is active, that is, $y_i = C_i(q) = 0$. Furthermore, $u_i \geq 0$ since the constraint forces will be always pushing in the direction of rendering y_i non-negative. This basic principle of handling geometric inequality constraints can be found e. g. in [26,18], and dates back to Fourier and Farkas.

The Hamiltonian $H(q,p)$ denotes the total energy, generally given as the sum of a kinetic energy $\frac{1}{2}p^T M^{-1}(q)p$ (where $M(q)$ denotes the mass matrix, depending on the configuration vector q) and a potential energy $V(q)$. The semi-explicit complementarity system (12) is called a Hamiltonian complementarity system, since the dynamics of every mode is Hamiltonian [28]. In particular, every mode is energy-conserving (since the constraint forces are workless); it should be noted though that the model is easily extended to mechanical systems with dissipation by replacing the second set of equations of (12) by

$$
\dot{p} = -\frac{\partial H}{\partial q}(q,p) - \frac{\partial R}{\partial \dot{q}}(\dot{q}) + \frac{\partial C^T}{\partial q}(q)u
\tag{13}
$$

where $R(\dot{q})$ denotes a Rayleigh dissipation function.

Example 3 (Variable-structure systems).
Consider a nonlinear input-output system of the form

$$
\dot{x} = f(x,\bar{u}), \quad \bar{y} = h(x,\bar{u})
\tag{14}
$$

in which the input and output variables are adorned with a bar for reasons that will become clear in a moment. Suppose that the system is in feedback coupling with a relay element given by

$$\begin{vmatrix} \bar{u} = 1, & \bar{y} \geq 0 \\ -1 \leq \bar{u} \leq 1, & \bar{y} = 0 \\ \bar{u} = -1, & \bar{y} \leq 0. \end{vmatrix} \qquad (15)$$

with | again denoting disjunction. Many of the systems considered by Filippov [12] can be rewritten in this form. At first sight, relay systems do not seem to fit in the complementarity framework. However, let us introduce new variables y_1, y_2, u_1, and u_2, together with the following new equations:

$$u_1 = \frac{1}{2}(1 - \bar{u})$$

$$u_2 = \frac{1}{2}(1 + \bar{u}) \qquad (16)$$

$$y = y_1 - y_2$$

Instead of considering (14) together with (15), we can also consider (14) together with the standard complementarity conditions for the vectors $y = \text{col}(y_1, y_2)$ and $u = \text{col}(u_1, u_2)$:

$$\begin{vmatrix} y_1 = 0, & u_1 \geq 0 \\ y_1 \geq 0, & u_1 = 0 \end{vmatrix}, \quad \begin{vmatrix} y_2 = 0, & u_2 \geq 0 \\ y_2 \geq 0, & u_2 = 0. \end{vmatrix} \qquad (17)$$

It can be easily verified that the trajectories of (14–16–17) are the same as those of (14–15). Note in particular that, although (17) in principle allows four modes, the conditions (16) imply that $u_1 + u_2 = 1$ so that the mode in which both u_1 and u_2 vanish is excluded, and the actual number of modes is three.

3.2   Linear complementarity systems

Linear complementarity systems are given by

$$\dot{x}(t) = Ax(t) + Bu(t) \qquad (18a)$$

$$y(t) = Cx(t) + Du(t) \qquad (18b)$$

$$y(t) \geq 0, \quad u(t) \geq 0, \quad y^{\top}(t)u(t) = 0. \qquad (18c)$$

The set of indices for which $y_i(t) = 0$ (we shall call this the *active index set*) need not be constant in time, so that the system may switch from one "operating mode" to another. To define the dynamics of (18) completely, we will have to specify when these mode switches occur, what their effect will be on the state variables, and how a new mode will be selected. A proposal for answering these questions (cf. [15]) will be explained below.

Let n denote the length of the vector $x(t)$ in the equations (18a–18b) and let k denote the number of inputs and outputs. There are then 2^k possible choices for the active index set. The equations of motion when the active index set is I are given by

$$
\begin{aligned}
\dot{x}(t) &= Ax(t) + Bu(t) \\
y(t) &= Cx(t) + Du(t) \\
y_i(t) &= 0, \quad i \in I \\
u_i(t) &= 0, \quad i \in I^c
\end{aligned}
\tag{19}
$$

where I^c denotes the index set that is complementary to I, that is, $I^c = \{i \in \{1, \ldots, k\} \mid i \notin I\}$. We shall say that the above equations represent the system in *mode I*. An equivalent and somewhat more explicit form is given by the (generalized) state equations

$$
\begin{aligned}
\dot{x}(t) &= Ax(t) + B_{\bullet I} u_I(t) \\
0 &= C_{I \bullet} x(t) + D_{II} u_I(t)
\end{aligned}
\tag{20}
$$

together with the output equations

$$
\begin{aligned}
y_{I^c}(t) &= C_{I^c \bullet} x(t) + D_{I^c I} u_I(t) \\
u_{I^c}(t) &= 0.
\end{aligned}
\tag{21}
$$

Here and below, the notation $M_{\bullet I}$ where M is a matrix of size $m \times k$ and I is a subset of $\{1, \ldots, k\}$ denotes the submatrix of M formed by taking the columns of M whose indices are in I. The notation $M_{I \bullet}$ denotes the submatrix obtained by taking the rows with indices in the index set I.

In order to formulate an event rule, we first need to introduce some concepts taken from the geometric theory of linear systems (see [37,3,20] for the general background). Denote by V_I the *consistent subspace* of mode I, i.e. the set of initial conditions x_0 for which there exist smooth functions $x(\cdot)$ and $u_I(\cdot)$, with $x(0) = x_0$, such that (20) is satisfied. The space V_I can be computed as the limit of the sequence defined by

$$
\begin{aligned}
V^0 &= \mathbb{R}^n \\
V^{i+1} &= \{x \in V^i \mid \exists u \in \mathbb{R}^{|I|} \text{ s.t. } Ax + B_{\bullet I} u \in V^i, \ C_{I \bullet} x + D_{II} u = 0\}.
\end{aligned}
\tag{22}
$$

There exists a linear mapping F_I such that (20) will be satisfied for $x_0 \in V_I$ by taking $u_I(t) = F_I x(t)$. The mapping F_I is uniquely determined, and more generally the function $u_I(\cdot)$ that satisfies (20) for given $x_0 \in V_I$ is uniquely determined, if the full-column-rank condition

$$
\ker \begin{bmatrix} B_{\bullet I} \\ D_{II} \end{bmatrix} = \{0\}
\tag{23}
$$

holds and moreover we have

$$
V_I \cap T_I = \{0\},
\tag{24}
$$

where T_I is the subspace that can be computed as the limit of the following sequence:

$$T^0 = \{0\}$$
$$T^{i+1} = \{x \in \mathbb{R}^n \mid \exists \tilde{x} \in T^i,\ \tilde{u} \in \mathbb{R}^{|I|}\ \text{s.t.} \tag{25}$$
$$x = A\tilde{x} + B_{\bullet I}\tilde{u},\ C_{I\bullet}\tilde{x} + D_{II}\tilde{u} = 0\}.$$

The subspace T_I is best thought of as the *jump space* associated to mode I, that is, as the space along which fast motions will occur that take an inconsistent initial state instantaneously to a point in the consistent space V_I; note that under the condition (24) this projection is uniquely determined. (The interpretation of T_I as a jump space can be made precise by introducing the class of *impulsive-smooth distributions* that was studied by Hautus [14]). The projection can be used to define a *jump rule*. However, there are 2^k possible projections, corresponding to all possible subsets of $\{1, \ldots, k\}$; which one of these to choose should be determined by a *mode selection rule*.

For the formulation of a mode selection rule we have to relate in some way index sets to continuous states. Such a relation can be established on the basis of the so-called *rational complementarity problem* (RCP). The RCP is defined as follows. Let a rational vector $q(s)$ of length k and a rational matrix $M(s)$ of size $k \times k$ be given. The rational complementarity problem is to find a pair of rational vectors $y(s)$ and $u(s)$ (both of length k) such that

$$y(s) = q(s) + M(s)u(s) \tag{26}$$

and moreover for all indices $1 \le i \le k$ we have either $y_i(s) = 0$ and $u_i(s) > 0$ for all sufficiently large s, or $u_i(s) = 0$ and $y_i(s) > 0$ for all sufficiently large s. The vector $q(s)$ and the matrix $M(s)$ are called the *data* of the RCP, and we write RCP($q(s)$,$M(s)$). We shall also consider an RCP with data consisting of a quadruple of constant matrices (A, B, C, D) (such as could be used to define (18a–18b)) and a constant vector x_0, namely by setting

$$q(s) = C(sI - A)^{-1}x_0 \quad \text{and} \quad M(s) = C(sI - A)^{-1}B + D.$$

We say that an index set $I \subset \{1, \ldots, k\}$ *solves* the RCP (26) if there exists a solution $(y(s), u(s))$ with $y_i(s) = 0$ for $i \in I$ and $u_i(s) = 0$ for $i \notin I$. The collection of index sets I that solve RCP($A, B, C, D; x_0$) will be denoted by $\mathcal{S}(A, B, C, D; x_0)$ or simply by $\mathcal{S}(x_0)$ if the quadruple (A, B, C, D) is given by the context.

The semantics of a linear complementarity system is now defined as follows. We assume that a quadruple (A, B, C, D) is given whose transfer matrix $G(s) = C(sI - A)^{-1}B + D$ is *totally invertible*, i.e. for each index set I the $k \times k$ matrix $G_{II}(s)$ is nonsingular. Under this condition, the two subspaces V_I and T_I as defined above form for all I a direct sum decomposition of the state space $\mathbb{R}^n$, so that the projection along T_I onto V_I is well-defined. We denote this projection by P_I. The linear complementarity system associated to the quadruple (A, B, C, D) is given by the event-flow formula

$$\dot{x} = Ax + Bu, \quad y = Cx + Du$$
$$u_I \geq 0, \quad y_I = 0, \quad u_{I^c} = 0, \quad y_{I^c} \geq 0 \tag{27}$$
$$I^\# \in S(x), \quad x^\# = P_{I^\#} x.$$

The expression in the second line should be read as a shorthand; for instance in case $k = 2$ the long form would be

$$\begin{vmatrix} I = \emptyset, & u_1 = 0, & u_2 = 0, & y_1 \geq 0, & y_2 \geq 0 \\ I = \{1\}, & y_1 = 0, & u_2 = 0, & y_2 \geq 0, & u_1 \geq 0 \\ I = \{2\}, & y_2 = 0, & u_1 = 0, & y_1 \geq 0, & u_2 \geq 0 \\ I = \{1,2\}, & y_1 = 0, & y_2 = 0, & u_1 \geq 0, & u_1 \geq 0. \end{vmatrix}$$

In a similar way the third line of (27) should be read as a disjunction with 2^k terms.

3.3   Existence and uniqueness of solutions of linear complementarity systems

Conditions for well-posedness of hybrid systems that are both necessary and sufficient have been given only for limited classes. One example appears in [28]. The statement in this paper concerns *bimodal* linear complementarity systems, so systems with only two modes ($k = 1$). Such a system of the form (18) has a transfer function $g(s) = C(sI - A)^{-1}B + D$ which is a rational function. The system is said to have *no feedthrough term* if the matrix D vanishes. The system is called *degenerate* if the transfer matrix $g(s)$ is of the form $g(s) = 1/q(s)$ where $q(s)$ is a polynomial; in this case the consistent subspace in the constrained mode is just the origin. The *Markov parameters* of the system are the coefficients of the expansion of $g(s)$ around infinity,

$$g(s) = g_0 + g_1 s^{-1} + g_2 s^{-2} + \cdots.$$

The *leading Markov parameter* is defined as the first parameter in the sequence $(g_0, g_1, \dots)$ that is nonzero. Having introduced this terminology, we can now formulate the following result [28, Thm. 4.8].

Theorem 1. *A nondegenerate bimodal linear complementarity system without feedthrough term and with nonzero transfer function is well-posed if and only if its leading Markov parameter is positive.*

It is typical to find that well-posedness of complementarity systems is linked to a positivity condition. If the number of pairs of complementary variables is larger than 1 an appropriate matrix version of the positivity condition has to be used. A relation of the rational complementarity problem with the linear complementarity problem of mathematical programming can be established in the following way [29,16].

Theorem 2. *For given* $q(s) \in \mathbb{R}^k(s)$ *and* $M(s) \in \mathbb{R}^{k \times k}(s)$, *the problem* $RCP(q(s), M(s))$ *is solvable if and only if there exists* $\mu \in \mathbb{R}$ *such that for all* $\lambda > \mu$ *the problem* $LCP(q(\lambda), M(\lambda))$ *is solvable. The same statement holds with "solvable" replaced by "uniquely solvable".*

The above theorem provides a convenient way of proving well-posedness for several classes of linear complementarity systems. The following example is taken from [16].

Example 4. A linear mechanical system may be described by equations of the form

$$M\ddot{q} + D\dot{q} + Kq = 0 \tag{28}$$

where q is the vector of generalized coordinates, M is the generalized mass matrix, D contains damping and gyroscopic terms, and K is the elasticity matrix. The mass matrix M is positive definite. Suppose now that we subject the above system to unilateral constraints of the form

$$Fq \geq 0 \tag{29}$$

where F is a given matrix. Under the assumption of inelastic collisions, the dynamics of the resulting system may be described by

$$M\ddot{q} + D\dot{q} + Kq = F^T u, \quad y = Fq \tag{30}$$

together with complementarity conditions between y and u. The associated RCP is the following:

$$y(s) = F(s^2 M + sD + K)^{-1}[(sM + D)q_0 + M\dot{q}_0] + \tag{31}$$
$$+ F(s^2 M + sD + K)^{-1}F^T u(s).$$

If F has full row rank, then the matrix $F(s^2 M + sD + K)^{-1}F^T$ is positive definite (although not necessarily symmetric) for all sufficiently large s because the term with s^2 becomes dominant. By combining the standard result on solvability of LCPs with Thm. 2, it follows that RCP is solvable and we can use this to prove the well-posedness of the constrained mechanical system; this provides some confirmation for the validity of the model that has been used, since physical intuition certainly suggests that a unique solution should exist.

One can easily imagine cases in which the matrix F does not have full row rank so that the fulfillment of some constraints already implies that some other constraints will also be satisfied; think for instance of a chair having four legs on the ground. In such cases the basic result on solvability of LCPs does not provide enough information, but there are alternatives available that make use of the special structure that is present in equations like (31). On the basis of this, one can still prove well-posedness; in particular the trajectories of the coordinate vector $q(t)$ are uniquely determined, even though the trajectories of the constraint force $u(t)$ are not.

A similar result can be obtained for linear RLC-circuits containing ideal diodes. Indeed, by extracting the diodes, and representing the remaining system with ports as a port-controlled Hamiltonian system with dissipation (cf. [32]) it follows by the positivity of the stored (quadratic) energy that the corresponding RCP is uniquely solvable. Furthermore, in this case (contrary to the mechanical case considered above) the system does not exhibit jumps in the continuous state variables.

3.4   Mechanical complementarity systems

As discussed in Example 2 mechanical systems with unilateral constraints can be represented as semi-explicit complementarity systems (12)–(13).

Assume that the unilateral constraints are *independent*, that is

$$\text{rank } \frac{\partial C^T}{\partial q}(q) = k, \text{ for all } q \text{ with } C(q) \geq 0. \tag{32}$$

Since the Hamiltonian is of the form (kinetic energy plus potential energy)

$$H(q,p) = \frac{1}{2}p^T M^{-1}(q)p + V(q), \quad M(q) = M^T(q) > 0 \tag{33}$$

where $M(q)$ is the generalized mass matrix, it follows that the system (12)–(13) has uniform relative degree 2 with decoupling matrix

$$D(q) = \left[\frac{\partial C^T}{\partial q}(q)\right]^T M^{-1}(q)\frac{\partial C^T}{\partial q}(q). \tag{34}$$

Hence, from $M(q) > 0$ and (32) it follows that $D(q)$ is *positive definite* for all q with $C(q) \geq 0$. It can be shown that this implies that the system has unique smooth solutions, [29,22].

A switch and jump rule for mechanical complementarity systems can be formulated as follows. Let us consider a mechanical system with n degrees of freedom $q = (q_1, \cdots, q_n)$ having kinetic energy $\frac{1}{2}\dot{q}^T M(q)\dot{q}$, where $M(q) > 0$ is the generalized mass matrix. Suppose the system is subject to k geometric inequality constraints

$$y_i = C_i(q) \geq 0, \quad i \in K = \{1, \cdots, k\} \tag{35}$$

If the i-th inequality constraint is *active*, that is $C_i(q) = 0$, then the system will experience a constraint force of the form $\frac{\partial C_i}{\partial q}(q)u_i$, where $\frac{\partial C_i}{\partial q}(q)$ is the column vector of partial derivatives of C_i and u_i a Lagrange multiplier.

Let us now consider an arbitrary initial continuous state $(q^-, \dot{q}^-)$. Define the vector of generalized velocities

$$v^- := \frac{\partial C_I}{\partial q}(q^-)\dot{q}^- \tag{36}$$

where I denotes the set of active indices at q^-. In order to describe the inelastic collision we consider the system of equalities and inequalities (in the unknowns v^+, λ)

$$v^+ = v^- + \frac{\partial C_I}{\partial q}(q^-)M^{-1}(q^-)\frac{\partial C_I^T}{\partial q}(q^-)\lambda$$
$$v^+ \geq 0, \ \lambda \geq 0, \ (v^+)^T\lambda = 0. \tag{37}$$

Here λ can be interpreted as a vector of *impulsive forces*. The system (37) is in the form of the *linear complementarity problem* (LCP). The general form of an LCP can be written as

$$y = x + Mu, \quad 0 \leq y \perp u \geq 0 \tag{38}$$

where the vector x and the square matrix M are given, and the vectors y and u are the unknowns. Recall that the LCP (38) has a unique solution (y, u) for each x if and only if the matrix M is a P-matrix, that is to say, if and only if all principal minors of the matrix M are positive. This is the case in particular if M is a positive definite matrix. Since $\frac{\partial C_I}{\partial q}(q^-)M^{-1}(q^-)\frac{\partial C_I^T}{\partial q}(q^-) > 0$, the LCP (37) indeed has a unique solution. The *jump rule* is now given by

$$(q^-, \dot{q}^-) \mapsto (q^+, \dot{q}^+), \quad q^+ = q^-, \quad \dot{q}^+ = \dot{q}^- + M^{-1}(q^-)\frac{\partial C_I^T}{\partial q}(q^-)\lambda. \tag{39}$$

The new velocity vector $\dot{q}^+$ may be equivalently characterized as the solution of the quadratic programming problem

$$\min_{\{\dot{q}^+ | C_I(q)\dot{q}^+ \geq 0\}} \frac{1}{2}(\dot{q}^+ - \dot{q}^-)^T M(q)(\dot{q}^+ - \dot{q}^-) \tag{40}$$

where $q := q^- = q^+$. This formulation is sometimes taken as the starting point for describing multiple inelastic collisions, see [8,25]. An appealing feature of the transition rule above is that the energy of the mechanical system will always decrease at the switching instant. One may take this as a starting point for stability analysis.

3.5   Relay systems

For piecewise linear relay systems of the form

$$\dot{x} = Ax + Bu, \quad y = Cx + Du, \quad u_i = -\text{sgn}(y_i) \ (i = 1, \ldots, k) \tag{41}$$

Thm. 2 can also be applied, but the application is somewhat less straightforward for the following reason. As noted above, it is possible to rewrite a relay system as a complementarity system (in several ways actually). Using the method (16), one arrives at a relation between the new inputs $\text{col}(u_1, u_2)$ and the new outputs $\text{col}(y_1, y_2)$ that may be written in the frequency domain as follows (ι denotes the vector all of whose entries are 1, and $G(s)$ denotes the transfer matrix $C(sI - A)^{-1}B + D$):

$$
\begin{bmatrix} u_1(s) \\ u_2(s) \end{bmatrix} = \begin{bmatrix} -G^{-1}(s)C(sI - A)^{-1}x_0 + s^{-1}\iota \\ G^{-1}(s)C(sI - A)^{-1}x_0 + s^{-1}\iota \end{bmatrix} +
$$
$$
+ \begin{bmatrix} G^{-1}(s) & -G^{-1}(s) \\ -G^{-1}(s) & G^{-1}(s) \end{bmatrix} \begin{bmatrix} y_1(s) \\ y_2(s) \end{bmatrix}. \quad (42)
$$

The matrix that appears on the right hand side is singular for all s and so the corresponding LCP does not always have a unique solution. However the vector that we find at the right hand side is of a special form and we only need to ensure existence of a unique solution for vectors of this particular form. On the basis of this observation, the following result is obtained.

Theorem 3. [21,16] *The piecewise relay system (41) is well-posed if the transfer matrix $G(s)$ is a P-matrix for all sufficiently large s.*

This result gives a criterion that is straightforward to verify (compute the determinants of all principal minors of $G(s)$, and check the signs of the leading Markov parameters), but that is restricted to piecewise linear systems. Filippov [12, §2.10] gives a criterion for well-posedness which works for general nonlinear systems, but needs to be verified on a point-by-point basis.

3.6   Discrete time complementarity systems and mixed logical dynamical systems

Linear complementarity systems may also be considered in *discrete time* by replacing the differential equations in (19) by difference equations. For *discrete time* hybrid dynamical systems with linear dynamics various other formalisms have been proposed in the literature; we mention *mixed logical dynamical systems* [4], *piecewise affine systems* [35], *extended linear complementarity systems* [33], and *max-min-plus-scaling systems* [34].

Recently in [17] it has been shown that all these formalisms are basically *equivalent*, and have more or less the same expressive power. This result of course enables the transfer of knowledge from one class of systems to another, and implies that for the study of a particular discrete time hybrid system one can choose the framework that is most suitable (for the goal one has in mind).

4   Conclusions

Clearly, only a few aspects of hybrid systems have been considered in this paper. Much more can be found in our book [30]. We hope we have convinced the reader that the area of hybrid systems constitutes a challenging research area, which is very well-motivated from applications. From a scientific point of view it seems clear that real progress can only be made by merging concepts and tools both from computer science and systems and control theory.

References

1. R. Alur, C. Courcoubetis, N. Halbwachs, T. A. Henzinger, P.-H. Ho, X. Nicollin, A. Olivero, J. Sifakis, and S. Yovine. The algorithmic analysis of hybrid systems. *Theoretical Computer Science*, 138:3–34, 1995.

2. P. J. Antsaklis, J. A. Stiver, and M. D. Lemmon. Hybrid system modeling and autonomous control systems. In R. L. Grossman, A. Nerode, A. P. Ravn, and H. Rischel, editors, *Hybrid Systems*, Lect. Notes Comp. Sci. 736, pages 366–392, New York, 1993. Springer.

3. G. Basile and G. Marro. *Controlled and Conditioned Invariants in Linear System Theory*. Prentice Hall, Englewood Cliffs, NJ, 1992.

4. A. Bemporad and M. Morari. Control of systems integrating logic, dynamics, and constraints. *Automatica*, 35:407–427, 1999.

5. A. Benveniste and P. Le Guernic. Hybrid dynamical systems theory and the SIGNAL language. *IEEE Trans. Automat. Contr.*, AC-35:535–546, 1990.

6. A. Benveniste. Compositional and uniform modelling of hybrid systems. *IEEE Trans. Automat. Contr.*, AC-43:579–584, 1998.

7. M. S. Branicky, V. S. Borkar, and S. K. Mitter. A unified framework for hybrid control. In *Proc. 33rd IEEE Conf. Dec. Contr.*, pages 4228–4234, 1994.

8. B. Brogliato. *Nonsmooth Impact Mechanics. Models, Dynamics and Control*. Lect. Notes Contr. Inform. Sci. 220. Springer, Berlin, 1996.

9. R. W. Cottle, J.-S. Pang, and R. E. Stone. *The Linear Complementarity Problem*. Academic Press, Boston, 1992.

10. M.H.A. Davis. *Markov Models and Optimization*. Chapman and Hall, London, 1993.

11. H. Elmqvist, F. Boudaud, J. Broenink, D. Brück, T. Ernst, P. Fritzon, A. Jeandel, K. Juslin, M. Klose, S. E. Mattsson, M. Otter, P. Sahlin, H. Tummescheit, and H. Vangheluwe. Modelica™ — a unified object-oriented language for physical systems modeling. http://www.Dynasim.se/Modelica/Modelica1.html.

12. A. F. Filippov. *Differential equations with discontinuous righthand sides*. Kluwer, Dordrecht, 1988.

13. N. Halbwachs, P. Caspi, P. Raymond, and D. Pilaud. The synchronous dataflow programming language LUSTRE. *Proc. IEEE*, 79:1305–1320, 1991.

14. M. L. J. Hautus. The formal Laplace transform for smooth linear systems. In G. Marchesini and S.K. Mitter, editors, *Mathematical Systems Theory*, Lect. Notes Econ. Math. Syst. 131, pages 29–47. Springer, New York, 1976.

15. W. P. M. H. Heemels, J. M. Schumacher, and S. Weiland. Linear complementarity systems. Internal Report 97 I/01, Dept. of EE, Eindhoven Univ. of Technol., July 1997. To appear in *SIAM J. Appl. Math*. www.cwi.nl/~jms/lcs.ps.Z www.cwi.nl/~jms/PUB/ARCH/lcs_revised.ps.Z (revised version).

16. W. P. M. H. Heemels, J. M. Schumacher, and S. Weiland. The rational complementarity problem. *Lin. Alg. Appl.* 294:93–135, 1999.

17. W. P. M. H. Heemels, B. De Schutter. On the equivalence of classes of hybrid systems: Mixed logical dynamical and complementarity systems. Internal Report 00 I/04, Dept. of EE, Eindhoven Univ. of Technol., June 2000.

18. C. W. Kilmister and J. E. Reeve. *Rational Mechanics*. Longmans, London, 1966.

19. S. Kowalewski, S. Engell, J. Preussig, and O. Stursberg. Verification of logic controllers for continuous plants using timed condition/event-system models. *Automatica*, 35:505–518, 1999.
20. M. Kuijper. *First-Order Representations of Linear Systems*. Birkhäuser, Boston, 1994.
21. Y.J. Lootsma, A.J. van der Schaft, and M.K. Çamlıbel. Uniqueness of solutions of relay systems. *Automatica*, 35:467–478, 1999.
22. P. Lötstedt. Mechanical systems of rigid bodies subject to unilateral constraints. *SIAM Journal of Applied Mathematics*, 42:281–296, 1982.
23. J. Lygeros, D.N. Godbole, S. Sastry. Verified hybrid controllers for automated vehicles. *IEEE Transactions on Automatic Control*, 43(4):522–539, 1998.
24. N. Lynch, R. Segala, F. Vaandrager, H.B. Weinberg. Hybrid I/O automata. *Hybrid Systems III: Verification and Control* (editors Alur R, Henzinger T, Sontag E). Lect. Notes Comp. Sci. 1066, pp. 496–510, Springer: New York, 1996.
25. J. J. Moreau. Liaisons unilatérales sans frottement et chocs inélastiques. *C. R. Acad. Sc. Paris*, 296:1473–1476, 1983.
26. J. Pérès. *Mécanique Générale*. Masson & Cie., Paris, 1953.
27. H. Samelson, R. M. Thrall, and O. Wesler. A partition theorem for Euclidean *n*-space. *Proc. Amer. Math. Soc.*, 9:805–807, 1958.
28. A. J. van der Schaft and J. M. Schumacher. The complementary-slackness class of hybrid systems. *Math. Contr. Signals Syst.*, 9:266–301, 1996.
29. A. J. van der Schaft and J. M. Schumacher. Complementarity modeling of hybrid systems. *IEEE Trans. Automat. Contr.*, AC-43:483–490, 1998.
30. A. J. van der Schaft and J. M. Schumacher. *An Introduction to Hybrid Dynamical Systems*. Springer, London, 2000.
31. A. J. van der Schaft and J. M. Schumacher. Compositionality issues in discrete, continuous, and hybrid systems. *Int. J. Robust & Nonlinear Control*, 2001, to appear.
32. A.J. van der Schaft. L_2-*Gain and Passivity Techniques in Nonlinear Control*, 2nd revised and enlarged edition. Springer-Verlag, Springer Communications and Control Engineering series, p. xvi+249, London, 2000 (first edition Lect. Notes in Control and Inf. Sciences, vol. 218, Springer-Verlag, Berlin, 1996).
33. B. De Schutter, B. De Moor. The extended linear complementarity problem and the modeling and analysis of hybrid systems. In *Hybrid Systems V* (editors P. Antsaklis, W. Kohn, M. Lemmon, A. Nerode, S. Sastry), Lecture Notes in Computer Science 1567, pp. 70–85, Springer, New York, 1999.
34. B. De Schutter, T. van den Boom. On model predictive control for max-min-plus-scaling discrete event systems. Technical report, ITS, Delft Univ. Techn., 2000.
35. E. Sontag. Nonlinear regulation: the piecewise linear approach. *IEEE Trans. Automat. Contr.*, 26:346–357, 1981.
36. J. C. Willems. Paradigms and puzzles in the theory of dynamical systems. *IEEE Trans. Automat. Control*, AC-36(3):259–294, 1991.
37. W. M. Wonham. *Linear Multivariable Control: A Geometric Approach*. Springer Verlag, New York, 3rd edition, 1985.

Part IV

Physics in Control

Introduction

Nonlinear systems and control theory has witnessed tremendous developments over the last three decades. In particular, the introduction of geometric tools like Lie brackets of vector fields on manifolds has greatly advanced the theory, and has enabled the proper generalization of many fundamental concepts known for linear control systems to the nonlinear world. While the emphasis in the eighties was primarily on the structural analysis of smooth nonlinear dynamical control systems, in the nineties this has been combined with analytic techniques for stability, stabilization and robust control, leading e.g. to backstepping techniques and nonlinear H_∞ control. Moreover, in the last decade the theory of passive systems, and its implications for regulation and tracking, has undergone a remarkable revival. This last development was also spurred by work in robotics on the possibilities of shaping, by feedback, the physical energy in such a way that it can be used as a suitable Lyapunov function for the control purpose at hand. This has led to what is sometimes called passivity–based control. Many other important developments have taken place, and much attention has been paid to special subclasses of systems like mechanical systems with nonholonomic constraints.

All this has resulted in very lively research in nonlinear control, with many actual and potential applications. The aim of the part "Physics in Control" is to stress the importance of physical modeling for nonlinear control. In one sense, this is common knowledge for everyday control engineering, but a general theoretical framework for modeling is also of utmost importance for the development of nonlinear control theory. Indeed, although in principle the same applies to linear control systems, in the latter case the relative ease of general linear control techniques may obviate the need for a representation of the system which makes explicit the physical characteristics of the system. On the other hand, the class of general nonlinear control systems is so overwhelmingly rich, that it cannot be expected that a single theory will cover the whole area, thus necessitating the exploitation of the inherent physical structure of many nonlinear control systems. In this Part some of the recent developments in systems and control theory of physical systems will be highlighted. This will include the Hamiltonian geometrization of network models of physical systems, and its implications for balance which is so prominent in Hamiltonian models and the theory of passive nonlinear systems, other physical balance relations such as mass balance can play a crucial role. Also the study of symmetries and its consequences for control substantially adds to this framework. Application areas of these approaches range from mechanical systems, robot manipulators, induction motors, power systems to (bio–) chemical processes.

On Modelling and Control of Mass Balance Systems

Georges Bastin

Centre for Systems Engineering and Applied Mechanics (CESAME), Université Catholique de Louvain, Louvain la Neuve, Belgium

1   Introduction

The aim of this chapter is to give a self content presentation of the modelling of engineering systems that are governed by a law of mass conservation and to briefly discuss some control problems regarding these systems.

A general state-space model of mass balance systems is presented. The equations of the model are shown to satisfy physical constraints of positivity and mass conservation. These conditions have strong structural implications that lead to particular Hamiltonian, Compartmental and Stoichiometric representations. The modelling of mass balance systems is illustrated with two simple industrial examples : a biochemical process and a grinding process.

In general, mass balance systems have multiple equilibria, one of them being the operating point of interest which is locally asymptotically stable. However if big enough disturbances occur, the process may be lead by accident to a behaviour which may be undesirable or even catastrophic. The control challenge is then to design a feedback controller which is able to prevent the process from such undesirable behaviours. Two solutions of this problem are briefly described for inflow controlled systems : (i) robust state feedback stabilisation of the total mass, (ii) output regulation for a class of minimum phase systems.

Some interesting stability properties of open loop mass balance systems are reviewed in Appendix.

2   Mass balance systems

In mass balance systems, each state variable x_i ($i = 1, \ldots, n$) represents an amount of some material (or some matter) inside the system, while each state equation describes a balance of flows as illustrated in Fig. 1 :

$$\dot{x}_i = r_i - q_i + p_i \tag{1}$$

where p_i represents the inflow rate, q_i the outflow rate and r_i an internal transformation rate. The flows p_i, q_i and r_i can be function of the state variables $x_1, \ldots x_n$ and possibly of control inputs $u_1, \ldots, u_m$. The state space model which is the natural behavioural representation of the system is therefore written in vector form :

$$\dot{x} = r(x, u) - q(x, u) + p(x, u) \qquad (2)$$

As a matter of illustration, some concrete examples of the phenomena that can be represented by the (p, q, r) flow rates in engineering applications are given in Table 1.

Transformations
   Physical : grinding, evaporation, condensation
   Chemical : reaction, catalysis, inhibition
   Biological : infection, predation, parasitism

Outflows
   Withdrawals, extraction
   Excretion, decanting, adsorption
   Emigration, mortality

Inflows
   Supply of raw material
   Feeding of nutrients
   Birth, immigration
     etc...etc...

Table 1.

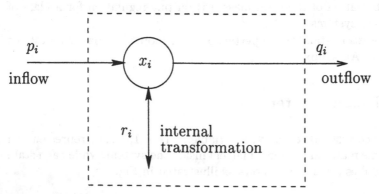

Fig. 1. Balance of flows

In this paper, we shall assume that the functions $p(x, u), q(x, u), r(x, u)$ are differentiable with respect to their arguments. The physical meaning of the model (2) implies that these functions must satisfy two kinds of conditions : positivity conditions and mass conservation conditions which are explicited hereafter.

3 Positivity

Since there cannot be negative masses, the model (2) makes sense only if the state variables $x_i(t)$ remain *non-negative* for all t :

$$x_i(t) \in R_+$$

where R_+ denotes the set of real non-negative numbers. It follows that :

$$x_i = 0 \implies \dot{x}_i \geq 0 \tag{3}$$

whatever the values of $x_j \in R_+$, $j \neq i$ and u_k. This requirement is satisfied if the functions $p(x, u), q(x, u), r(x, u)$ have the following properties :

1. The inflow and outflow functions are defined to be non-negative :

$$\left. \begin{array}{c} p(x, u) \\ q(x, u) \end{array} \right\} : R_+^n \times R^m \to R_+^n$$

2. There cannot be an outflow if there is no material inside the system :

$$x_i = 0 \implies q_i(x, u) = 0 \tag{4}$$

3. The transformation rate $r_i(x, u) : R_+^n \times R^m \to R$ may be positive or negative but it must be defined to be positive when x_i is zero :

$$x_i = 0 \implies r_i(x, u) \geq 0 \tag{5}$$

4 Conservation of mass

Provided the quantities x_i are expressed in appropriate normalized units, the total mass contained in the system may be expressed as[1] :

$$M = \sum_i x_i$$

When the system is *closed* (neither inflows nor outflows), the dynamics of M are written :

$$\dot{M} = \sum_i r_i(x, u)$$

---

[1] To simplify the notations, it will be assumed throughout the paper that the summation $\sum_i$ is taken over all possible values of i (here $i = 1, \ldots, n$) and $\sum_{i \neq j}$ over all possible values of i except j.

It is obvious that the total mass inside a closed system must be conserved ($\dot{M} = 0$), which implies that the transformation functions $r_i(x, u)$ satisfy the condition :

$$\sum_i r_i(x, u) = 0 \tag{6}$$

The positivity conditions (4)- (5) and the mass conservation condition (6) have strong structural implications that are now presented.

5 Hamiltonian representation

A necessary consequence of the mass conservation condition (6) is that $n(n - 1)$ functions $r_{ij}(x, u)$ ($i = 1, \ldots, n$; $j = 1, \ldots, n$; $i \neq j$) may be selected such that :

$$r_i(x, u) = \sum_{j \neq i} r_{ji}(x, u) - \sum_{j \neq i} r_{ij}(x, u) \tag{7}$$

(note the indices !). Indeed, the summation over i of the right hand sides of (7) equals zero. It follows that any mass balance system (2) can be written under the form of a so-called *port-controlled Hamiltonian representation* (see [10], [11]) :

$$\dot{x} = [F(x, u) - D(x, u)] \left(\frac{\partial M}{\partial x} \right)^T + p(x, u) \tag{8}$$

where the storage function is the total mass $M(x) = \sum_i x_i$. The matrix $F(x, u)$ is skew-symmetric :

$$F(x, u) = -F^T(x, u)$$

with off-diagonal entries $f_{ij}(x, u) = r_{ji}(x, u) - r_{ij}(x, u)$. The matrix $D(x, u)$ represents the natural damping or dissipation provided by the outflows. It is diagonal and positive :

$$D(x, u) = \text{diag} \left(q_i(x, u) \right) \geq 0$$

The last term $p(x, u)$ in (8) obviously represents a supply of mass to the system from the outside.

6 Compartmental representation

There is obviously an infinity of ways of defining the r_{ij} functions in (7). We assume that they are selected to be non-negative :

$$r_{ij}(x, u) : R_+^n \times R^m \rightarrow R_+$$

and differentiable since $r_i(x, u)$ is required to be differentiable.

Then condition (5) is satisfied if :

$$x_i = 0 \Rightarrow r_{ij}(x, u) = 0 \tag{9}$$

Now, it is a well known fact (see e.g. [7], page 67) that if $r_{ij}(x, u)$ is differentiable and if condition (9) holds, then $r_{ij}(x, u)$ may be written as :

$$r_{ij} = x_i \bar{r}_{ij}(x, u)$$

for some appropriate function $\bar{r}_{ij}(x, u)$ which is defined on $R_+^n \times R^m$, non-negative and at least continuous. Obviously, the same is true for $q_i(x, u)$ due to condition (4) :

$$q_i(x, u) = x_i \bar{q}_i(x, u)$$

The functions $\bar{r}_{ij}$ and $\bar{q}_i$ are called fractional rates. It follows that the mass balance system (2) is then written under the following alternative representation :

$$x = G(x, u)x + p(x, u) \tag{10}$$

where $G(x, u)$ is a so-called *compartmental matrix* with the following properties :

1. $G(x, u)$ is a Metzler matrix with non-negative off-diagonal entries :

$$g_{ij}(x, u) = \bar{r}_{ji}(x, u) \geq 0 \quad i \neq j$$

(note the inversion of indices !)

2. The diagonal entries of $G(x, u)$ are non-positive :

$$g_{ii}(x, u) = -\bar{q}_i(x, u) - \sum_{j \neq i} \bar{r}_{ij}(x, u) \leq 0$$

3. The matrix $G(x, u)$ is diagonally dominant :

$$|g_{ii}(x, u)| \geq \sum_{j \neq i} g_{ji}(x, u)$$

The term *compartmental* is motivated by the fact that a mass balance system may be represented by a network of conceptual reservoirs called compartments. Each quantity (state variable) x_i is supposed to be contained in a compartment which is represented by a box in the network (see Fig. 2). The internal transformation rates are represented by directed arcs : there is an arc from compartment i to compartment j when there is a non-zero entry $g_{ji} = \bar{r}_{ij}$ in the compartmental matrix G. These arcs are labeled with the fractional rates $\bar{r}_{ij}$. Additional arcs, labeled respectively with fractional outflow rates $\bar{q}_i$ and inflow rates p_i are used to represent inflows and outflows. Concrete examples of compartmental networks will be given in Fig.4 and Fig.6.

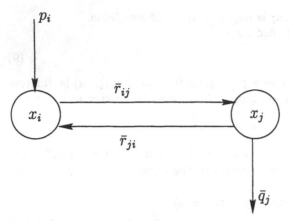

Fig. 2. Network of compartments

A compartment is said to be *outflow connected* if there is a path from that compartment to a compartment from which there is an outflow arc. The system is said to be *fully outflow connected* if all compartments are outflow connected. As stated in the following property, the non singularity of a compartmental matrix can be checked directly on the network.

Property 1. For a given value of $(x, u) \in R_+^n \times R^m$, the compartmental matrix $G(x, u)$ of a mass balance system (10) is non singular if and only if the system is fully outflow connected.   ■

A proof of this property can be found e.g. in [7].

7   Stoichiometric representation

In many cases the transformation rates $r_i(x, u), i = 1, n$ can be expressed as linear combinations of a smaller set of non-negative and differentiable basis functions $\rho_1(x, u), \rho_2(x, u), ..., \rho_k(x, u)$ $(k < n)$:

$$r_i(x, u) = \sum_j c_{ij} \rho_j(x, u)$$

This situation typically arises in chemical systems where the non-zero coefficients c_{ij} are the stoichiometric coefficients of the underlying reaction network and the functions $\rho_j(x, u)$ are the reaction rates. The matrix $C = [c_{ij}]$ is therefore called *stoichiometric* and by defining the vector $\rho(x, u) = (\rho_1(x, u), \rho_2(x, u), ..., \rho_k(x, u))^T$ we have :

$$r(x, u) = C\rho(x, u)$$

As we will see in the examples, this stoichiometric representation is also relevant in many other physical and biological systems, As stated in the following property, the mass conservation condition (6) can easily be checked from the stoichiometric matrix C independently of the rate functions $\rho_j(x, u)$.

Property 2. The mass conservation condition $\sum_i r_i(x, u) = 0$ is satisfied if the sum of the entries of each column of C is zero :

$$\sum_i c_{ij} = 0 \quad \forall j$$

or equivalently if the vector $\varepsilon = (1, 1, \ldots, 1)^T$ belongs to the kernel of the transpose of the stoichiometric matrix : $\varepsilon^T C = 0$. ∎

8   Examples of mass-balance systems

8.1   A biochemical process

A continuous stirred tank reactor is represented in Fig.3. The following biochemical reactions take place in the reactor :

$$A \xrightarrow[X]{} B$$
$$B \xrightarrow[X]{} X$$

where X represents a microbial population and A, B organic matters. The first reaction represents the hydrolysis of species A into species B, catalysed by cellular enzymes. The second reaction represents the growth of microorganisms on substrate B. It is obviously an auto-catalytic reaction. Assuming

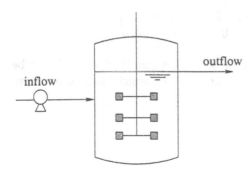

Fig. 3. Stirred tank reactor

mass action kinetics, the dynamics of the reactor may be described by the model :

$$\dot{x}_1 = +k_1 x_1 x_2 - u x_1$$
$$\dot{x}_2 = -k_1 x_1 x_2 + k_2 x_1 x_3 - u x_2$$
$$\dot{x}_3 = -k_2 x_1 x_3 - u x_3 + u x_3^{in}$$

with the following notations and definitions :

x_1 = concentration of species X in the reactor
x_2 = concentration of species B in the reactor
x_3 = concentration of species A in the reactor
x_3^{in} = concentration of species A in the influent
u = dilution rate (control input)
k_1, k_2 = rate constants.

This could be for instance the model of a biological depollution process where $u x_3^{in}$ is the pollutant inflow while $u(x_2 + x_3)$ is the residual pollution outflow. It is readily seen to be a mass-balance model with the following definitions :

$$r(x,u) = \begin{pmatrix} +k_1 x_1 x_2 \\ -k_1 x_1 x_2 + k_2 x_1 x_3 \\ -k_2 x_1 x_3 \end{pmatrix} \quad q(x,u) = \begin{pmatrix} u x_1 \\ u x_2 \\ u x_3 \end{pmatrix} \quad p(x,u) = \begin{pmatrix} 0 \\ 0 \\ u x_3^{in} \end{pmatrix}$$

The Hamiltonian representation is :

$$F(x,u) = \begin{pmatrix} 0 & k_1 x_1 x_2 & 0 \\ -k_1 x_1 x_2 & 0 & k_2 x_1 x_3 \\ 0 & -k_2 x_1 x_3 & 0 \end{pmatrix} \quad D(x,u) = \begin{pmatrix} u x_1 & 0 & 0 \\ & u x_2 & 0 \\ 0 & 0 & u x_3 \end{pmatrix}$$

The compartmental matrix is :

$$G(x,u) = \begin{pmatrix} -u & k_1 x_1 & 0 \\ 0 & -u - k_1 x_1 & k_2 x_1 \\ 0 & 0 & -u - k_2 x_1 \end{pmatrix}$$

The compartmental network of the process is shown in Fig.4 where it can be seen that the system is fully outflow connected. The stoichiometric represen-

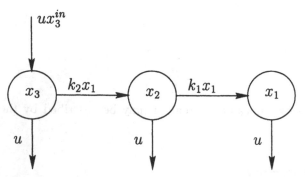

Fig. 4. Compartmental network of the biochemical process model

tation is :

$$C = \begin{pmatrix} 1 & 0 \\ -1 & 1 \\ 0 & -1 \end{pmatrix} \quad p(x) = \begin{pmatrix} k_1 x_1 x_2 \\ k_2 x_1 x_3 \end{pmatrix}$$

8.2   A grinding process

An industrial grinding circuit, as represented in Fig.5 is made up of the interconnection of a mill and a separator. The mill is fed with raw material. After grinding, the material is introduced in a separator where it is separated in two classes : fine particles which are given off and oversize particles which are recycled to the mill. A simple dynamical model has been proposed for this system in [6]:

$$\dot{x}_1 = -\gamma_1 x_1 + (1 - \alpha)\phi(x_3)$$
$$\dot{x}_2 = -\gamma_2 x_2 + \alpha\phi(x_3)$$
$$\dot{x}_3 = \gamma_2 x_2 - \phi(x_3) + u$$
$$\phi(x_3) = k_1 x_3 e^{-k_2 x_3}$$

with the following notations and definitions :

x_1 = hold-up of fine particles in the separator
x_2 = hold-up of oversize particles in the separator
x_3 = hold-up of material in the mill
u = inflow rate
$\gamma_1 x_1$ = outflow rate of fine particles
$\gamma_2 x_2$ = flowrate of recycled particles
$\phi(x_3)$ = outflowrate from the mill = grinding function
α = separation constant $(0 < \alpha < 1)$
$\gamma_1, \gamma_2, k_1, k_2$ = characteristic positive constant parameters

This model is readily seen to be a mass-balance system with the following definitions :

$$r(x, u) = \begin{pmatrix} (1 - \alpha)\phi(x_3) \\ -\gamma_2 x_2 + \alpha\phi(x_3) \\ \gamma_2 x_2 - \phi(x_3) \end{pmatrix} \quad q(x, u) = \begin{pmatrix} -\gamma_1 x_1 \\ 0 \\ 0 \end{pmatrix} \quad p(x, u) = \begin{pmatrix} 0 \\ 0 \\ u \end{pmatrix}$$

The Hamiltonian representation is :

$$F(x, u) = \begin{pmatrix} 0 & 0 & (1 - \alpha)\phi(x_3) \\ 0 & 0 & -\gamma_2 x_2 + \alpha\phi(x_3) \\ -(1 - \alpha)\phi(x_3) & \gamma_2 x_2 - \alpha\phi(x_3) & 0 \end{pmatrix}$$

$$D(x, u) = \begin{pmatrix} \gamma_1 x_1 & 0 & 0 \\ 0 & 0 & 0 \\ 0 & 0 & 0 \end{pmatrix}$$

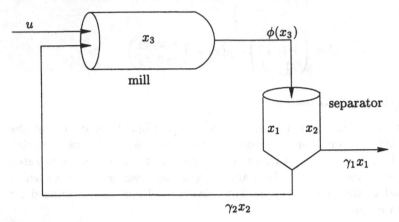

Fig. 5. Grinding circuit

The compartmental matrix is :

$$G(x,u) = \begin{pmatrix} -\gamma_1 & 0 & (1-\alpha)k_1e^{-k_2x_3} \\ 0 & -\gamma_2 & \alpha k_1e^{-k_2x_3} \\ 0 & +\gamma_2 & -k_1e^{-k_2x_3} \end{pmatrix}$$

The compartmental network of the process is shown in Fig.6 where it can be seen that the system is fully outflow connected.

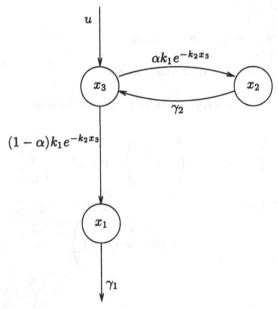

Fig. 6. Compartmental network of the grinding process model

The stoichiometric representation is :

$$C = \begin{pmatrix} 1-\alpha & 0 \\ \alpha & -1 \\ -1 & 1 \end{pmatrix} \quad \rho(x) = \begin{pmatrix} \phi(x_3) \\ \gamma_2 x_2 \end{pmatrix}$$

9   A fundamental control problem

Let us consider a mass-balance system with constant inputs denoted $\bar{u}$:

$$\dot{x} = r(x,\bar{u}) - q(x,\bar{u}) + p(x,\bar{u}) \tag{11}$$

An equilibrium of this system is a state vector $\bar{x}$ which satisfies the equilibrium equation :

$$r(\bar{x},\bar{u}) - q(\bar{x},\bar{u}) + p(\bar{x},\bar{u}) = 0$$

In general, mass balance systems (11) have multiple equilibria. One of these equilibria is the operating point of interest. It is generally locally asymptotically stable. This means that an open loop operation may be acceptable in practice. But if big enough disturbances occur, it may arise that the system is driven too far from the operating point towards a region of the state space which is outside of its basin of attraction. From time to time, the process may therefore be lead by accident to a behaviour which may be undesirable or even catastrophic. We illustrate the point with our two examples.

Example 1 : The biochemical process
     For a constant inflow rate $\bar{u} < k_1 x_3^{in}$, the biochemical process has three equilibria (see Fig.7). Two of these equilibria (E_1, E_2) are solutions of the following equations :

$$\bar{x}_2 = \frac{\bar{u}}{k_1} \quad \bar{x}_1 + \bar{x}_3 = x_3^{in} - \frac{\bar{u}}{k_1} \quad \bar{x}_3(\bar{u} + k_2\bar{x}_1) = \bar{u}x_3^{in}$$

The third equilibrium $(E3)$ is

$$\bar{x}_1 = 0 \quad \bar{x}_2 = 0 \quad \bar{x}_3 = x_3^{in}$$

As we shall see later on, this system is globally stable in the sense that all trajectories are bounded independently of $\bar{u}$. Furthermore, by computing the Jacobian matrix, it can be easily checked that $E1$ and $E3$ are asymptotically stable while $E2$ is unstable.
     $E1$ is the normal operating point corresponding to a high conversion of substrate x_3 into product x_1. It is stable and the process can be normally operated at this point. But there is another stable equilibrium $E3$ called "wash-out steady state" which is highly undesirable because it corresponds to a complete loss of productivity : $\bar{x}_1 = 0$. The pollutant just goes through the tank without any degradation.

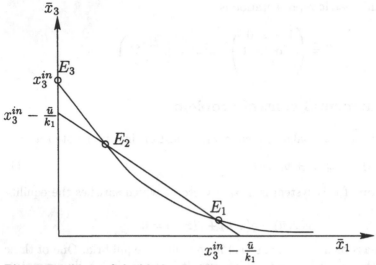

Fig. 7. Equilibria of the biochemical process

The problem is that an intermittent disturbance (like for instance a pulse of toxic matter) may irreversibly drive the process to this wash-out steady-state, making the process totally unproductive.

Example 2 : The grinding process

The equilibria of the grinding process $(\bar{x}_1, \bar{x}_2, \bar{x}_3)$ are parametrized by a constant input flowrate $\bar{u}$ as follows :

$$\bar{x}_1 = \frac{\gamma_1}{\bar{u}} \qquad \bar{x}_2 = \frac{\alpha\bar{u}}{\gamma_2(1-\alpha)} \qquad \phi(\bar{x}_3) = \frac{\bar{u}}{(1-\alpha)}$$

In view of the shape of $\phi(x_3)$ as illustrated in Fig.8, there are two distinct equilibria if :

$$\bar{u} < (1-\alpha)\phi_{max}$$

The equilibrium $E1$ on the left of the maximum is stable and the other one $E2$ is unstable. Furthermore, for any value of $\bar{u}$, the trajectories become unstable as soon as the state enters the set D defined by :

$$D\begin{cases} (1-\alpha)\phi(x_3) < \gamma_1 x_1 < \bar{u} \\ \alpha\phi(x_3) < \gamma_2 x_2 \\ \partial\phi/\partial x_3 < 0 \end{cases}$$

Indeed, it can be shown that this set D is forward invariant and if $x(0) \in D$ then $x_1 \to 0$ $x_2 \to 0$ $x_3 \to \infty$. In some sense, the system is *Bounded Input - Unbounded State (BIUS)*. This means that there can be an irreversible accumulation of material in the mill with a decrease of the production to

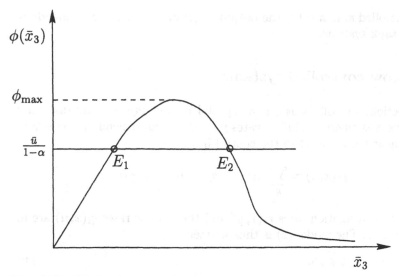

Fig. 8. Equilibria of the grinding process

zero. In the industrial jargon, this is called *mill plugging*. In practice, the state may be lead to the set D by intermittent disturbances like variations of hardness of the raw material.   ■

In both examples we thus have a stable open loop operating point with a potential process destabilisation which can take two forms :

- drift of the state x towards another (unproductive) equilibrium
- unbounded increase of the total mass $M(x)$

The control challenge is then to design a feedback controller which is able to prevent the process from such undesirable behaviours.

Ideally a good control law should meet the following specifications :

S1. The feedback control action is bounded;
S2. The closed loop system has a single equilibrium in the positive orthant which is globally asymptotically stable;
S3. The single closed-loop equilibrium may be assigned by an appropriate set point.

Moreover, it could be desirable that the feedback stabilisation be robust against modelling uncertainties regarding $r(x)$ which is the most uncertain term of the model in many applications.

This is indeed a vast problem which is far to be completely explored. Hereafter, we limit ourselves to the presentation of two specific solutions of this problem namely (i) the state feedback stabilisation of the total mass in inflow controlled systems; (ii) the output regulation with state boundedness in stirred tank systems.

10   Inflow controlled systems

In this section, we will focus on the special case of *inflow-controlled* mass-balance systems where the inflow rates $p_i(x, u)$ do not depend on the state x and are linear with respect to the control inputs u_k :

$$p_i(x, u) = \sum_k b_{ik} u_k \quad b_{ik} \geq 0 \quad u_k \geq 0$$

while the transformation rates $r_i(x, u)$ and the outflow rates $q_i(x, u)$ are independent of u. The model (2) is thus written as :

$$\dot{x} = r(x) - q(x) + Bu \tag{12}$$

with B the $n \times m$ matrix with entries b_{ik}.

The Hamiltonian representation specializes as :

$$\dot{x} = [F(x) - D(x)] \left(\frac{\partial M}{\partial x} \right)^T + Bu \tag{13}$$

and the compartmental representation as :

$$\dot{x} = G(x)x + Bu \tag{14}$$

with appropriate definitions of the matrices $F(x), D(x)$ and $G(x)$.

The grinding process model presented in the previous section is an example of an inflow-controlled mass balance system.

10.1   Bounded input - (un)bounded state

Obviously, the state x of any mass-balance system is bounded if and only if the total mass $M(x) = \sum_i x_i$ is itself bounded. In an inflow-controlled system, the dynamics of the total mass are written as :

$$\dot{M} = -\sum_i q_i(x) + \sum_{i,k} b_{ik} u_k \tag{15}$$

From this expression, a natural condition for state boundedness is clearly that the total outflow $\sum_i q_i(x)$ should exceed the total inflow $\sum_{i,k} b_{ik} u_k$ when the total mass $M(x)$ is big enough (in order to make the right hand side of (15) negative). This intuitive condition is made technically precise as follows.

Property 3. Assume that :

(A1) the input $u(t)$ is bounded :

$$0 \leq u_k(t) \leq u_k^{\max} \quad \forall t \; \forall k = 1, \ldots, m$$

(A2) There exists a constant M_0 such that

$$\sum_i q_i(x) \geq \sum_{i,k} b_{ik} u_k^{\max}$$

when $M(x) \geq M_0$

Then, the state of the system (12) is bounded and the simplex

$$\Delta = \{x \in R_+^n : M(x) \leq M_0\}$$

is forward invariant.

The system is BIBS if condition (A2) holds for any $u^{\max}$, for example if each $q_i(x) \to \infty$ as $x_i \to \infty$. ∎

As a matter of illustration, it is readily checked that inflow-controlled systems with linear outflows in all compartments i.e. $q_i(x) = a_i x_i, a_i > 0, \forall i$ are necessarily BIBS. Indeed in this case we have

$$\sum_i q_i(x) = \sum_i a_i x_i \geq \min_i(a_i) M(x)$$

and therefore $M_0 = \frac{\sum_k b_{ik} u_k^{max}}{\min_i(a_i)}$

In contrast, as we have seen in the previous section, the grinding process of Example 2 is *not* BIBS. Even worse, the state variable x_3 may be unbounded for any value of $u^{\max} > 0$. This means that the process is *globally unstable* for any bounded input.

10.2   Systems without inflows

Consider the case of systems *without inflows* $u = 0$ which are written in compartmental form

$$\dot{x} = G(x)x \tag{16}$$

Obviously, the origin $x = 0$ is an equilibrium of the system.

Property 4. If the compartmental matrix $G(x)$ is full rank for all $x \in R_+^n$ (equivalently if the system is fully outflow connected), then the origin $x = 0$ is a globally asymptotically stable (GAS) equilibrium of the unforced system $\dot{x} = G(x)x$ in the non negative orthant, with the total mass $M(x) = \sum_i x_i$ as Lyapunov function. ∎

Indeed, for such systems, the total mass can only decrease along the system trajectories since there are outflows but no inflows :

$$\dot{M} = -\sum_i q_i(x)$$

Property 4 says that the total mass $M(x)$ and the state x will decrease until the system is empty if there are no inflows and the compartmental matrix is nonsingular for all x. A proof of this property and other related results can be found in [2].

10.3   Robust state feedback stabilisation of the total mass

We now consider a single-input inflow-controlled mass-balance system of the form :

$$\dot{x}_i = r_i(x) - q_i(x) + b_i u \quad i = 1, \ldots, n \tag{17}$$

with $b_i \geq 0 \ \forall i, \sum_i b_i > 0$

This system may be globally unstable (bounded input/unbounded state). The symptom of this instability is an unbounded accumulation of mass inside the system like for instance in the case of the grinding process of Example 2.

One way of approaching the problem is to consider that the control objective is to globally stabilise the total mass $M(x)$ at a given set point $M^* > 0$ in order to prevent the unbounded mass accumulation.

In order to achieve this control objective, the following *positive* control law is proposed in [1] :

$$u(x) = \max(0, \tilde{u}(x)) \tag{18}$$

$$\tilde{u}(x) = \left(\sum_i b_i\right)^{-1} \left[\sum_i q_i(x) + \lambda(M^* - M(x))\right] \tag{19}$$

where $\lambda > 0$ is an arbitrary design parameter. The stabilising properties of this control law are as follows.

Property 5. If the system (17) is fully outflow connected, then the closed loop system (17)-(18)-(19) has the following properties for any initial condition $x(0) \in R_+^n$:

1. the set $\Omega = \{x \in R_+^n : M(x) = M^*\}$ is forward invariant
2. the state $x(t)$ is bounded for all $t \geq 0$ and $\lim_{t \to \infty} M(x) = M^*$.

■

The proof of this property can be found in [1]. It is worth noting that the control law (18)-(19) is independent from the internal transformation term $r(x)$. This means that the feedback stabilisation is robust against a full modelling uncertainty regarding $r(x)$ provided it satisfies the conditions of positivity and mass conservativity.

The application of this control law to the example of the grinding process is studied in [1] where it is shown that the closed loop system has indeed a single globally stable equilibrium (although the open loop may have 0, 1, or 2 equlibria).

10.4   Output regulation for a class of BIBS systems

In order to avoid undesirable equilibria, a possible solution is to regulate some output variable at a set point y^* which uniquely assigns the equilibrium of interest. Here is an example of such a solution. We consider the class of single-input BIBS mass-balance systems of the form :

$$\dot{x}_i = r_i(x) - a_i x_i \quad i = 1, \ldots, n-1$$
$$\dot{x}_n = r_n(x) - a_n x_n + u$$

with $a_i > 0 \ \forall i$. We assume that the measured output $y = x_n$ is the state of an *initial* compartment. The species x_n can only be consumed inside the system but not produced. In other terms, in the compartmental graph of the system, there are several arcs going from compartment n to other compartments but absolutely *no* arcs coming from other compartments. Then, with the notations :

$$\xi = (x_1, \ldots, x_{n-1})^T \quad y = x_n$$

and appropriate definitions of φ and ψ, the system is rewritten as :

$$\dot{\xi} = \varphi(\xi, y) \tag{20}$$
$$\dot{y} = -(\psi(\xi, y) + a_n)y + u \tag{21}$$

and the function $\psi(\xi, y)$ is non-negative.

The goal is to regulate the measured output y at a given set point $y^* > 0$. In order to achieve this objective, the following control law is considered :

$$u = (\psi(\xi, y) + a_n)[(1 - \lambda)y + \lambda y^*] \tag{22}$$

where λ is a design parameter such that :

$$0 < \lambda < 1$$

With this control law, the closed loop system is written as :

$$\dot{\xi} = \varphi(\xi, y) \tag{23}$$
$$\dot{y} = -(\psi(\xi, y) + a_n)\lambda(y^* - y) \tag{24}$$

The stabilisation properties of this control law are analysed under the following assumptions :

A1. The state is initialised in the non negative orthant with $0 \le y(0) \le y^{max}$ for some arbitrary $y^{max} > y^*$.

A2. The function $\psi(\xi, y)$ is bounded :

$$0 \le \psi(\xi, y) \le \psi^{max} \quad \forall (\xi, y) \in R^n_+$$

A3. The zero dynamics $\dot{\xi} = \varphi(\xi, y^*)$ have a single equilibrium $\bar{\xi} \in R^{n-1}_+$ which is GAS in the non negative orthant.

Assumption A3 is a standard global minimum phase assumption.

Property 6 Under Assumptions A1, A2 and A3, the closed loop system (23)-(24) has the following properties :

1. The control input is positive and bounded :

$$0 \le u(t) \le (\psi^{max} + a_n)[(1 - \lambda)y^{max} + \lambda y^*]$$

2. The state is bounded
3. The regulation error converges to zero : $(y^* - y) \to 0$ as $t \to \infty$.
4. The closed loop system has a single equilibrium $(\bar{\xi}, y^*)$ which is GAS in the non negative orthant.

∎

Again the important point is that the closed loop system is guaranteed to have a single GAS equilibrium although the open loop system may have several equilibria as we have seen above.

11   Mass balance systems in stirred tanks

In many engineering applications, the system under consideration takes place in liquid phase in a stirred tank with a constant volume as represented in Fig.3. The state variables x_i represent the concentrations of various species in the tank. We consider the very common case of stirred tank mass balance systems with the volumetric flow rate as single control input. In such systems, both the mass inflow rates $p_i(x, u)$ and the mass outflow rates $q_i(x, u)$ linearly depend on the input u :

$$p_i(x, u) = u x_i^{in} \quad q_i(x, u) = u x_i \tag{25}$$

while the transformation rates $r_i(x, u)$ are independent of u. $x_i^{in} \ge 0$ denotes the constant concentration of the i-th species in the influent stream. Obviously, $x_i^{in} = 0$ for those species which are not fed to the tank but are only produced inside the system. The consistency of the model also requires that the control input be non negative : $u(t) \ge 0 \ \forall t$. The general mass-balance (2) is thus written as :

$$\dot{x} = r(x) + u(x^{in} - x)$$

with x^{in} the $n \times 1$ vector with entries x_i^{in}. The stoichiometric representation specializes as :

$$\dot{x} = C\rho(x) + u(x^{in} - x) \tag{26}$$

The biochemical process model presented above is an example of a stirred tank mass balance system.

State boudedness

For a stirred tank system, the dynamics of the total mass $M(x) = \sum_i x_i$ are written as :

$$\dot{M} = u\left(\sum_i x_i^{in} - M\right) \tag{27}$$

which implies that $M(x)$ and therefore x are bounded *independently of the control input u*. Furthermore, the simplex

$$\Delta = \left\{x_i \geq 0 : \sum_i (x_i^{in} - x_i) \geq 0\right\}$$

is forward invariant. A weaker but more explicit consequence is that if x is initialised in Δ, then each state variable is bounded as :

$$0 \leq x_i(t) \leq \sum_i x_i^{in} \quad \forall t$$

Stoichiometric invariants

From equation (27) we see also that the set $\Omega = \{x \in R_+^n : \sum_i (x_i - x_i^{in}) = 0\}$ is forward invariant. This is a typical special case of *stoichiometric invariants* which are classically considered in the Chemical Engineering literature (see e.g. [3]). For any non-zero vector $\lambda^T = (\lambda_1, \ldots, \lambda_n)$ such that $\lambda^T C = 0$ (the vector λ is in the kernel of the transpose of the stoichiometric matrix C), a stoichiometric invariant is defined as the set

$$\Omega = \{x \in R_+^n : \lambda^T(x - x^{in}) = 0\}$$

It is indeed easy to check that this set is forward invariant along the trajectories of the stirred tank system (26).

The nonlinear control of mass balance systems in stirred tank reactors is discussed e.g. in [8] (see also [9] for related results).

Summary

In this chapter a general state-space model of mass balance systems has been presented and illustrated with two simple industrial examples : a biochemical process and a grinding process. In general, mass balance systems have multiple equilibria, one of them being the operating point of interest which is locally asymptotically stable. However if big enough disturbances occur, the process may be lead by accident to a behaviour which may be undesirable or even catastrophic. The control challenge is then to design a feedback controller which is able to prevent the process from such undesirable behaviours. We have presented two very specific solutions for single input systems. But it is obvious that the fundamental control problem formulated in this chapter is far from being solved and deserves deeper investigations. In particular a special interest should be devoted to control design methodologies which explicitely account for the structural specificities (Hamiltonian, Compartmental, Stoichiometric) of mass balance systems and rely on the construction of physically based control laws.

Acknowledgements

This paper presents research results of the Belgian Program on Interuniversity Attraction Poles, Prime Minister's Office, Science Policy Programming.

Appendix : stability conditions

In this appendix some interesting stability results for mass balance systems with constant inputs are collected. These results can be useful for Lyapunov control design or for the stability analysis of zero-dynamics.

Compartmental Jacobian matrix
   We consider the general case of inflow controlled mass balance systems with constant inflows :

$$\dot{x} = r(x) - q(x) + p(\bar{u})$$

The Jacobian matrix of the system is defined as :

$$J(x) = \frac{\partial}{\partial x}[r(x) - q(x)]$$

When this matrix has a compartmental structure, we have the following stability result.

Property A1

   a) If $J(x)$ is a compartmental matrix $\forall\ x \in R_+^n$, then all bounded orbits tend to an equilibrium in R_+^n.

b) If there is a *bounded* closed convex set $D \subseteq R_+^n$ which is *forward invariant* and if $J(x)$ is a *non singular* compartmental matrix $\forall x \in D$, then there is a *unique* equilibrium $\bar{x} \in D$ which is GAS in D with Lyapunov function $V(x) = \sum_i |r_i(x) - q_i(x) + p_i(\bar{u})|$.

∎

A proof of part a) can be found in [7] Appendix 4 while part b) is a concise reformulation of a theorem by Rosenbrock [12].

The assumption that $J(x)$ is compartmental $\forall x \in R_+^n$ is fairly restrictive. For instance, this assumption is *not* satisfied neither for the grinding process nor for the biochemical processes that we have used as examples in this paper. A simple sufficient condition to have $J(x)$ compartmental for all x is as follows.

Property A2 The Jacobian matrix $J(x) = \frac{\partial}{\partial x}[r(x) - q(x)]$ is compartmental $\forall x \in R_+^n$ if the functions $r(x)$ and $q(x)$ satisfy the following *monotonicity* conditions :

1) $\dfrac{\partial q_i}{\partial x_i} \geq 0 \quad \dfrac{\partial q_i}{\partial x_k} = 0 \ \ k \neq i$

2) $\dfrac{\partial r_{ij}}{\partial x_i} \geq 0 \quad \dfrac{\partial r_{ij}}{\partial x_j} \leq 0 \quad \dfrac{\partial r_{ij}}{\partial x_k} = 0 \ \ k \neq i \neq j$

∎

In the next two sections, we describe two examples of systems that have a single GAS equilibrium in the nonnegative orthant although their Jacobian matrix is not compartmental.

The Gouzé's condition

We consider a class of stirred tank mass-balance systems of the form :

$$\dot{x}_i = \sum_{j \neq i}[r_{ji}(x_j) - r_{ij}(x_i)] + \bar{u}(x_i^{in} - x_i) \tag{28}$$

where the transformation rates $r_{ij}(x_i)$ depend on x_i only.

For example this can be the model of a stirred tank chemical reactor with monomolecular reactions as explained in [5] (see also [13]).

The set $\Omega = \{x \in R_+^n : M(x) = \sum_i x_i^{in}\}$ is bounded, convex, compact and invariant. By the Brouwer fixed point theorem, it contains at least an equilibrium point $\bar{x} = (\bar{x}_1, \bar{x}_2, \ldots, \bar{x}_n)$ which satisfies the set of algebraic equations :

$$\sum_{j \neq i}[r_{ji}(\bar{x}_j) - r_{ij}(\bar{x}_i)] + \bar{u}(x_i^{in} - \bar{x}_i) = 0$$

The following property then gives a condition for this equilibrium to be unique and GAS in the non negative orthant.

Property A3 If $(r_{ij}(x_i) - r_{ij}(\bar{x}_i))(x_i - \bar{x}_i) \geq$ $\forall x_i \geq 0$, then the equilibrium $(\bar{x}_1, \ldots, \bar{x}_n)$ of the system (28) is GAS in the non negative orthant with Lyapunov function.

$$V(x) = \sum_i |x_i - \bar{x}_i|$$

■

The proof of this property is given in [5]. The interesting feature is that the rate functions $r_{ij}(x_i)$ can be *non-monotonic* (which makes the Jacobian matrix non-compartmental) in contrast with the assumptions of Property A2.

Conservative Lotka-Volterra systems

We consider now a class of Lotka-Volterra ecologies of the form :

$$\dot{x}_i = x_i \left(\sum_{j \neq i} a_{ij} x_j - a_{i0} \right) + \bar{u}_i \ \ i = 1, \ldots, n \tag{29}$$

with $a_{i0} > 0$ the natural mortality rates;
$a_{ij} = -a_{ij} \ \forall i \neq j$ the predation coefficients (i.e. $A = [a_{ij}]$ is skew symmetric);
$\bar{u}_i \geq 0$ the feeding rate of species x_i with $\sum_i \bar{u}_i > 0$.

This is a mass balance system with a bilinear Hamiltonian representation :

$$F(x) = [a_{ij} x_i x_j] \quad D(x) = (\text{diag } a_{i0} x_i)$$

Assume that the system has an equilibrium in the positive orthant $\text{int}\{R_+^n\}$ i.e. there is a strictly positive solution $(\bar{x}_1, \bar{x}_2, \ldots, \bar{x}_n)$ to the set of algebraic equations :

$$a_{i0} = \sum_{j \neq i} a_{ij} \bar{x}_j + \frac{\bar{u}_i}{\bar{x}_i} \ \ i = 1, \ldots, n$$

Assume that this equilibrium $(\bar{x}_1, \bar{x}_2, \ldots, \bar{x}_n)$ is the only trajectory in the set :

$$D = \{x \in \text{int}\{R_+^n\} : \bar{u}_i(x_i - \bar{x}_i) = 0 \forall i\}$$

Then we have the following stability property.

Property A4 The equilibrium $(\bar{x}_1, \bar{x}_2, \ldots, \bar{x}_n)$ of the Lotka-Volterra system (29) is unique and GAS in the positive orthant with Lyapunov function

$$V(x) = \sum_i (x_i - \bar{x}_i ln x_i)$$

■

The proof is established, as usual, by using the time derivative of V:

$$\dot{V}(x) = -\sum_i \left[\frac{\bar{u}_i \bar{x}_i}{x_i} \left(1 - \frac{x_i}{\bar{x}_i} \right)^2 \right]$$

and the La Salle's invariance principle (see also [4] for related results).

References

1. Bastin G. and L. Praly, "Feedback stabilisation with positive control of a class of mass-balance systems", Proceedings 14th IFAC World Congress, Volume C, Paper 2a-03-2, pp. 79–84, July 1999, Beijing, July 1999.
2. Eisenfeld J., "On washout in nonlinear compartmental systems", Mathematical Biosciences, Vol. 58, pp. 259–275, 1982.
3. Fjeld M., O.A. Asbjornsen, K.J. Astrom, "Reaction invariants and their importance in the analysis of eigenvectors, state observability and controllability of the continuous stirred tank reactor", Chemical Engineering science, Vol. 29, pp. 1917–1926, 1974.
4. Gouzé J-L, "Global behaviour of n-dimensional Lotka-Volterra systems", Mathematicam Biosciences, Vol. 113, pp. 231–243, 1993.
5. Gouzé J-L, "Stability of a class of stirred tank reactors", CD-Rom Proceedings ECC97, Paper FR–A–C7, Brussels, July 1995.
6. Grognard F., F. Jadot, L. Magni, G. Bastin, R. Sepulchre, V. Wertz, "Robust stabilisation of a nonlinear cement mill model", to be published in IEEE Transactions of Automatic Control, March 2001.
7. Jacquez J.A. and C.P. Simon, "Qualitative theory of compartmental systems", SIAM Review, Vol. 35(1), pp. 43–79, 1993.
8. Jadot F., "Dynamics and robust nonlinear PI control of stirred tank reactors", PhD Thesis, CESAME, University of Louvain la Neuve, Belgium, June 1996.
9. Jadot F., G. Bastin and F. Viel, "Robust global stabilisation of stirred tank reactors by saturated output feedback", European Journal of Control, Vol.5 , pp. 361–371, 1999.
10. Maschke B. M. and A. J. van der Schaft, "Port controlled Hamiltonian systems : modeling origins and system theoretic properties", Proc. IFAC NOLCOS 92, pp. 282–288, Bordeaux, June 1992.
11. Maschke B. M., R. Ortega and A. J. van der Schaft, "Energy based Lyapunov functions for Hamiltonian systems with dissipation", Proc. IEEE Conf. Dec. Control, Tampa, December 1998.
12. Rosenbrock H., "A Lyapounov function with applications to some nonlinear physical problems", Automatica, Vol. 1, pp. 31–53, 1962.
13. Rouchon P., "Remarks on some applications of nonlinear control techniques to chemical processes", Proc. IFAC NOLCOS 92, pp. 37–42, Bordeaux, June 1992.

Network Modelling of Physical Systems: a Geometric Approach*

Arjan van der Schaft[1], Bernhard Maschke[2], and Romeo Ortega[3]

[1] Fac. of Mathematical Sciences, University of Twente, Enschede,
The Netherlands
[2] Laboratoire d'Automatique et de Genie des Procedes, Université Claude
Bernard – Lyon 1, Villeurbanne Cedex, France
[3] Lab. des Signaux et Systèmes, CNRS–Supélec, Gif–sur–Yvette, France

Abstract. It is discussed how *network modeling* of lumped-parameter physical systems naturally leads to a geometrically defined class of systems, called *port-controlled Hamiltonian systems (with dissipation)*. The structural properties of these systems are investigated, in particular the existence of Casimir functions and their implications for stability. It is shown how the power-conserving interconnection with a controller system which is also a port-controlled Hamiltonian system defines a closed-loop port-controlled Hamiltonian system; and how this may be used for control by shaping the internal energy. Finally, extensions to implicit system descriptions (constraints, no a priori input-output structure) are discussed.

1 Introduction

Nonlinear systems and control theory has witnessed tremendous developments over the last three decades, see for example the textbooks [12,25]. Especially the introduction of geometric tools like Lie brackets of vector fields on manifolds has greatly advanced the theory, and has enabled the proper generalization of many fundamental concepts known for *linear control systems* to the nonlinear world. While the emphasis in the eighties has been primarily on the *structural* analysis of smooth nonlinear dynamical control systems, in the nineties this has been combined with analytic techniques for stability, stabilization and robust control, leading e.g. to backstepping techniques and nonlinear H_∞- control. Moreover, in the last decade the theory of *passive* systems, and its implications for regulation and tracking, has undergone a remarkable revival. This last development was also spurred by work in robotics on the possibilities of *shaping* by feedback the *physical energy* in such a way that it can be used as a suitable Lyapunov function for the control purpose at hand, see e.g. the influential paper [42]. This has led to what is called *passivity-based control*, see e.g. [26,32,13].

In this lecture we want to stress the importance of *modelling* for nonlinear control. Of course, this is well-known for (nonlinear) control applications, but

---

* This paper is an adapted and expanded version of [33]. Part of this material can be also found in [32].

in our opinion also the development of nonlinear control *theory* for physical systems should be integrated with a theoretical framework for modelling. We discuss how *network modelling* of (lumped-parameter) physical systems naturally leads to a geometrically defined class of systems, called *port-controlled Hamiltonian systems with dissipation* (PCHD systems). This provides a unified mathematical framework for the description of physical systems stemming from different physical domains, such as mechanical, electrical, thermal, as well as mixtures of them.

Historically, the Hamiltonian approach has its roots in analytical mechanics and starts from the principle of least action, via the Euler-Lagrange equations and the Legendre transform, towards the Hamiltonian equations of motion. On the other hand, the network approach stems from electrical engineering, and constitutes a cornerstone of systems theory. While most of the *analysis* of physical systems has been performed within the Lagrangian and Hamiltonian framework, the network modelling point of view is prevailing in *modelling* and *simulation* of (complex) physical systems. The framework of PCHD systems *combines* both points of view, by associating with the interconnection structure ("generalized junction structure" in bond graph terminology) of the network model a *geometric structure* given by a *Poisson structure*, or more generally a *Dirac structure*. The Hamiltonian dynamics is then defined with respect to this Poisson (or Dirac) structure *and* the Hamiltonian given by the total stored energy, as well as the energy-dissipating elements and the ports of the system.

Dirac structures encompass the "canonical" structures which are classically being used in the geometrization of mechanics, since they also allow to describe the geometric structure of systems with *constraints* as arising from the interconnection of sub-systems. Furthermore, Dirac structures allow to extend the Hamiltonian description of *distributed parameter systems* to include variable boundary conditions, leading to port-controlled distributed parameter Hamiltonian systems with boundary ports, see [17].

The structural properties of PCHD systems can be investigated through geometric tools stemming from the theory of Hamiltonian systems. We shall indicate how the *interconnection* of PCHD systems leads to another PCHD system, and how this may be exploited for control and design. In particular, we investigate the existence of Casimir functions for the feedback interconnection of a plant PCHD system and a controller PCHD system, leading to a reduced PCHD system on invariant manifolds with *shaped* energy. We thus provide an interpretation of *passivity-based control* from an *interconnection* point of view. This point of view can be further extended to what has been recently called Interconnection-Damping Assignment Passivity-Based Control (IDA-PBC).

2   Port-controlled Hamiltonian systems

2.1   From the Euler-Lagrange and Hamiltonian equations to port-controlled Hamiltonian systems

Let us briefly recall the standard Euler-Lagrange and Hamiltonian equations of motion. The standard *Euler-Lagrange equations* are given as

$$\frac{d}{dt}\left(\frac{\partial L}{\partial \dot{q}}(q,\dot{q})\right) - \frac{\partial L}{\partial q}(q,\dot{q}) = \tau, \tag{1}$$

where $q = (q_1,\dots,q_k)^T$ are generalized configuration coordinates for the system with k degrees of freedom, the Lagrangian L equals the difference $K - P$ between kinetic energy K and potential energy P, and $\tau = (\tau_1,\dots,\tau_k)^T$ is the vector of generalized forces acting on the system. Furthermore, $\frac{\partial L}{\partial \dot{q}}$ denotes the column-vector of partial derivatives of $L(q,\dot{q})$ with respect to the generalized velocities $\dot{q}_1,\dots,\dot{q}_k$, and similarly for $\frac{\partial L}{\partial q}$. In standard mechanical systems the kinetic energy K is of the form

$$K(q,\dot{q}) = \frac{1}{2}\dot{q}^T M(q)\dot{q} \tag{2}$$

where the $k \times k$ inertia (generalized mass) matrix $M(q)$ is symmetric and positive definite for all q. In this case the vector of generalized *momenta* $p = (p_1,\dots,p_k)^T$, defined for any Lagrangian L as $p = \frac{\partial L}{\partial \dot{q}}$, is simply given by

$$p = M(q)\dot{q}, \tag{3}$$

and by defining the state vector $(q_1,\dots,q_k,p_1,\dots,p_k)^T$ the k second-order equations (1) transform into $2k$ first-order equations

$$\dot{q} = \frac{\partial H}{\partial p}(q,p) \quad (= M^{-1}(q)p)$$

$$\dot{p} = -\frac{\partial H}{\partial q}(q,p) + \tau \tag{4}$$

where

$$H(q,p) = \frac{1}{2}p^T M^{-1}(q)p + P(q) \quad (= \frac{1}{2}\dot{q}^T M(q)\dot{q} + P(q) \ ) \tag{5}$$

is the total energy of the system. The equations (4) are called the *Hamiltonian equations* of motion, and H is called the *Hamiltonian*. The following *energy balance* immediately follows from (4):

$$\frac{d}{dt}H = \frac{\partial^T H}{\partial q}(q,p)\dot{q} + \frac{\partial^T H}{\partial p}(q,p)\dot{p} = \frac{\partial^T H}{\partial p}(q,p)\tau = \dot{q}^T\tau, \tag{6}$$

expressing that the increase in energy of the system is equal to the supplied work (*conservation of energy*).

If the potential energy is *bounded from below*, that is $\exists C > -\infty$ such that $P(q) \geq C$, then it follows that (4) with inputs $u = \tau$ and outputs $y = \dot{q}$ is a *passive* (in fact, a *lossless*) state space system with storage function $H(q, p) - C \geq 0$ (see e.g. [43,11,32] for the general theory of passive and dissipative systems). Since the energy is only defined up to a constant, we may as well take as potential energy the function $P(q) - C \geq 0$, in which case the total energy $H(q, p)$ becomes nonnegative and thus itself is the storage function.

System (4) is an example of a *Hamiltonian system* with collocated inputs and outputs, which more generally is given in the following form

$$\dot{q} = \frac{\partial H}{\partial p}(q, p), \qquad (q, p) = (q_1, \ldots, q_k, p_1, \ldots, p_k)$$

$$\dot{p} = -\frac{\partial H}{\partial q}(q, p) + B(q)u, \qquad u \in \mathbb{R}^m, \tag{7}$$

$$y = B^T(q)\frac{\partial H}{\partial p}(q, p) \quad (= B^T(q)\dot{q}), \qquad y \in \mathbb{R}^m,$$

Here $B(q)$ is the input force matrix, with $B(q)u$ denoting the generalized forces resulting from the control inputs $u \in \mathbb{R}^m$. The state space of (7) with local coordinates (q, p) is usually called the *phase space*. Normally $m < k$, in which case we speak of an *underactuated* system.

Because of the form of the output equations $y = B^T(q)\dot{q}$ we again obtain the energy balance

$$\frac{dH}{dt}(q(t), p(t)) = u^T(t)y(t) \tag{8}$$

and if H is bounded from below, any Hamiltonian system (7) is a lossless state space system. For a system-theoretic treatment of Hamiltonian systems (7), we refer to e.g. [4,29,30,6,25].

A major generalization of the class of Hamiltonian systems (7) is to consider systems which are described in local coordinates as

$$\dot{x} = J(x)\frac{\partial H}{\partial x}(x) + g(x)u, \qquad x \in \mathcal{X}, u \in \mathbb{R}^m$$

$$y = g^T(x)\frac{\partial H}{\partial x}(x), \qquad y \in \mathbb{R}^m \tag{9}$$

Here $J(x)$ is an $n \times n$ matrix with entries depending smoothly on x, which is assumed to be *skew-symmetric*

$$J(x) = -J^T(x), \tag{10}$$

and $x = (x_1, \ldots, x_n)$ are local coordinates for an n-dimensional state space manifold $\mathcal{X}$. Because of (10) we easily recover the energy-balance $\frac{dH}{dt}(x(t)) =$

$u^T(t)y(t)$, showing that (9) is lossless if $H \geq 0$. We call (9) with J satisfying (10) a *port-controlled Hamiltonian (PCH) system* with *structure matrix $J(x)$* and *Hamiltonian H* ([21,16,15]).

As an important mathematical note, we remark that in many examples the structure matrix J will satisfy the "*integrability*" conditions

$$\sum_{l=1}^{n} \left[J_{lj}(x)\frac{\partial J_{ik}}{\partial x_l}(x) + J_{li}(x)\frac{\partial J_{kj}}{\partial x_l}(x) + J_{lk}(x)\frac{\partial J_{ji}}{\partial x_l}(x) \right] = 0$$

$$i, j, k = 1, \ldots, n \qquad (11)$$

In this case we may find, by Darboux's theorem (see e.g. [14]) around any point x_0 where the rank of the matrix $J(x)$ is constant, local coordinates $\tilde{x} = (q, p, s) = (q_1, \ldots, q_k, p_1, \ldots, p_k, s_1, \ldots s_l)$, with $2k$ the rank of J and $n = 2k + l$, such that J in these coordinates takes the form

$$J = \begin{bmatrix} 0 & I_k & 0 \\ -I_k & 0 & 0 \\ 0 & 0 & 0 \end{bmatrix} \qquad (12)$$

The coordinates (q, p, s) are called *canonical* coordinates, and J satisfying (10) and (11) is called a *Poisson structure matrix*. In such canonical coordinates the equations (9) are very close to the standard Hamiltonian form (7).

PCH systems arise systematically from *network-type models* of physical systems as formalized within the (generalized) bond graph language ([28,3]). Indeed, the structure matrix $J(x)$ and the input matrix $g(x)$ may be directly associated with the network interconnection structure given by the bond graph, while the Hamiltonian H is just the sum of the energies of all the energy-storing elements; see our papers [16,21,18,22,35,36,23,31]. This is most easily exemplified by electrical circuits.

Example 1 (LCTG circuits) *Consider a controlled LC-circuit consisting of two parallel inductors with magnetic energies $H_1(\varphi_1), H_2(\varphi_2)$ (φ_1 and φ_2 being the magnetic flux linkages), in parallel with a capacitor with electric energy $H_3(Q)$ (Q being the charge). If the elements are linear then $H_1(\varphi_1) = \frac{1}{2L_1}\varphi_1^2, H_2(\varphi_2) = \frac{1}{2L_2}\varphi_2^2$ and $H_3(Q) = \frac{1}{2C}Q^2$. Furthermore let $V = u$ denote a voltage source in series with the first inductor. Using Kirchhoff's laws one immediately arrives at the dynamical equations*

$$\begin{bmatrix} \dot{Q} \\ \dot{\varphi}_1 \\ \dot{\varphi}_2 \end{bmatrix} = \underbrace{\begin{bmatrix} 0 & 1 & -1 \\ -1 & 0 & 0 \\ 1 & 0 & 0 \end{bmatrix}}_{J} \begin{bmatrix} \frac{\partial H}{\partial Q} \\ \frac{\partial H}{\partial \varphi_1} \\ \frac{\partial H}{\partial \varphi_2} \end{bmatrix} + \begin{bmatrix} 0 \\ 1 \\ 0 \end{bmatrix} u \qquad (13)$$

$$y = \frac{\partial H}{\partial \varphi_1} \qquad (= \text{current through first inductor})$$

with $H(Q, \varphi_1, \varphi_2) := H_1(\varphi_1) + H_2(\varphi_2) + H_3(Q)$ the total energy. Clearly the matrix J is skew-symmetric, and since J is constant it trivially satisfies (11). In [22] it has been shown that in this way every LC-circuit with independent elements can be modelled as a port-controlled Hamiltonian system, with the constant skew-symmetric matrix J being solely determined by the network topology (i.e., Kirchhoff's laws). Furthermore, also any LCTG-circuit with independent elements can be modelled as a PCH system, with J determined by Kirchhoff's laws and the constitutive relations of the transformers T and gyrators G. □

Another important class of PCH systems are mechanical systems as arising from *reduction by a symmetry group*, such as Euler's equations for a rigid body.

2.2   Basic properties of port-controlled Hamiltonian systems

Recall that a port-controlled Hamiltonian system is defined by a state space manifold $\mathcal{X}$ endowed with a *triple* (J, g, H). The pair $(J(x), g(x)), x \in \mathcal{X}$, captures the *interconnection structure* of the system, with $g(x)$ modeling in particular the *ports* of the system. Independently from the interconnection structure, the function $H : \mathcal{X} \to \mathbb{R}$ defines the total stored *energy* of the system.

PCH systems are intrinsically *modular* in the sense that any power-conserving interconnection of a number of PCH systems again defines a PCH system, with its overall interconnection structure determined by the interconnection structures of the composing individual PCH systems together with their power-conserving interconnection, and the Hamiltonian just the sum of the individual Hamiltonians (see [36,31,7]). The only thing which needs to be taken into account is the fact that a general power-conserving interconnection of PCH systems not always leads to a PCH system with respect to a Poisson structure $J(x)$ and input matrix $g(x)$ as above, since the interconnection may introduce *algebraic constraints* between the state variables of the individual sub-systems. Nevertheless, also in this case the resulting system still can be seen as a PCH system, which now, however, is defined with respect to a *Dirac structure*, generalizing the notion of a Poisson structure. The resulting class of *implicit* PCH systems, see e.g. [36,31,7], will be discussed in Section 4.

From the structure matrix $J(x)$ of a port-controlled Hamiltonian system one can directly extract useful information about the dynamical properties of the system. Since the structure matrix is directly related to the modeling of the system (capturing the interconnection structure) this information usually has a direct physical interpretation. A very important property is the possible existence of dynamical invariants *independent* of the Hamiltonian H. Consider the set of p.d.e.'s

$$\frac{\partial^T C}{\partial x}(x) J(x) = 0, \qquad x \in \mathcal{X}, \tag{14}$$

in the unknown (smooth) function $C : \mathcal{X} \to \mathbb{R}$. If (14) has a solution C then it follows that the time-derivative of C along the port-controlled Hamiltonian system (9) satisfies

$$\begin{aligned} \frac{dC}{dt} &= \frac{\partial^T C}{\partial x}(x)J(x)\frac{\partial H}{\partial x}(x) + \frac{\partial^T C}{\partial x}(x)g(x)u \\ &= \frac{\partial^T C}{\partial x}(x)g(x)u \end{aligned} \tag{15}$$

Hence, for the input $u = 0$, or for *arbitrary* input functions if additionally $\frac{\partial^T C}{\partial x}(x)g(x) = 0$, the function $C(x)$ *remains constant* along the trajectories of the port-controlled Hamiltonian system, *irrespective* of the precise form of the Hamiltonian H. A function $C : \mathcal{X} \to \mathbb{R}$ satisfying (14) is called a *Casimir function* (of the structure matrix $J(x)$).

It follows that the level sets $L_C := \{x \in \mathcal{X} | C(x) = c\}, c \in \mathbb{R}$, of a Casimir function C are *invariant* sets for the autonomous Hamiltonian system $\dot{x} = J(x)\frac{\partial H}{\partial x}(x)$, while the dynamics *restricted* to any level set L_C is given as the *reduced* Hamiltonian dynamics

$$\dot{x}_C = J_C(x_C)\frac{\partial H_C}{\partial x}(x_C) \tag{16}$$

with H_C and J_C the *restriction* of H, respectively J, to L_C. The existence of Casimir functions has immediate consequences for stability analysis of (9) for $u = 0$. Indeed, if $C_1, \cdots, C_r$ are Casimirs, then by (14) not only $\frac{dH}{dt} = 0$ for $u = 0$, but

$$\frac{d}{dt}\left(H + H_a(C_1, \cdots, C_r)\right)(x(t)) = 0 \tag{17}$$

for *any* function $H_a : \mathbb{R}^r \to \mathbb{R}$. Hence, if H is not positive definite at an equilibrium $x^* \in \mathcal{X}$, then $H + H_a(C_1, \cdots, C_r)$ may be rendered positive definite at x^* by a proper choice of H_a, and thus may serve as a Lyapunov function. This method for stability analysis is called the *Energy-Casimir method*, see e.g. [14].

Example 2 (Example 1 continued) *The quantity* $\phi_1 + \phi_2$ *is a Casimir function.*

2.3 Port-controlled Hamiltonian systems with dissipation

Energy-dissipation is included in the framework of port-controlled Hamiltonian systems (9) by terminating some of the ports by resistive elements. In the sequel we concentrate on PCH systems with *linear* resistive elements $u_R = -Sy_R$ for some positive semi-definite symmetric matric $S = S^T \geq 0$, where u_R and y_R are the power variables at the resistive ports. This leads to models of the form

$$\begin{aligned} \dot{x} &= [J(x) - R(x)]\frac{\partial H}{\partial x}(x) + g(x)u \\ y &= g^T(x)\frac{\partial H}{\partial x}(x) \end{aligned} \tag{18}$$

where $R(x)$ is a positive semi-definite symmetric matrix, depending smoothly on x. In this case the energy-balancing property (7) takes the form

$$
\frac{dH}{dt}(x(t)) = u^T(t)y(t) - \frac{\partial^T H}{\partial x}(x(t))R(x(t))\frac{\partial H}{\partial x}(x(t))
$$
$$
\leq u^T(t)y(t). \tag{19}
$$

showing passivity if the Hamiltonian H is bounded from below. We call (18) a *port-controlled Hamiltonian system with dissipation* (PCHD system). Note that in this case *two* geometric structures play a role: the internal power-conserving interconnection structure given by $J(x)$, and an additional resistive structure given by $R(x)$.

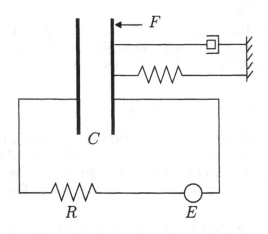

Fig. 1. Capacitor microphone

Example 3 *([24]) Consider the capacitor microphone depicted in Figure 1. Here the capacitance $C(q)$ of the capacitor is varying as a function of the displacement q of the right plate (with mass m), which is attached to a spring (with spring constant $k > 0$) and a damper (with constant $c > 0$), and affected by a mechanical force F (air pressure arising from sound). Furthermore, E is a voltage source. The dynamical equations of motion can be written* *as the PCHD system*

$$\begin{bmatrix} \dot{q} \\ \dot{p} \\ \dot{Q} \end{bmatrix} = \left(\begin{bmatrix} 0 & 1 & 0 \\ -1 & 0 & 0 \\ 0 & 0 & 0 \end{bmatrix} - \begin{bmatrix} 0 & 0 & 0 \\ 0 & c & 0 \\ 0 & 0 & \frac{1}{R} \end{bmatrix} \right) \begin{bmatrix} \frac{\partial H}{\partial q} \\ \frac{\partial H}{\partial p} \\ \frac{\partial H}{\partial Q} \end{bmatrix} + \begin{bmatrix} 0 \\ 1 \\ 0 \end{bmatrix} F + \begin{bmatrix} 0 \\ 0 \\ \frac{1}{R} \end{bmatrix} E$$

$$y_1 = \frac{\partial H}{\partial p} = \dot{q} \tag{20}$$

$$y_2 = \frac{1}{R}\frac{\partial H}{\partial Q} = I$$

with p the momentum, R the resistance of the resistor, I the current through the voltage source, and the Hamiltonian H being the total energy

$$H(q,p,Q) = \frac{1}{2m}p^2 + \frac{1}{2}k(q-\bar{q})^2 + \frac{1}{2C(q)}Q^2, \tag{21}$$

with $\bar{q}$ denoting the equilibrium position of the spring. Note that $F\dot{q}$ is the mechanical power, and EI the electrical power applied to the system. In the application as a microphone the voltage over the resistor will be used (after amplification) as a measure for the mechanical force F.

A rich class of examples of PCHD systems is provided by electro-mechanical systems such as induction motors, see e.g. [27]. In some examples the interconnection structure $J(x)$ is actually *varying*, depending on the mode of operation of the system, as is the case for power converters (see e.g. [9]) or for mechanical systems with variable constraints.

3   Control of port-controlled Hamiltonian systems with dissipation

The aim of this section is to discuss a general methodology for controlling PCH or PCHD systems which exploits their Hamiltonian properties in an intrinsic way. Since this exposition is based on ongoing recent research (see e.g. [19,39,20,27,32]) we only try to indicate its potential. An expected benefit of such a methodology is that it leads to physically interpretable controllers, which possess inherent robustness properties. Future research is aimed at corroborating these claims.

We have already seen that PCH or PCHD systems are *passive* if the Hamiltonian H is bounded from below. Hence in this case we can use all the results from the theory of passive systems, such as asymptotic stabilization by the insertion of *damping* by negative output feedback, see e.g. [32]. The emphasis in this section is however on the somewhat complementary aspect of *shaping the energy* of the system, which directly involves the Hamiltonian structure of the system, as opposed to the more general passivity structure.

3.1 Control by interconnection

Consider a port-controlled Hamiltonian system with dissipation (18) regarded as a plant system to be controlled. Recall the well-known result that the standard feedback interconnection of two passive systems again is a passive system; a basic fact which can be used for various stability and control purposes ([11,26,32]). In the same vein we consider the interconnection of the plant (18) with *another* port-controlled Hamiltonian system with dissipation

$$
C : \quad
\begin{aligned}
\dot{\xi} &= [J_C(\xi) - R_C(\xi)]\tfrac{\partial H_C}{\partial \xi}(\xi) + g_C(\xi)u_C \\
y_C &= g_C^T(\xi)\tfrac{\partial H_C}{\partial \xi}(\xi)
\end{aligned}
\qquad \xi \in \mathcal{X}_C
\tag{22}
$$

regarded as the *controller* system, via the standard feedback interconnection

$$
\begin{aligned}
u &= -y_C + e \\
u_C &= y + e_C
\end{aligned}
\tag{23}
$$

with e, e_C external signals inserted in the feedback loop. The closed-loop system takes the form

$$
\begin{bmatrix} \dot{x} \\ \dot{\xi} \end{bmatrix} = (\underbrace{\begin{bmatrix} J(x) & -g(x)g_C^T(\xi) \\ g_C(\xi)g^T(x) & J_C(\xi) \end{bmatrix}}_{J_{cl}(x,\xi)} - \underbrace{\begin{bmatrix} R(x) & 0 \\ 0 & R_C(\xi) \end{bmatrix}}_{R_{cl}(x,\xi)}) \begin{bmatrix} \frac{\partial H}{\partial x}(x) \\ \frac{\partial H_C}{\partial \xi}(\xi) \end{bmatrix}
$$

$$
+ \begin{bmatrix} g(x) & 0 \\ 0 & g_C(\xi) \end{bmatrix} \begin{bmatrix} e \\ e_C \end{bmatrix}
\tag{24}
$$

$$
\begin{bmatrix} y \\ y_C \end{bmatrix} = \begin{bmatrix} g(x) & 0 \\ 0 & g_C(\xi) \end{bmatrix} \begin{bmatrix} \frac{\partial H}{\partial x}(x) \\ \frac{\partial H_C}{\partial \xi}(\xi) \end{bmatrix}
$$

which again is a port-controlled Hamiltonian system with dissipation, with state space given by the product space $\mathcal{X} \times \mathcal{X}_C$, total Hamiltonian $H(x) + H_C(\xi)$, inputs (e, e_C) and outputs (y, y_C). Hence the feedback interconnection of any two PCHD systems results in another PCHD system; just as in the case of passivity. This is a special case of a theorem ([32]), which says that any regular power-conserving interconnection of PCHD systems defines another PCHD system.

It is of interest to investigate the *Casimir functions* of the closed-loop system, especially those relating the state variables ξ of the controller system to the state variables x of the plant system. Indeed, from a control point of view the Hamiltonian H is *given* while H_C can be *assigned*. Thus if we can find Casimir functions $C_i(\xi, x), i = 1, \cdots, r$, relating ξ to x then by the Energy-Casimir method the Hamiltonian $H + H_C$ of the closed-loop system may be replaced by the Hamiltonian $H + H_C + H_a(C_1, \cdots, C_r)$, thus creating the possibility of obtaining a suitable Lyapunov function for the closed-loop system.

Example 4 *[38] Consider the "plant" system*

$$\begin{bmatrix} \dot{q} \\ \dot{p} \end{bmatrix} = \begin{bmatrix} 0 & 1 \\ -1 & 0 \end{bmatrix} \begin{bmatrix} \frac{\partial H}{\partial q} \\ \frac{\partial H}{\partial p} \end{bmatrix} + \begin{bmatrix} 0 \\ 1 \end{bmatrix} u$$

$$y = \begin{bmatrix} 0 & 1 \end{bmatrix} \begin{bmatrix} \frac{\partial H}{\partial q} \\ \frac{\partial H}{\partial p} \end{bmatrix}$$

(25)

with q the position and p being the momentum of the mass m, in feedback interconnection ($u = -y_C + e, u_C = y$) with the controller system (see Figure 2)

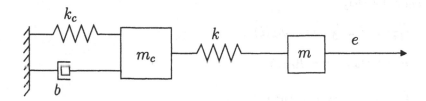

Fig. 2. Controlled mass

$$\begin{bmatrix} \Delta \dot{q}_c \\ \dot{p}_c \\ \Delta \dot{q} \end{bmatrix} = \begin{bmatrix} 0 & 1 & 0 \\ -1 & -b & 1 \\ 0 & -1 & 0 \end{bmatrix} \begin{bmatrix} \frac{\partial H_C}{\partial \Delta q_c} \\ \frac{\partial H_C}{\partial p_c} \\ \frac{\partial H_C}{\partial \Delta q} \end{bmatrix} + \begin{bmatrix} 0 \\ 0 \\ 1 \end{bmatrix} u_C$$

$$y_C = \frac{\partial H_C}{\partial \Delta q}$$

where Δq_c is the displacement of the spring k_c, Δq is the displacement of the spring k, and p_c is the momentum of the mass m_c. The plant Hamiltonian is $H(p) = \frac{1}{2m}p^2$, and the controller Hamiltonian is given as $H_C(\Delta q_c, p_c, \Delta q) = \frac{1}{2}(\frac{p_c^2}{m_c} + k(\Delta q)^2 + k_c(\Delta q_c)^2)$. The variable $b > 0$ is the damping constant, and e is an external force. The closed-loop system possesses the Casimir function

$$C(q, \Delta q_c, \Delta q) = \Delta q - (q - \Delta q_c), \tag{26}$$

implying that along the solutions of the closed-loop system

$$\Delta q = q - \Delta q_c + c \tag{27}$$

with c a constant depending on the initial conditions. With the help of LaSalle's Invariance principle it can be shown that restricted to the invariant manifolds (27) the system is asymptotically stable for the equilibria $q = \Delta q_c = p = p_c = 0$. □

As a special case (see [32] for a more general discussion) let us consider Casimir functions for (24) of the form

$$\xi_i - G_i(x) \quad , \quad i = 1, \ldots, \dim \mathcal{X}_C = n_C \tag{28}$$

That means that we are looking for solutions of the p.d.e.'s (with e_i denoting the i-th basis vector)

$$\left[-\frac{\partial^T G_i}{\partial x}(x) \; e_i^T \right] \begin{bmatrix} J(x) - R(x) & -g(x)g_C^T(\xi) \\ g_C(\xi)g^T(x) & J_C(\xi) - R_C(\xi) \end{bmatrix} = 0$$

for $i = 1, \ldots, n_C$, relating all the controller state variables $\xi_1, \ldots, \xi_{n_C}$ to the plant state variables x. Denoting $G = (G_1, \ldots, G_{n_C})^T$ this means ([32]) that G should satisfy

$$\frac{\partial^T G}{\partial x}(x) J(x) \frac{\partial G}{\partial x}(x) = J_C(\xi)$$

$$R(x) \frac{\partial G}{\partial x}(x) = 0 = R_C(\xi) \tag{29}$$

$$\frac{\partial^T G}{\partial x}(x) J(x) = g_C(\xi) g^T(x)$$

In this case the reduced dynamics on any multi-level set

$$L_C = \{(x, \xi) | \xi_i = G_i(x) + c_i, i = 1, \ldots n_C\} \tag{30}$$

can be immediately recognized ([32]) as the PCHD system

$$\dot{x} = [J(x) - R(x)] \frac{\partial H_s}{\partial x}(x), \tag{31}$$

with the same interconnection and dissipation structure as before, but with *shaped* Hamiltonian H_s given by

$$H_s(x) = H(x) + H_C(G(x) + c). \tag{32}$$

In the context of actuated mechanical systems this amounts to the shaping of the *potential energy* as in the classical paper [42], see [32].

A direct interpretation of the shaped Hamiltonian H_s in terms of *energy-balancing* is obtained as follows. Since $R_C(\xi) = 0$ by (29) the controller Hamiltonian H_C satisfies $\frac{dH_C}{dt} = u_C^T y_C$. Hence along any multi-level set L_C given by (30) $\frac{dH_s}{dt} = \frac{dH}{dt} + \frac{dH_C}{dt} = \frac{dH}{dt} - u^T y$, since $u = -y_C$ and $u_C = y$. Therefore, up to a constant,

$$H_s(x(t)) = H(x(t)) - \int_0^t u^T(\tau) y(\tau) d\tau, \tag{33}$$

and the shaped Hamiltonian H_s is the original Hamiltonian H *minus the energy* supplied to the plant system (18) by the controller system (22). From a stability analysis point of view (33) can be regarded as an effective way of *generating* candidate Lyapunov functions H_s from the Hamiltonian H.

3.2 Passivity-based control of port-controlled Hamiltonian systems with dissipation

In the previous section we have seen how under certain conditions the feedback interconnection of a PCHD system having Hamiltonian H (the "plant") with another PCHD system with Hamiltonian H_C (the "controller") leads to a reduced dynamics given by (31) for the shaped Hamiltonian H_s. From a *state feedback* point of view the dynamics (31) could have been directly obtained by a state feedback $u = \alpha(x)$ such that

$$g(x)\alpha(x) = [J(x) - R(x)]\frac{\partial H_C(G(x) + c)}{\partial x} \tag{34}$$

Indeed, such an $\alpha(x)$ is given in explicit form as

$$\alpha(x) = -g_C^T(G(x) + c)\frac{\partial H_C}{\partial \xi}(G(x) + c) \tag{35}$$

The state feedback $u = \alpha(x)$ is customarily called a *passivity-based control law*, since it is based on the passivity properties of the original plant system (18) and transforms (18) into *another* passive system with *shaped* storage function (in this case H_s).
Seen from this perspective we have shown in the previous section that the passivity-based state feedback $u = \alpha(x)$ satisfying (34) *can be derived* from the interconnection of the PCHD plant system (18) with a PCHD controller system (22). This fact has some favorable consequences. Indeed, it implies that the passivity-based control law defined by (34) can be equivalently *generated* as the feedback interconnection of the passive system (18) with another passive system (22). In particular, this implies an inherent *invariance* property of the controlled system: the plant system (18), the controller system (32), as well as any other passive system interconnected to (18) in a power-conserving fashion, may change in any way as long as they remain passive, and for any perturbation of this kind the controlled system will remain stable. For a further discussion of passivity-based control from this point of view we refer to [27].

3.3 Interconnection and damping assignment passivity-based control

A further generalization of the previous subsection is to use state feedback in order to *change* the interconnection structure and the resistive structure of the plant system, and thereby to create more flexibility to shape the storage function for the (modified) port-controlled Hamiltonian system to a desired form. This methodology has been called Interconnection-Damping Assignment Passivity-Based Control (IDA-PBC) in [27], and has been succesfully applied to a number of applications. The method is especially attractive if the newly assigned interconnection and resistive structures are judiciously chosen on the basis of physical considerations, and represent some "ideal" interconnection and resistive structures for the physical plant. For an extensive treatment of IDA-PBC we refer to [27].

4   Physical systems with algebraic constraints

From a general modeling point of view physical systems are, at least in first instance, often described by DAE's, that is, a mixed set of differential and *algebraic* equations. This stems from the fact that in many modelling approaches the system under consideration is naturally regarded as obtained from interconnecting simpler sub-systems. These interconnections in general, give rise to algebraic constraints between the state space variables of the sub-systems; thus leading to implicit systems. While in the linear case one may argue that it is often relatively straightforward to eliminate the algebraic constraints, and thus to reduce the system to an *explicit* form without constraints, in the nonlinear case such a conversion from implicit to explicit form is usually fraught with difficulties. Indeed, if the algebraic constraints are nonlinear then they need not be analytically solvable (locally or globally). More importantly perhaps, even if they are analytically solvable, then often one would prefer *not* to eliminate the algebraic constraints, because of the complicated and physically not easily interpretable expressions for the reduced system which may arise.

4.1   Power-conserving interconnections

In order to geometrically describe network models of physical systems we first consider the notion of a *Dirac structure*, formalizing the concept of a power-conserving interconnection. Let $\mathcal{F}$ be an ℓ-dimensional linear space, and denote its dual (the space of linear functions on $\mathcal{F}$) by $\mathcal{F}^*$. The product space $\mathcal{F} \times \mathcal{F}^*$ is considered to be the space of power variables, with *power* intrinsically defined by

$$P = <f^*|f>, \quad (f, f^*) \in \mathcal{F} \times \mathcal{F}^*, \tag{36}$$

where $<f^*|f>$ denotes the duality product, that is, the linear function $f^* \in \mathcal{F}^*$ acting on $f \in \mathcal{F}$. Often we call $\mathcal{F}$ the space of *flows* f, and $\mathcal{F}^*$ the space of *efforts* e, with the power of an element $(f, e) \in \mathcal{F} \times \mathcal{F}^*$ denoted as $<e|f>$.

Remark 1 *If $\mathcal{F}$ is endowed with an inner product structure $<,>$, then $\mathcal{F}^*$ can be naturally identified with $\mathcal{F}$ in such a way that $<e|f> = <e, f>$, $f \in \mathcal{F}$, $e \in \mathcal{F}^* \simeq \mathcal{F}$.*

Example 5 *Let $\mathcal{F}$ be the space of generalized velocities, and $\mathcal{F}^*$ be the space of generalized forces, then $< e|f >$ is mechanical power. Similarly, let $\mathcal{F}$ be the space of currents, and $\mathcal{F}^*$ be the space of voltages, then $< e|f >$ is electrical power.*

There exists on $\mathcal{F} \times \mathcal{F}^*$ a canonically defined symmetric bilinear form

$$< (f_1, e_1), (f_2, e_2) >_{\mathcal{F} \times \mathcal{F}^*} := < e_1|f_2 > + < e_2|f_1 > \tag{37}$$

for $f_i \in \mathcal{F}$, $e_i \in \mathcal{F}^*, i = 1, 2$. Now consider a linear subspace $S \subset \mathcal{F} \times \mathcal{F}^*$, and its orthogonal complement with respect to the bilinear form $<, >_{\mathcal{F} \times \mathcal{F}^*}$ on $\mathcal{F} \times \mathcal{F}^*$, denoted as $S^\perp \subset \mathcal{F} \times \mathcal{F}^*$. Clearly, if S has dimension d, then the subspace $S^\perp$ has dimension $2\ell - d$. (Since dim $(\mathcal{F} \times \mathcal{F}^*) = 2\ell$, and $<, >_{\mathcal{F} \times \mathcal{F}^*}$ is a non-degenerate form.)

Definition 1 *[5,8,7] A constant Dirac structure on $\mathcal{F}$ is a linear subspace $\mathcal{D} \subset \mathcal{F} \times \mathcal{F}^*$ such that*

$$\mathcal{D} = \mathcal{D}^\perp \tag{38}$$

It immediately follows that the dimension of any Dirac structure $\mathcal{D}$ on an ℓ-dimensional linear space is equal to ℓ. Furthermore, let $(f, e) \in \mathcal{D} = \mathcal{D}^\perp$. Then by (37)

$$0 = < (f, e), (f, e) >_{\mathcal{F} \times \mathcal{F}^*} = 2 < e|f > . \tag{39}$$

Thus for all $(f, e) \in \mathcal{D}$ we obtain $< e|f >= 0$; and hence any Dirac structure $\mathcal{D}$ on $\mathcal{F}$ defines a power-conserving relation between the power variables $(f, e) \in \mathcal{F} \times \mathcal{F}^*$.

Remark 2 *The property dim $\mathcal{D} =$ dim $\mathcal{F}$ is intimately related to the usually expressed statement that a physical interconnection can not determine at the same time both the flow and effort (e.g. current and voltage, or velocity and force).*

Constant Dirac structures admit different *matrix representations*. Here we just list three of them, without giving proofs and algorithms to convert one representation into another, see e.g. [7].
Let $\mathcal{D} \subset \mathcal{F} \times \mathcal{F}^*$, with dim $\mathcal{F} = \ell$, be a constant Dirac structure. Then $\mathcal{D}$ can be represented as

1. (*Kernel and Image representation*, [7,35]).

$$\mathcal{D} = \{(f, e) \in \mathcal{F} \times \mathcal{F}^* | Ff + Ee = 0\} \tag{40}$$

   for $\ell \times \ell$ matrices F and E satisfying

   (i) $EF^T + FE^T = 0$

   (ii) rank $[F \vdots E] = \ell$
$$\tag{41}$$

Equivalently,

$$\mathcal{D} = \{(f, e) \in \mathcal{F} \times \mathcal{F}^* | f = E^T \lambda, \ e = F^T \lambda, \ \lambda \in \mathbb{R}^\ell\} \tag{42}$$

2. (*Constrained input-output representation*, [7]).

$$\mathcal{D} = \{(f, e) \in \mathcal{F} \times \mathcal{F}^* | f = -Je + G\lambda, \ G^T e = 0\} \tag{43}$$

for an $\ell \times \ell$ skew-symmetric matrix J, and a matrix G such that $\mathrm{Im} G = \{f | (f, 0) \in \mathcal{D}\}$. Furthermore, $\mathrm{Ker} J = \{e | (0, e) \in \mathcal{D}\}$.

3. (*Canonical coordinate representation*, [5]).
There exist linear coordinates (q, p, r, s) for $\mathcal{F}$ such that in these coordinates and dual coordinates for $\mathcal{F}^*$, $(f, e) = (f_q, f_p, f_r, f_s, e_q, e_p, e_r, e_s) \in \mathcal{D}$ if and only if

$$\begin{cases} f_q = e_p, \ f_p = -e_q \\ \\ f_r = 0, \ e_s = 0 \end{cases} \tag{44}$$

Example 6 *Kirchhoff's laws are a special case of (40). By taking $\mathcal{F}$ the space of currents and $\mathcal{F}^*$ the space of voltages, Kirchhoff's current laws determine a subspace V of $\mathcal{F}$, while Kirchhoff's voltage laws determine the orthogonal subspace V^{orth} of $\mathcal{F}^*$. Hence, the Dirac structure determined by Kirchhoff's laws is given as $V \times V^{orth} \subset \mathcal{F} \times \mathcal{F}^*$, with kernel representation of the form*

$$\mathcal{D} = \{(f, e) \in \mathcal{F} \times \mathcal{F}^* | Ff = 0, Ee = 0\}, \tag{45}$$

for suitable matrices F and E (consisting only of elements $+1$, -1 and 0), such that $\mathrm{Ker}\, F = V$ and $\mathrm{Ker}\, E = V^{orth}$. In this case the defining property $\mathcal{D} = \mathcal{D}^\perp$ of the Dirac structure amounts to Tellegen's theorem.

Example 7 *Any skew-symmetric map $J : \mathcal{F}^* \to \mathcal{F}$ defines the Dirac structure*

$$\mathcal{D} = \{(f, e) \in \mathcal{F} \times \mathcal{F}^* | f = -Je\}, \tag{46}$$

as a special case of (43). Furthermore, any interconnection structure (J, g) with J skew-symmetric defines a Dirac structure given in hybrid input-output representation as

$$\begin{bmatrix} f_S \\ e_P \end{bmatrix} = \begin{bmatrix} -J & -g \\ g^T & 0 \end{bmatrix} \begin{bmatrix} e_S \\ f_P \end{bmatrix} \tag{47}$$

Given a Dirac structure $\mathcal{D}$ on $\mathcal{F}$, the following subspaces of $\mathcal{F}$, respectively $\mathcal{F}^*$, will shown to be of importance in the next section

$$G_1 := \{f \in \mathcal{F} \mid \exists e \in \mathcal{F}^* \text{ s.t. } (f, e) \in \mathcal{D}\} \tag{48}$$

$$P_1 := \{e \in \mathcal{F}^* \mid \exists f \in \mathcal{F} \text{ s.t. } (f, e) \in \mathcal{D}\}$$

The subspace G_1 expresses the set of admissible flows, and P_1 the set of admissible efforts. In the image representation (42) they are given as

$$G_1 = \mathrm{Im}\, E^T, \qquad P_1 = \mathrm{Im}\, F^T. \tag{49}$$

4.2   Implicit port-controlled Hamiltonian systems

From a network modeling perspective, see e.g. [28,3], a (lumped-parameter) physical system is directly described by a set of (possibly multi-dimensional) *energy-storing* elements, a set of *energy-dissipating* or *resistive* elements, and a set of *ports* (by which interaction with the environment can take place), interconnected to each other by a *power-conserving interconnection*, see Figure 3. Associated with the energy-storing elements are energy-variables

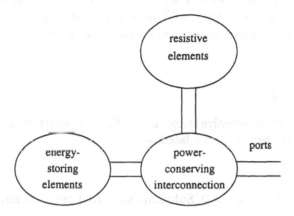

Fig. 3. Network model of physical systems

$x_1, \cdots, x_n$, being coordinates for some n-dimensional state space manifold $\mathcal{X}$, and a total energy $H : \mathcal{X} \to \mathbb{R}$. The power-conserving interconnection also includes power-conserving elements like (in the electrical domain) transformers, gyrators, or (in the mechanical domain) transformers, kinematic pairs and kinematic constraints. In first instance (see later on for the non-constant case) the power-conserving interconnection will be formalized by a constant Dirac structure on a finite-dimensional linear space $\mathcal{F} := \mathcal{F}_S \times \mathcal{F}_R \times \mathcal{F}_P$, with $\mathcal{F}_S$ denoting the space of flows f_S connected to the energy-storing elements, $\mathcal{F}_R$ denoting the space of flows f_R connected to the dissipative (resistive) elements, and $\mathcal{F}_P$ the space of external flows f_P which can be connected to the environment. Dually, we write $\mathcal{F}^* = \mathcal{F}_S^* \times \mathcal{F}_R^* \times \mathcal{F}_P^*$, with $e_S \in \mathcal{F}_S^*$ the efforts connected to the energy-storing elements, $e_R \in \mathcal{F}_R^*$ the efforts connected to the resistive elements, and $e_P \in \mathcal{F}_P^*$ the efforts to be connected to the environment of the system.

In kernel representation, the Dirac structure on $\mathcal{F} = \mathcal{F}_S \times \mathcal{F}_R \times \mathcal{F}_P$ is given as

$$\mathcal{D} = \{(f_S, f_R, f_P, e_S, e_R, e_P) \,| $$

$$F_S f_S + E_S e_S + F_R f_R + E_R e_R + F_P f_P + E_P e_P = 0\} \tag{50}$$

for certain matrices $F_S, E_S, F_R, E_R, F_P, E_P$ satisfying

(i) $\quad E_S F_S^T + F_S E_S^T + E_R F_R^T + F_R E_R^T + E_P F_P^T + F_P E_P^T = 0$

$$(51)$$

(ii) $\quad \text{rank} \left[F_S \vdots F_R \vdots F_P \vdots E_S \vdots E_R \vdots E_P \right] = \dim \mathcal{F}$

The flow variables of the energy-storing elements are given as $\dot{x}(t) = \frac{dx}{dt}(t), t \in \mathbb{R}$, and the effort variables of the energy-storing elements as $\frac{\partial H}{\partial x}(x(t))$ (implying that $< \frac{\partial H}{\partial x}(x(t))|\dot{x}(t) >= \frac{dH}{dt}(x(t))$ is the increase in energy). In order to have a consistent sign convention for energy flow we put

$$f_S = -\dot{x}$$
$$(52)$$
$$e_S = \frac{\partial H}{\partial x}(x)$$

Restricting to *linear* resistive elements, the flow and effort variables connected to the resistive elements are related as

$$f_R = -S e_R \qquad (53)$$

for some matrix $S = S^T \geq 0$. Substitution of (52) and (53) into (50) yields

$$-F_S \dot{x}(t) + E_S \frac{\partial H}{\partial x}(x(t)) - F_R S e_R + E_R e_R + F_P f_P + E_P e_P = 0 \qquad (54)$$

with $F_S, E_S, F_R, E_R, F_P, E_P$ satisfying (51). We call (54) an *implicit port-controlled Hamiltonian system with dissipation*, defined with respect to the constant Dirac structure $\mathcal{D}$, the Hamiltonian H, and the resistive structure S.

Actually, for many purposes this definition of an implicit PCHD system is not general enough, since often the Dirac structure is not constant, but *modulated* by the state variables x. In this case the matrices $F_S, E_S, F_R, E_R, F_P, E_P$ depend (smoothly) on x, leading to the implicit PCHD system

$$-F_S(x(t))\dot{x}(t) + E_S(x(t))\frac{\partial H}{\partial x}(x(t)) - F_R(x(t))S e_R(t)$$
$$(55)$$
$$+E_R(x(t))e_R(t) + F_P(x(t))f_P(t) + E_P(x(t))e_P(t) = 0, \quad t \in \mathbb{R}$$

with

$$E_S(x)F_S^T(x) + F_S(x)E_S^T(x) + E_R(x)F_R^T(x) + F_R(x)E_R^T(x)$$
$$+ E_P(x)F_P^T(x) + F_P(x)E_P^T(x) = 0, \quad \forall x \in \mathcal{X}$$
$$(56)$$

$$\text{rank} \left[F_S(x) \vdots F_R(x) \vdots F_P(x) \vdots E_S(x) \vdots E_R(x) \vdots E_P(x) \right] = \dim \mathcal{F}$$

Remark 3 *Strictly speaking the flow and effort variables $\dot{x}(t) = -f_S(t)$, respectively $\frac{\partial H}{\partial x}(x(t)) = e_S(t)$, are not living in the constant linear space $\mathcal{F}_S$, respectively $\mathcal{F}_S^*$, but instead in the tangent spaces $T_{x(t)}\mathcal{X}$, respectively cotangent spaces $T_{x(t)}^*\mathcal{X}$, to the state space manifold $\mathcal{X}$. This is formalized in the definition of a non-constant Dirac structure on a manifold; see [5,8,7,32].*

By the power-conservation property of a Dirac structure (cf. (39)) it follows directly that any implicit PCHD system satisfies the energy-inequality

$$\frac{dH}{dt}(x(t)) = \;< \frac{\partial H}{\partial x}(x(t))|\dot{x}(t) >=$$

$$= -e_R^T(t)Se_R(t) + e_P^T(t)f_P(t) \le e_P^T(t)f_P(t), \tag{57}$$

showing passivity if $H \ge 0$. The *algebraic constraints* that are present in the implicit system (55) are expressed by the subspace P_1, and the Hamiltonian H. In fact, since the Dirac structure $\mathcal{D}$ is modulated by the x-variables, also the subspace P_1 is modulated by the x-variables, and thus the effort variables e_S, e_R and e_P necessarily satisfy $(e_S, e_R, e_P) \in P_1(x)$, $x \in \mathcal{X}$, and thus, because of (49),

$$e_S \in \text{Im } F_S^T(x), e_R \in \text{Im } F_R^T(x), e_P \in \text{Im } F_P^T(x). \tag{58}$$

The second and third inclusions entail the expression of e_R and e_P in terms of the other variables, while the first inclusion determines, since $e_S = \frac{\partial H}{\partial x}(x)$, the following algebraic constraints on the state variables

$$\frac{\partial H}{\partial x}(x) \in \text{Im } F_S^T(x). \tag{59}$$

The *Casimir functions* $C : \mathcal{X} \to \mathbb{R}$ of the implicit system (55) are determined by the subspace $G_1(x)$. Indeed, necessarily $(f_S, f_R, f_P) \in G_1(x)$, and thus by (49)

$$f_S \in \text{Im } E_S^T(x), f_R \in \text{Im } E_R^T(x), f_P \in \text{Im } E_P^T(x). \tag{60}$$

Since $f_S = -\dot{x}(t)$, the first inclusion yields the *flow constraints* $\dot{x}(t) \in \text{Im } E_S^T(x(t))$, $t \in \mathbb{R}$. Thus $C : \mathcal{X} \to \mathbb{R}$ is a Casimir function if $\frac{dC}{dt}(x(t)) = \frac{\partial^T C}{\partial x}(x(t))\dot{x}(t) = 0$ for all $\dot{x}(t) \in \text{Im } E_S^T(x(t))$. Hence $C : \mathcal{X} \to \mathbb{R}$ is a Casimir of the implicit PCHD system (54) if it satisfies the set of p.d.e.'s

$$\frac{\partial C}{\partial x}(x) \in \text{Ker } E_S(x) \tag{61}$$

Remark 4 *Note that $C : \mathcal{X} \to \mathbb{R}$ satisfying (61) is a Casimir function of (54) in a strong sense: it is a dynamical invariant $\left(\frac{dC}{dt}(x(t)) = 0\right)$ for every port behavior and every resistive relation (53).*

Example 8 *[7,36,35] Consider a mechanical system with k degrees of freedom, locally described by k configuration variables $q = (q_1, \ldots, q_k)$. Suppose that there are constraints on the generalized velocities $\dot{q}$, described as $A^T(q)\dot{q} = 0$, with $A(q)$ a $r \times k$ matrix of rank r everywhere (that is, there are r independent kinematic constraints). This leads to the following constrained Hamiltonian equations*

$$\dot{q} = \frac{\partial H}{\partial p}(q, p)$$

$$\dot{p} = -\frac{\partial H}{\partial q}(q, p) + A(q)\lambda + B(q)u$$

$$y = B^T(q)\frac{\partial H}{\partial p}(q, p) \tag{62}$$

$$0 = A^T(q)\frac{\partial H}{\partial p}(q, p)$$

where $B(q)u$ are the external forces (controls) applied to the system, for some $k \times m$ matrix $B(q)$, while $A(q)\lambda$ are the constraint forces. The Lagrange multipliers $\lambda(t)$ are uniquely determined by the requirement that the constraints $A^T(q(t))\dot{q}(t) = 0$ have to be satisfied for all t. One way of proceeding with these equations is to eliminate the constraint forces, and to reduce the equations of motion to the constrained state space $\mathcal{X}_c = \{(q, p) \mid A^T(q)\frac{\partial H}{\partial p}(q, p) = 0\}$, thereby obtaining an (explicit) port-controlled Hamiltonian system; see [34]. An alternative, and more direct, approach is to view the constrained Hamiltonian equations (62) as an implicit port-controlled Hamiltonian system with respect to the Dirac structure $\mathcal{D}$, given in constrained input-output representation (43) by

$$\mathcal{D} = \{(f_S, f_P, e_S, e_P)|0 = A^T(q)e_S, \ e_P = B^T(q)e_S,$$

$$-f_S = \begin{bmatrix} 0 & I_k \\ -I_k & 0 \end{bmatrix} e_S + \begin{bmatrix} 0 \\ A(q) \end{bmatrix} \lambda + \begin{bmatrix} 0 \\ B(q) \end{bmatrix} f_P, \ \lambda \in \mathbb{R}^r\} \tag{63}$$

In this case, the algebraic constraints on the state variables (q, p) are given as $A^T(q)\frac{\partial H}{\partial p}(q, p) = 0$, while the Casimir functions C are determined by the equations

$$\frac{\partial^T C}{\partial q}(q)\dot{q} = 0, \quad \text{for all } \dot{q} \text{ satisfying } A^T(q)\dot{q} = 0. \tag{64}$$

Hence, finding Casimir functions amounts to integrating the kinematic constraints $A^T(q)\dot{q} = 0$.

Remark 5 *For a proper notion of integrability of non-constant Dirac structures, generalizing the Jacobi identity for the structure matrix $J(x)$, we refer e.g. to [7]. For example, the Dirac structure (63) is integrable if and only if the kinematic constraints are* holonomic.

In principle, the theory presented in Section 3 for stabilization of *explicit* port-controlled Hamiltonian systems can be directly extended, mutatis mutandis, to *implicit* port-controlled Hamiltonian system. In particular, the standard feedback interconnection of an implicit port-controlled Hamiltonian system P with port variables f_P, e_P (the "plant") with another implicit port-controlled Hamiltonian system with port variables f_P^C, e_P^C (the "controller"), via the interconnection relations

$$f_P = -e_P^C + f^{\text{ext}}$$

$$f_P^C = e_P + e^{\text{ext}} \tag{65}$$

is readily seen to result in a closed-loop implicit port-controlled Hamiltonian system with port variables $f^{\text{ext}}, e^{\text{ext}}$. Furthermore, as in the explicit case, the Hamiltonian of this closed-loop system is just the *sum* of the Hamiltonian of the plant PCHD system and the Hamiltonian of the controller PCHD system. Finally, the Casimir analysis for the closed-loop system can be performed along the same lines as before.

5   Conclusions and future research

We have shown how network modelling of (lumped-parameter) physical systems, e.g. using bond graphs, leads to a mathematically well-defined class of open dynamical systems, which are called port-controlled Hamiltonian systems (with dissipation). Furthermore, we have tried to emphasize that this definition is completely *modular*, in the sense that any power-conserving interconnection of these systems defines a system in the same class, with overall interconnection structure defined by the individual interconnection structures, together with the power-conserving interconnection.

Clearly, the theory presented in this paper opens up the way for many other control and design problems than the stabilization problem as briefly discussed in the present paper. Its potential for set-point regulation has already received some attention (see [19,20,27,32]), while the extension to *tracking problems* is wide open. In this context we also like to refer to some recent work concerned with the shaping of the *Lagrangian*, see e.g. [2]. Also, the control of mechanical systems with nonholonomic kinematic constraints can be fruitfully approached from this point of view, see e.g. [10], as well as the modelling and control of multi-body systems, see [18,23,40]. The framework of PCHD systems seems perfectly suited to theoretical investigations on the topic of *impedance control*; see already [38] for some initial results in this direction. Also the connection with multi-modal (*hybrid*) systems, corresponding to PCHD systems with varying interconnection structure [9], needs further investigations. Finally, our current research is concerned with the formulation of *distributed parameter systems* as port-controlled Hamiltonian systems, see [17], and applications in tele-manipulation [41] and smart structures [37].

References

1. A.M. Bloch & P.E. Crouch, "Representations of Dirac structures on vector spaces and nonlinear *LC* circuits", Proc. Symposia in Pure Mathematics, Differential Geometry and Control Theory, G. Ferreyra, R. Gardner, H. Hermes, H. Sussmann, eds., Vol. 64, pp. 103-117, AMS, 1999.

2. A. Bloch, N. Leonard & J.E. Marsden, "Matching and stabilization by the method of controlled Lagrangians", in Proc. 37th IEEE Conf. on Decision and Control, Tampa, FL, pp. 1446-1451, 1998.

3. P.C. Breedveld, *Physical systems theory in terms of bond graphs*, PhD thesis, University of Twente, Faculty of Electrical Engineering, 1984

4. R.W. Brockett, "Control theory and analytical mechanics", in *Geometric Control Theory*, (eds. C. Martin, R. Hermann), Vol. VII of Lie Groups: History, Frontiers and Applications, Math. Sci. Press, Brookline, pp. 1-46, 1977.

5. T.J. Courant, "Dirac manifolds", *Trans. American Math. Soc.*, 319, pp. 631-661, 1990.

6. P.E. Crouch & A.J. van der Schaft, *Variational and Hamiltonian Control Systems*, Lect. Notes in Control and Inf. Sciences 101, Springer-Verlag, Berlin, 1987.

7. M. Dalsmo & A.J. van der Schaft, "On representations and integrability of mathematical structures in energy-conserving physical systems", *SIAM J. Control and Optimization*, 37, pp. 54-91, 1999.

8. I. Dorfman, *Dirac Structures and Integrability of Nonlinear Evolution Equations*, John Wiley, Chichester, 1993.

9. G. Escobar, A.J. van der Schaft & R. Ortega, "A Hamiltonian viewpoint in the modelling of switching power converters", *Automatica*, Special Issue on Hybrid Systems, 35, pp. 445-452, 1999.

10. K. Fujimoto, T. Sugie, "Stabilization of a class of Hamiltonian systems with nonholonomic constraints via canonical transformations", Proc. European Control Conference '99, Karlsruhe, 31 August - 3 September 1999.

11. D.J. Hill & P.J. Moylan, "Stability of nonlinear dissipative systems," *IEEE Trans. Aut. Contr.*, AC-21, pp. 708-711, 1976.

12. A. Isidori, *Nonlinear Control Systems* (2nd Edition), Communications and Control Engineering Series, Springer-Verlag, London, 1989, 3rd Edition, 1995.

13. R. Lozano, B. Brogliato, O. Egeland and B. Maschke, *Dissipative systems*, Communication and Control Engineering series, Springer, London, March 2000.

14. J.E. Marsden & T.S. Ratiu, *Introduction to Mechanics and Symmetry*, Texts in Applied Mathematics 17, Springer-Verlag, New York, 1994.

15. B.M. Maschke, *Interconnection and structure of controlled Hamiltonian systems: a network approach*, (in French), Habilitation Thesis, No.345, Dec. 10, 1998, University of Paris-Sud , Orsay, France.

16. B.M. Maschke, A.J. van der Schaft, "An intrinsic Hamiltonian formulation of network dynamics: non-standard Poisson structures and gyrators", J. Franklin Institute, vol. 329, no.5, pp. 923-966, 1992.

17. B.M. Maschke, A.J. van der Schaft, "Port controlled Hamiltonian representation of distributed parameter systems", Proc. IFAC Workshop on Lagrangian and Hamiltonian methods for nonlinear control, Princeton University, March 16-18, pp. 28-38, 2000.

18. B.M. Maschke, C. Bidard & A.J. van der Schaft, "Screw-vector bond graphs for the kinestatic and dynamic modeling of multibody systems", in Proc. ASME Int. Mech. Engg. Congress, 55-2, Chicago, U.S.A., pp. 637-644, 1994.
19. B.M. Maschke, R. Ortega & A.J. van der Schaft, "Energy-based Lyapunov functions for forced Hamiltonian systems with dissipation", in Proc. 37th IEEE Conference on Decision and Control, Tampa, FL, pp. 3599-3604, 1998.
20. B.M. Maschke, R. Ortega, A.J. van der Schaft & G. Escobar, "An energy-based derivation of Lyapunov functions for forced systems with application to stabilizing control", in Proc. 14th IFAC World Congress, Beijing, Vol. E, pp. 409-414, 1999.
21. B.M. Maschke & A.J. van der Schaft, "Port-controlled Hamiltonian systems: Modelling origins and system-theoretic properties", in Proc. 2nd IFAC NOL-COS, Bordeaux, pp. 282-288, 1992.
22. B.M. Maschke, A.J. van der Schaft & P.C. Breedveld, "An intrinsic Hamiltonian formulation of the dynamics of LC-circuits, *IEEE Trans. Circ. and Syst.*, CAS-42, pp. 73-82, 1995.
23. B.M. Maschke & A.J. van der Schaft, "Interconnected Mechanical Systems, Part II: The Dynamics of Spatial Mechanical Networks", in *Modelling and Control of Mechanical Systems*, (eds. A. Astolfi, D.J.N. Limebeer, C. Melchiorri, A. Tornambe, R.B. Vinter), pp. 17-30, Imperial College Press, London, 1997.
24. J.I. Neimark & N.A. Fufaev, *Dynamics of Nonholonomic Systems*, Vol. 33 of Translations of Mathematical Monographs, American Mathematical Society, Providence, Rhode Island, 1972.
25. H. Nijmeijer & A.J. van der Schaft, *Nonlinear Dynamical Control Systems*, Springer-Verlag, New York, 1990.
26. R. Ortega, A. Loria, P.J. Nicklasson & H. Sira-Ramirez, *Passivity-based Control of Euler-Lagrange Systems*, Springer-Verlag, London, 1998.
27. R. Ortega, A.J. van der Schaft, B.M. Maschke & G. Escobar, "Interconnection and damping assignment passivity-based control of port-controlled Hamiltonian systems", 1999, submitted for publication.
28. H. M. Paynter, *Analysis and design of engineering systems*, M.I.T. Press, MA, 1960.
29. A.J. van der Schaft, *System theoretic properties of physical systems*, CWI Tract 3, CWI, Amsterdam, 1984.
30. A.J. van der Schaft, "Stabilization of Hamiltonian systems", *Nonl. An. Th. Math. Appl.*, 10, pp. 1021-1035, 1986.
31. A.J. van der Schaft, "Interconnection and geometry", in *The Mathematics of Systems and Control, From Intelligent Control to Behavioral Systems* (eds. J.W. Polderman, H.L. Trentelman), Groningen, 1999.
32. A.J. van der Schaft, *L₂-Gain and Passivity Techniques in Nonlinear Control*, 2nd revised and enlarged edition, Springer-Verlag, Springer Communications and Control Engineering series, p. xvi+249, London, 2000 (first edition Lect. Notes in Control and Inf. Sciences, vol. 218, Springer-Verlag, Berlin, 1996).
33. A.J. van der Schaft, "Port-controlled Hamiltonian systems: Towards a theory for control and design of nonlinear physical systems", J. of the Society of Instrument and Control Engineers of Japan (SICE), vol. 39, no.2, pp. 91-98, 2000.
34. A.J. van der Schaft & B.M. Maschke, "On the Hamiltonian formulation of nonholonomic mechanical systems", *Rep. Math. Phys.*, 34, pp. 225-233, 1994.

35. A.J. van der Schaft & B.M. Maschke, "The Hamiltonian formulation of energy conserving physical systems with external ports", *Archiv für Elektronik und Übertragungstechnik*, 49, pp. 362-371, 1995.

36. A.J. van der Schaft & B.M. Maschke, "Interconnected Mechanical Systems, Part I: Geometry of Interconnection and implicit Hamiltonian Systems", in *Modelling and Control of Mechanical Systems*, (eds. A. Astolfi, D.J.N. Limebeer, C. Melchiorri, A. Tornambe, R.B. Vinter), pp. 1-15, Imperial College Press, London, 1997.

37. K. Schlacher, A. Kugi, "Control of mechanical structures by piezoelectric actuators and sensors". In *Stability and Stabilization of Nonlinear Systems*, eds. D. Aeyels, F. Lamnabhi-Lagarrigue, A.J. van der Schaft, Lecture Notes in Control and Information Sciences, vol. 246, pp. 275-292, Springer-Verlag, London, 1999.

38. S. Stramigioli, *From Differentiable Manifolds to Interactive Robot Control*, PhD Dissertation, University of Delft, Dec. 1998.

39. S. Stramigioli, B.M. Maschke & A.J. van der Schaft, "Passive output feedback and port interconnection", in Proc. 4th IFAC NOLCOS, Enschede, pp. 613-618, 1998.

40. S. Stramigioli, B.M. Maschke, C. Bidard, "A Hamiltonian formulation of the dynamics of spatial mechanism using Lie groups and screw theory", to appear in Proc. Symposium Commemorating the Legacy, Work and Life of Sir R.S. Ball, J. Duffy and H. Lipkin organizers, July 9-11, 2000, University of Cambridge, Trinity College, Cambridge, U.K..

41. S. Stramigioli, A.J. van der Schaft, B. Maschke, S. Andreotti, C. Melchiorri, "Geometric scattering in tele-manipulation of port controlled Hamiltonian systems", 39th IEEE Conf. Decision & Control, Sydney, 2000.

42. M. Takegaki & S. Arimoto, "A new feedback method for dynamic control of manipulators", *Trans. ASME, J. Dyn. Systems, Meas. Control*, 103, pp. 119-125, 1981.

43. J.C. Willems, "Dissipative dynamical systems - Part I: General Theory", *Archive for Rational Mechanics and Analysis*, 45, pp. 321-351, 1972.

Energy Shaping Control Revisited

Romeo Ortega[1], Arjan J. van der Schaft[2], Iven Mareels[3], and Bernhard Maschke[4]

[1] Lab. des Signaux et Systèmes, CNRS-SUPELEC, Gif–sur–Yvette, France
[2] Fac. of Mathematical Sciences, University of Twente, Enschede,
   The Netherlands
[3] Dept. Electrical and Computer Engineering, University of Melbourne, Australia
[4] Automatisme Industriel, Paris, France

1 Introduction

Energy is one of the fundamental concepts in science and engineering practice, where it is common to view dynamical systems as energy–transformation devices. This perspective is particularly useful in studying complex *nonlinear* systems by decomposing them into simpler subsystems which, upon interconnection, add up their energies to determine the full system's behavior. The action of a controller may be also understood in energy terms as another dynamical system—typically implemented in a computer—interconnected with the process to modify its behavior. The control problem can then be recast as finding a dynamical system and an interconnection pattern such that the overall energy function takes the desired form. This "energy shaping" approach is the essence of passivity based control (PBC), a controller design technique that is very well-known in mechanical systems.

Our objectives in this article are threefold: First, to call attention to the fact that PBC does not rely on some particular structural properties of mechanical systems, but hinges on the more fundamental (and universal) property of energy balancing. Second, to identify the physical obstacles that hamper the use of "standard" PBC in applications other than mechanical systems. In particular, we will show that "standard" PBC is stymied by the presence of unbounded energy dissipation, hence it is applicable only to systems that are stabilizable with passive controllers. Third, to revisit a PBC theory that has been recently developed to overcome the dissipation obstacle as well as to make the incorporation of process prior knowledge more systematic. These two important features allow us to design energy based controllers for a wide range of physical systems.

Intelligent Control Paradigm

Control design problems have traditionally been approached from a signal–processing viewpoint; that is, the plant to be controlled and the controller are viewed as signal–processing devices that transform certain input signals into outputs. The control objectives are expressed in terms of keeping some error signals small and reducing the effect of certain disturbance inputs on the given regulated outputs, despite the presence of some unmodeled dynamics. To make the problem mathematically tractable, the admissible disturbances and unmodeled dynamics are assumed to be norm–bounded, and consequently, the indicators of performance are the size of the gains of the operators that map these various signals. In the case of linear time–invariant systems, this "intelligent control paradigm" (paraphrasing Willems [1]) has been very successful, essentially because disturbances and unmodeled dynamics can be discriminated, via filtering, using frequency–domain considerations. The problem of reducing the gains of *nonlinear operators* can also be expressed in a clear, analytic way [2]. There are, however, two fundamental differences with respect to the linear time–invariant case: first, the solution involves some far from obvious computations. Second, and perhaps more important, since nonlinear systems "mix" the frequencies, it is not clear how to select the most likely disturbances, and we have to "crank up" the gain to quench the (large set of) undesirable signals and meet the specifications. Injecting high gains in the loop, besides being intrinsically conservative—hence yielding below–par performance—brings along a series of well–known undesirable features (e.g., noise amplification, actuator wear, and high energy consumption).

There are many practical control problems where we have available *structural* information about the plant. In these cases, it is reasonable to expect that the conservatism mentioned above could be reduced if we could incorporate this prior information in the controller design. Unfortunately, a procedure to systematically carry out this objective does not seem to be available. [The typical approach is to classify the nonlinearities according to the role they play in the derivative of a Lyapunov function candidate. This test has very little to do with the physics of the system. It is obviously tied up with the particular choice of the Lyapunov function, which stemming from our linear inheritance, is systematically taken to be a quadratic function in the "errors."] It is our contention that the inability to incorporate prior knowledge is inherent to the signal–processing viewpoint of the intelligent control paradigm and is therefore independent of the particular design technique. In the authors' opinion, this situation has stymied communication between practitioners and control theorists, seriously jeopardizing the future of modern model–based nonlinear control systems design.

The purpose of this article is to contribute, if modestly, to the reversal of this trend calling attention to the importance of incorporating energy principles in control. To achieve our objective, we propose to abandon the intelligent control paradigm and instead adopt the behavioral framework proposed by Willems [1]. In Willems's far–reaching interpretation of control, we start from a mathematical model obtained from first principles, say, a set of higher order differential equations and some algebraic equations. Among the vector of time trajectories satisfying these equations are components that are available for interconnection. The controller design then reduces to defining an additional set of equations for these interconnection variables to impose a desired behavior on the controlled system. We are interested here in the incorporation into this paradigm of the essential energy component. Therefore, we view dynamical systems (plants and controllers) as *energy-transformation* devices, which we interconnect (in a power-preserving manner) to achieve the desired behavior. More precisely, we are interested in lumped-parameter systems that satisfy an *energy balancing* principle, where the interconnection with the environment is established through *power port variables*. The power port variables are conjugated, and their product has units of power, for instance, currents and voltages in electrical circuits or forces and velocities in mechanical systems. This is the scenario that arises from any form of physical network modeling.

Our first control objective is to regulate the static behavior (i.e., the equilibria), which is determined by the shape of the energy function. It is therefore natural to recast our control problem in terms of finding a dynamical system and an interconnection pattern such that the overall energy function takes the desired form. There are at least two important advantages of adopting such an "energy shaping" perspective of control:

1. The energy function determines not just the static behavior, but also, via the energy transfer between subsystems, its transient behavior. Focusing our attention on the systems energy, we can then aim, not just at stabilization, but also at *performance* objectives that can, in principle, be expressed in terms of "optimal" energy transfer. Performance and not stability is, of course, the main concern in applications.
2. Practitioners are familiar with energy concepts, which can serve as a *lingua franca* to facilitate communication with control theorists, incorporating prior knowledge and providing physical interpretations of the control action.

Background

The idea of energy shaping has its roots in the work of Takegaki and Arimoto [3] in robot manipulator control, a field where it is very well known and highly successful. Simultaneously and independently of [3] the utilization of these ideas for a large class of Euler-Lagrange systems was suggested in [4]. [See also Slotine's innovative paper [5] and the related view on the control of physical systems by Hogan [6].] Using the fundamental notion of *passivity*, the principle was later formalized in [7], where the term passivity-based control (PBC) was coined to define a controller design methodology whose aim is to render the closed-loop system passive with a given storage function. The importance of linking passivity to energy shaping can hardly be overestimated. On the one hand, viewing the control action in terms of interconnections of passive systems provides an *energy balancing* interpretation of the stabilization mechanism. More precisely, we have defined in [8] a class of systems (which includes mechanical systems) such that the application of PBC yields a closed–loop energy that is equal to the difference between the stored and the supplied energies. For obvious reasons, we call this special class of PBC energy balancing PBC. On the other hand, showing that the approach does not rely on some particular structural properties of mechanical systems, but hinges instead on the more fundamental (and universal) property of passivity, it can be extended to cover a wide range of applications.

In carrying out this extension, two approaches have been pursued:

• The first approach is similar to classical Lyapunov–based design, where we first select the storage function to be assigned and then design the controller that ensures this objective. Extensive applications of this line of research may be found in [9] (see also [10]–[15]) and are not reviewed in the present work. [It should be noted that in this approach, the desired storage function—typically quadratic in the increments—does not qualify as an *energy* function in any meaningful physical sense. Actually, it has been shown that the stabilization mechanism is akin to systems inversion instead of energy shaping [9], hence a stable invertibility assumption is usually required.]

• The second, newer approach stems from the energy balancing view of mechanical systems discussed above. The closed–loop storage function—which is now a *bona fide* energy function—is not postulated *a priori*, but is instead obtained *as a result* of our choice of desired subsystems interconnections and damping. This idea was first advanced for stability analysis in [16]; the extension for controller design was then reported in [17] and [8]; since then many successful applications, including mass–balance systems [18], electrical machines [19], power systems [20], magnetic levitation systems [21], and power converters [22], have been reported.

The aim of the present work is to provide a new energy balancing perspective of PBC that embraces and unifies its classical and modern versions. To enhance readability and widen our target audience, we strip away as much as possible the mathematical details and concentrate instead on the basic underlying principles and limitations. To underscore the fact that the principles are universal, we present them in a very general circuit–theoretic framework, without any additional mathematical structure attached to the system models. Particular emphasis is given to exhibiting the physical interpretation of the concepts, for instance, the central role played by dissipation. Toward this end, we illustrate our main points with simple physical examples.

The remainder of the article is organized as follows. First, we review the basic notions of passivity and stabilization via energy shaping. Next, we define the concept of energy balancing PBC and prove that this principle is applicable to all mechanical systems. Later we show that systems which extract an infinite amount of energy from the controller (i.e., systems with unbounded dissipation) cannot be stabilized with energy balancing PBCs. To characterize the class of systems that are stabilizable with energy balancing PBCs and eventually extend PBC to systems with unbounded dissipation, we propose to adopt Willems's "control–as–interconnection" viewpoint, a perspective that naturally provides a geometric interpretation to the notion of energy shaping. Then, after identifying a class of "admissible dissipations," we view the control action as the interconnection of the system with a passive controller. To stabilize systems with unbounded dissipations, we propose to model the action of the control as a state–modulated power–preserving interconnection of the plant with an infinite energy source system. These developments, which lead to the definition of a new class of PBCs called *interconnection and damping assignment* PBC, are presented for the so–called port–controlled Hamiltonian systems. Finally, we detail the application of interconnection and damping assignment PBC to a physical example, and then we present some concluding remarks.

Notation All vectors in the article, including the gradient, are defined as column vectors. Also, we use throughout κ to denote a generic positive constant.

2   Passivity and Energy Shaping

We are interested here in lumped–parameter systems interconnected to the external environment through some *port power variables* $u \in \mathcal{R}^m$ and $y \in \mathcal{R}^m$, which are conjugated in the sense that their product has units of power (e.g., currents and voltages in electrical circuits, or forces and velocities in mechanical systems). We assume the system satisfies the *energy–balance equation*

$$\underbrace{H[x(t)] - H[x(0)]}_{stored\ energy} = \underbrace{\int_0^t u^\top(s)y(s)ds}_{supplied} - \underbrace{d(t)}_{dissipated} \tag{1}$$

where $x \in \mathcal{R}^n$ is the state vector, $H(x)$ is the *total energy* function, and $d(t)$ is a nonnegative function that captures the dissipation effects (e.g., due to resistances and frictions). Energy balancing is, of course, a universal property of physical systems; therefore, our class, which is nothing other than the well–known *passive* systems, captures a very broad range of applications that include nonlinear and time–varying dynamics.

Two important corollaries follow from (1)

- The energy of the uncontrolled system (i.e., with $u \equiv 0$) is nonincreasing (that is, $H[x(t)] \leq H[x(0)]$), and it will actually decrease in the presence of dissipation. If the energy function is bounded from below, the system will eventually stop at a point of minimum energy. Also, as expected, the rate of convergence of the energy function is increased if we extract energy from the system, for instance, setting $u = -K_{di}y$, with $K_{di} = K_{di}^\top > 0$ a so–called damping injection gain.

- Given that

$$- \int_0^t u^\top(s)y(s)ds \leq H[x(0)] < \infty \tag{2}$$

the total amount of energy that can be extracted from a passive system is bounded. [This property, which (somehow misleadingly) is often stated with the inequality inversed, will be instrumental in identifying the class of systems that are stabilizable with energy balancing PBC.]

2.1 Standard Formulation of PBC

The point where the open–loop energy is minimal (which typically coincides with the zero state) is usually not the one of practical interest, and control is introduced to operate the system around some nonzero equilibrium point, say x_*. In the standard formulation of PBC, we label the port variables as inputs and outputs (say u and y, respectively) and pose the stabilization problem in a classical way. [We consider first static state feedback control laws and postpone the case of dynamic controllers to the section on admissible dissipations. Also, we refer the reader to [8] and references therein for further details on the dynamic and output feedback cases.]

- Select a control action $u = \beta(x) + v$ so that the closed–loop dynamics satisfies the new energy balancing equation

$$H_d[x(t)] - H_d[x(0)] = \int_0^t v^\top(s)z(s)ds - d_d(t)$$

where $H_d(x)$, the desired total energy function, has a strict minimum at x_*, z (which may be equal to y) is the new passive output, and we have replaced the natural dissipation term by some function $d_d(t) \geq 0$ to increase the convergence rate. Assigning an energy function with a minimum at the desired value is usually referred to as *energy shaping* while the modification of the dissipation is called *damping injection*.

Later, we will show that this classical distinction between inputs and outputs is restrictive, and the "control–as–interconnection" perspective of Willems is needed to cover a wider range of applications.

2.2 Discussion

Remark 1. For simplicity, we have treated *all* the components of the vector u as manipulated variables. In many practical cases, this vector contains some external (non–manipulated) variables such as disturbances or sources (see [8], [19] for some examples). Furthermore, there are some applications where the control action does not enter at all in u, for instance, in switched devices [22].

The analysis we will present in the sequel applies as well—*mutatis mutandi*—to those cases.

Remark 2. The choice of the desired dissipation in the damping injection stage is far from obvious. For instance, contrary to conventional wisdom, and except for the highly unusual case where we can achieve *exponential* stability, performance is not necessarily improved by adding positive damping, but it can actually be degraded as illustrated in [22], [23]. Furthermore, as shown in [18], [21] there are cases in which shuffling the damping between the channels can be beneficial for our control objective; this will be illustrated in the last section of this paper.

Remark 3. It is well known that solving the stabilization problem via passivation automatically ensures some robustness properties. Namely, stability will be preserved for all passive unmodeled dynamics between the port variables u, z. When $z = y$, these correspond to phenomena such as frictions and parasitic resistances.

Remark 4. It is clear also that if the dissipation is such that the passivity property is strengthened to output strict passivity, that is,

$$\int_0^t v^\top(s)z(s)ds \geq \delta \int_0^t |z(s)|^2 ds - \kappa$$

for some $\delta, \kappa > 0$, then we can show (with a simple completion of the squares argument) that the map $v \mapsto z$ has gain smaller than $\frac{1}{\delta}$. Consequently, we can reduce the amplification factor of the energy of the input noise by increasing the damping. See, however, Remark 2 above.

Remark 5. Passivity can be used for stabilization independently of the notion of energy shaping. In fact, it suffices to find an output $z = h(x)$ such that z square integrable implies $x(t) \to x_*$ as $t \to \infty$. Stabilization via passivation for general nonlinear systems, which has its roots in [24], [25], is one of the most active current research areas in nonlinear control, and some constructive results are available for systems in special forms [26], [27]. The energy shaping approach is a reasonable way to incorporate the information about the energy functions that is available in physical systems to simplify the passivation problem. Besides making the procedure more systematic, it usually yields physically interpretable controllers, considerably simplifying their commissioning stage. See [9] for an extensive discussion on these issues, including a detailed historical review, and the application of PBC to many practical examples.

3   Stabilization via Energy Balancing

There is a class of systems, which interestingly enough includes *mechanical* systems, for which the solution to the problem posed above is very simple, and it reduces to being able to find a function $\beta(x)$ such that the energy supplied by the controller can be expressed as a function of the state. Indeed, from (1) we see that *if* we can find a function $\beta(x)$ such that

$$-\int_0^t \beta^\top[x(s)]y(s)ds = H_a[x(t)] + \kappa \tag{3}$$

for some function $H_a(x)$, then the control $u = \beta(x) + v$ will ensure that the map $v \mapsto y$ is passive with *new energy function*

$$H_d(x) \overset{\triangle}{=} H(x) + H_a(x). \tag{4}$$

If, furthermore, $H_d(x)$ has a minimum at the desired equilibrium x_*, then it will be stable. Notice that the closed–loop energy is equal to the difference between the stored and the supplied energies. Therefore, we refer to this particular class of PBCs as *energy balancing* PBCs.

3.1 Mechanical Systems

Let us look at the classical example of position regulation of fully actuated *mechanical systems* with generalized coordinates $q \in \mathcal{R}^{n/2}$ and total energy

$$H(q, \dot{q}) = \frac{1}{2}\dot{q}^\top D(q)\dot{q} + V(q)$$

where $D(q) = D^\top(q) > 0$ is the generalized mass matrix and $V(q)$ is the systems potential energy, which is also bounded from below. It has been shown in [7] that for these systems, the passive outputs are the generalized velocities (that is, $y = \dot{q}$). The simplest way to satisfy condition (3) and shape the energy is by choosing

$$\beta(q) = \frac{\partial V}{\partial q}(q) - K_p(q - q_*)$$

where q_* is the desired constant position and $K_p = K_p^\top > 0$ is a proportional gain. Indeed, replacing the expression above and $y = \dot{q}$ in (3) we get

$$-\int_0^t \beta^\top[q(s)]\dot{q}(s)ds = -V[q(t)] + \frac{1}{2}[q(t) - q_*]^\top K_p[q(t) - q_*] + \kappa$$

and the new total energy for the passive closed–loop map $v \mapsto \dot{q}$ is

$$H_d(q, \dot{q}) = \frac{1}{2}\dot{q}^\top D(q)\dot{q} + \frac{1}{2}(q - q_*)^\top K_p(q - q_*),$$

which has a minimum in $(q_*, 0)$, as desired. To ensure that the trajectories actually converge to this minimum (i.e., that the equilibrium is asymptotically stable), we add some damping $v = -K_{di}\dot{q}$, as discussed above.

Of course, the controller presented above is the very well–known PD+gravity compensation of [3]. The purpose of the exercise is to provide a new interpretation for the action of this controller, underscoring the fact that the storage function that is assigned to the closed loop is (up to an integration constant) precisely the difference between the stored and the supplied energies (i.e., $H(x) - \int_0^t u^\top(s)y(s)ds$). Hence application of PBC for position regulation of mechanical systems yields energy balancing PBCs.

Remark 6. With the elementary procedure described above, it is possible to re–derive most of the energy balancing PBCs (e.g., with saturated inputs, output feedback) reported for position regulation of robot manipulators. This usually requires some ingenuity to find out the "right" energy function to be assigned. It is clear, however, that the technique is restricted to potential energy shaping. Later we present a new methodology that allows us also to shape the kinetic energy, which is required for some underactuated mechanical devices (see [28], [29], [30]).

Remark 7. In the underactuated case, when the number of control actions is smaller than the number of degrees of freedom, we find that $y = M^\top \dot{q}$, with M the input matrix for the force (or torque) vector u. As shown in [9], the energy shaping procedure still applies in these cases, provided some dissipation propagation condition is satisfied.

3.2 General (f, g, h) Systems

Energy–balancing stabilization can, in principle, be applied to general (f, g, h) nonlinear passive systems of the form

$$\Sigma : \begin{cases} \dot{x} = f(x) + g(x)u \\ y = h(x). \end{cases} \tag{5}$$

From the celebrated nonlinear version of the Kalman–Yakubovich–Popov lemma [31], we know that for this class of systems, passivity is *equivalent* to the existence of a nonnegative scalar function $H(x)$ such that

$$\left(\frac{\partial H}{\partial x}(x)\right)^\top f(x) \leq 0$$

$$h(x) = g^\top(x)\frac{\partial H}{\partial x}(x).$$

We have the following simple proposition.

Proposition 1. Consider the passive system (5) with storage function $H(x)$ and an admissible equilibrium x_*. If we can find a vector function $\beta(x)$ such that the partial differential equation

$$\left(\frac{\partial H_a}{\partial x}(x)\right)^\top [f(x) + g(x)\beta(x)] = -h^\top(x)\beta(x) \tag{6}$$

can be solved for $H_a(x)$, and the function $H_d(x)$ defined as (4) has a minimum at x_*, then $u = \beta(x) + v$ is an energy balancing PBC. Consequently, setting $v \equiv 0$, we have that x_* is a stable equilibrium with the difference between the stored and the supplied energies constituting a Lyapunov function.

The proof follows immediately, noting that the left–hand side of (6) equals $\dot{H}_a$ while the right–hand side is $-y^\top u$, and then integrating from 0 to t.

Caveat emptor. This result, although quite general, is of limited interest. First of all, (f, g, h) models do not reveal the role played by the energy function in the system dynamics. Hence it is difficult to incorporate prior information to select a $\beta(x)$ to solve the PDE (6). A more practical and systematic result will be presented later for a more suitable class of models, namely, the so–called port–controlled Hamiltonian systems. Second, we will show below that, beyond the realm of mechanical systems, the applicability of energy balancing PBC is severely restricted by the system's natural dissipation.

4   Dissipation Obstacle

To investigate the conditions under which the PDE (6) is solvable we make the following observation

Fact A necessary condition for the *global* solvability of the PDE (6) is that $h^\top(x)\beta(x)$ vanishes at all the zeros of $f(x) + g(x)\beta(x)$, that is,

$$f(\bar{x}) + g(\bar{x})\beta(\bar{x}) = 0 \Rightarrow h^\top(\bar{x})\beta(\bar{x}) = 0.$$

Now $f(x) + g(x)\beta(x)$ is obviously zero at the equilibrium x_*, hence the right–hand side $-y^\top u$, which is the *power* extracted from the controller, should also be zero at the equilibrium. This means that energy balancing PBC is applicable only if the energy dissipated by the system is bounded, and consequently if it can be stabilized *extracting a finite amount of energy* from the controller. This is indeed the case in regulation of mechanical systems where the extracted power is the product of force and velocity and we want to drive the velocity to zero. Unfortunately, it is no longer the case for most electrical or electromechanical systems where power involves the product of voltages and currents and the latter may be nonzero for nonzero equilibria.

Let us illustrate this point with simple linear time–invariant RLC circuits. First, we prove that the series RLC circuit is stabilizable with an energy balancing PBC. Then we move the resistance to a parallel connection and show that, since for this circuit the power at any nonzero equilibrium is nonzero, energy balancing stabilization is no longer possible.

4.1   Finite Dissipation Example

Consider the series RLC circuit of Fig. 1, where the port power variables are the input voltage and the current. The "natural" state variables for this circuit are the charge in the capacitor and the flux in the inductance $x \triangleq [q_C, \phi_L]^\top$, and the total energy function is

$$H(x) = \frac{1}{2C}x_1^2 + \frac{1}{2L}x_2^2. \tag{7}$$

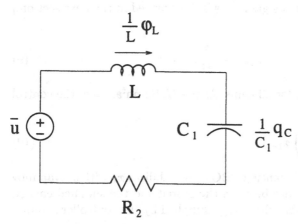

Fig. 1. Series RLC circuit

The dynamic equations are given by

$$\Sigma : \begin{cases} \dot{x}_1 = \frac{1}{L}x_2 \\ \dot{x}_2 = -\frac{1}{C}x_1 - \frac{R}{L}x_2 + u \\ y = \frac{1}{L}x_2 \end{cases} \tag{8}$$

The circuit clearly satisfies (1) with $d(t) = R \int_0^t [\frac{1}{L}x_2(s)]^2 ds$ (i.e., the energy dissipated in the resistor).

We are given an equilibrium x_* that we want to stabilize. It is clear from (8) that the admissible equilibria are of the form $x_* = [x_{1*}, 0]^\top$. It is important to note that the extracted power at any admissible equilibrium is zero.

To design our energy balancing PBC, we look for a solution of the PDE (6), which in this case takes the form

$$\left(\frac{1}{L}x_2\right)\frac{\partial H_a}{\partial x_1}(x) - \left[\frac{1}{C}x_1 + \frac{R}{L}x_2 - \beta(x)\right]\frac{\partial H_a}{\partial x_2}(x) = -\frac{1}{L}x_2\beta(x).$$

Notice that the energy function $H(x)$ already "has a minimum" at $x_2 = 0$; thus we only have to "shape" the x_1 component, so we look for a function of the form $H_a = H_a(x_1)$. In this case, the PDE reduces to

$$\beta(x_1) = -\frac{\partial H_a}{\partial x_1}(x_1),$$

which, for any given $H_a(x_1)$, defines the control law as $u = \beta(x_1)$. To shape the energy $H_d(x)$, we add a quadratic term and complete the squares (in the increments $x - x_*$) by proposing

$$H_a(x_1) = \frac{1}{2C_a}x_1^2 - \left(\frac{1}{C} + \frac{1}{C_a}\right)x_{1*}x_1 + \kappa.$$

[The particular notation for the gain $\frac{1}{C_a}$ will be clarified in the next section.] Replacing in (4), yields

$$H_d(x) = \frac{1}{2}\left(\frac{1}{C} + \frac{1}{C_a}\right)(x_1 - x_{1*})^2 + \frac{1}{2L}x_2^2 + \kappa, \tag{9}$$

which has a minimum at x_* for all gains $C_a > -C$. Summarizing, the control law

$$u = -\frac{x_1}{C_a} + \left(\frac{1}{C} + \frac{1}{C_a}\right)x_{1*} \tag{10}$$

with $C_a > -C$ is an energy balancing PBC that stabilizes x_* with a Lyapunov function equal to the difference between the stored and the supplied energy. Finally, it is easy to verify that the energy supplied by the controller is finite.

4.2 Infinite Dissipation Example

Even though in the previous example we could find a very simple energy balancing solution to our stabilization problem, it is easy to find systems that are not stabilizable with energy balancing PBCs. For instance, consider a parallel RLC circuit. With the same definitions as before, the dynamic equations are now

$$\Sigma : \begin{cases} \dot{x}_1 = -\frac{1}{RC}x_1 + \frac{1}{L}x_2 \\ \dot{x}_2 = -\frac{1}{C}x_1 + u \\ y = \frac{1}{L}x_2 \end{cases} \tag{11}$$

Notice that only the dissipation structure has changed, but the admissible equilibria are now of the form $x_* = [Cu_*, \frac{L}{R}u_*]^\top$ for any u_*. The problem is that the power at any equilibrium except the trivial one is nonzero, and consequently any stabilizing controller will yield $\lim_{t\to\infty} |\int_0^t u(s)y(s)ds| = \infty$ (we will eventually run down the battery!).

We will not elaborate further here on the infinite dissipation problem. A precise characterization, within the context of port–controlled Hamiltonian systems, will be given in the next section.

Remark 8. The well–known analogies between electrical and mechanical systems might lead us to conclude that, with another choice of states, we could overcome the infinite dissipation obstacle for energy balancing PBC. The obstacle is, however, "coordinate–free." The point is that in the mechanical case the dissipation only comes in at the momentum level, where also the input is appearing. This eliminates the possibility of infinite dissipation.

Remark 9. In the linear time–invariant case, we can design an energy balancing PBC working on incremental states. However, this procedure is usually not feasible, and despite its popularity is actually unnatural, for the general nonlinear case. The PBC design procedure we will present later to handle the infinite dissipation case does not rely on the generation of incremental dynamics. Furthermore, except for the linear case, the resulting energy functions will be nonquadratic.

5   Admissible Dissipations for Energy Balancing PBC

In the previous section we showed that energy balancing PBC is applicable only to systems with finite dissipation—obviously including conservative systems that have no dissipation at all. We have also shown that this class of systems contains all mechanical systems, as well as some electrical circuits with dissipation. A natural question then is how to characterize the "admissible dissipations" for energy balancing PBC. To provide an answer to this question, we find it convenient to adopt a variation of Willems's "control as interconnection" viewpoint. This perspective is also used in the next section, where viewing the action of the controller as an infinite energy source with a state modulated interconnection to the plant, we extend PBC to systems with infinite dissipation.

5.1   Passive Controllers

As shown in Fig. 2, we view the controller, Σ_c, as a one–port system that will be coupled with the plant to be controlled, Σ, via a two–port interconnection subsystem, Σ_I. We need the following definition.

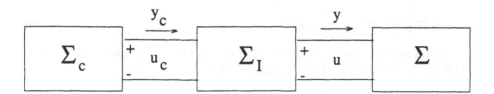

Fig. 2. Control as interconnection.

Definition 1. The interconnection of Fig. 2 is said to be *power preserving* if the two–port subsystem Σ_I is lossless; that is, if it satisfies

$$\int_0^t [y^\top(s), y_c^\top(s)] \begin{bmatrix} u(s) \\ u_c(s) \end{bmatrix} ds = 0.$$

We now make the following important observation.

Proposition 2. Consider the interconnection of Fig. 2 with some external inputs (v, v_c) as

$$\begin{bmatrix} u \\ u_c \end{bmatrix} = \Sigma_I \begin{bmatrix} y \\ y_c \end{bmatrix} + \begin{bmatrix} v \\ v_c \end{bmatrix}.$$

Assume Σ_I is power preserving and Σ, Σ_c are passive systems with states $x \in \mathcal{R}^n$, $\zeta \in \mathcal{R}^{n_c}$, and energy functions $H(x)$, $H_c(\zeta)$, respectively. Then $[v^\top, v_c^\top]^\top \mapsto [y^\top, y_c^\top]^\top$ is also a passive system with *new energy function* $H(x) + H_c(\zeta)$.

This fundamental property is proven with the following simple calculations:

$$\int_0^t [v^\top(s), v_c^\top(s)] \begin{bmatrix} y(s) \\ y_c(s) \end{bmatrix} ds = \int_0^t u^\top(s)y(s)ds + \int_0^t u_c^\top(s)y_c(s)ds$$
$$\geq H[x(t)] + H_c[\zeta(t)] - H[x(0)] - H_c[\zeta(0)] \tag{12}$$

where the first equation follows from the lossless property of Σ_I and the last inequality is obtained from the passivity of each subsystem.

5.2   Invariant Functions Method

From Proposition 2, we conclude that passive controllers and power–preserving interconnections can, in principle, be used to "shape" the closed–loop total energy. However, although $H_c(\zeta)$ can be freely assigned, the systems energy function $H(x)$ is given, and it is not clear how we can effectively shape the overall energy. The central idea of the invariant functions method [32], [33] is to restrict the motion of the closed–loop system to a certain subspace of the extended state space (x, ζ), say

$$\Omega \triangleq \{(x, \zeta) | \zeta = F(x) + \kappa\} \tag{13}$$

In this way, we have a functional relationship between x and ζ, and we can express the closed–loop total energy as a function of x only, namely

$$H_d(x) \triangleq H(x) + H_c[F(x) + \kappa], \tag{14}$$

[Notice that $H_c[F(x) + \kappa]$ plays the same role as $H_a(x)$ in (4).] This function can now be shaped with a suitable selection of the controller energy $H_c(\zeta)$. The problem then translates into finding a function $F(\cdot)$ that renders Ω *invariant*. [Recall that a set $\Omega \subset \mathcal{R}^n$ is invariant if the following implication holds: $x(0) \in \Omega \Rightarrow x(t) \in \Omega, \forall t \geq 0$.]

Let us illustrate this idea of generation of invariant subspaces to design stabilizing PBCs with the simple series RLC circuit example described by

(8). Following Proposition 2 we consider passive controllers with state ζ and energy function $H_c(\zeta)$ to be defined. Since, as discussed above, we only need to modify the first coordinate, we propose to take ζ a scalar. Furthermore, for simplicity, we choose the dynamics of the controller to be a simple integrator; that is

$$\Sigma_c : \begin{cases} \dot{\zeta} = u_c \\ y_c = \frac{\partial H_c}{\partial \zeta}(\zeta) \end{cases} . \tag{15}$$

Notice that if $H_c(\zeta)$ is bounded from below, then $u_c \mapsto y_c$ is indeed passive.

We already know that this system is stabilizable with an energy balancing PBC; therefore, we interconnect the circuit and the controller with the standard negative feedback interconnection

$$\begin{bmatrix} u \\ u_c \end{bmatrix} = \begin{bmatrix} 0 & -1 \\ 1 & 0 \end{bmatrix} \begin{bmatrix} y \\ y_c \end{bmatrix} . \tag{16}$$

To establish a relationship between x_1 and ζ, of the form $\zeta = F(x_1) + \kappa$, we define an invariant function candidate

$$C(x_1, \zeta) \triangleq F(x_1) - \zeta \tag{17}$$

and look for an $F(\cdot)$ such that $\frac{d}{dt}C \equiv 0$. Some simple calculations with (8), (15), (16), and (17) yield

$$\frac{d}{dt}C = \frac{1}{L}x_2 \left(\frac{\partial F}{\partial x_1}(x_1) - 1 \right),$$

from which we conclude that we should take $F(x_1) = x_1$, and the invariant subspaces are the linear spaces $\Omega = \{(x_1, x_2, \zeta) | \zeta = x_1 + \kappa\}$.

We now have to select the energy function of the controller such that in these invariant subspaces, the total energy function $H(x) + H_c(\zeta)$ has a minimum at x_*. Following the same rationale as in the previous section we aim at a quadratic function (in the increments $x - x_*$); hence we fix

$$H_c(\zeta) = \frac{1}{2C_a}\zeta^2 - \left(\frac{1}{C} + \frac{1}{C_a} \right) x_{1*}\zeta$$

where C_a is a design parameter. As expected, the closed-loop energy, which results from (14) with $F(x_1) = x_1$, coincides with (9). [Notice that we have taken $\kappa = 0$. This is without loss of generality, because κ is determined by the controller's initial conditions].

Remark 10. One important feature of PBC is that we can usually give a physical interpretation to the action of the controller. Indeed, a physical realization of the energy balancing PBC (15) consists of a constant voltage source in series with a capacitor C_a. Notice, however, that the control action can be implemented *without* the addition of dynamics. Indeed, the input–output relationship of the controller dynamics (15) together with the interconnection (16), reduces to the static state feedback (10).

Remark 11. For simplicity, we have assumed above that $n_c = n$. In [17] we consider the more general case when $n_c \neq n$, and not all the controller states are related with the plant state variables.

Remark 12. Even though stabilization is ensured for all values of the added capacitance such that $C_a > -C$, it is clear that the system Σ_c is passive only for positive values of C_a.

Remark 13. The problem of finding a function $F(\cdot)$ that renders Ω invariant involves, of course, the solution of a PDE which is, in general, difficult to find. (In the simple case above, this is the trivial equation $\frac{\partial F}{\partial x_1}(x_1) = 1$.) One of the main messages we want to convey in this article is that the search for a solution of the PDE can be made systematic by incorporation of additional structure to the problem—starting with the choice of a suitable system representation. We will further elaborate this point in the next subsection.

5.3   Energy Balancing PBC of Port–Controlled Hamiltonian Systems

To characterize a class of (finite dissipation) systems stabilizable with energy balancing PBC and simplify the solution of the PDE discussed above, we need to incorporate more structure into the system dynamics, in particular, making explicit the damping terms and the dependence on the energy function. Toward this end, we consider port–controlled Hamiltonian models that encompass a very large class of physical nonlinear systems. They result from the network modeling of energy–conserving lumped–parameter physical systems with independent storage elements, and have been advocated as an alternative to more classical Euler–Lagrange (or standard Hamiltonian) models in a series of recent papers (see [2] for a list of references). These models are of the form

$$\Sigma : \begin{cases} \dot{x} = [J(x) - \mathcal{R}(x)]\frac{\partial H}{\partial x}(x) + g(x)u \\ y = g^\top(x)\frac{\partial H}{\partial x}(x) \end{cases} \tag{18}$$

where $H(x)$ is the energy function, $J(x) = -J^\top(x)$ captures the interconnection structure, and $\mathcal{R}(x) = \mathcal{R}^\top(x) \geq 0$ is the dissipation matrix. Clearly these systems satisfy the energy balancing equation (1).

Motivated by Proposition 2, we consider port–controlled Hamiltonian controllers of the form

$$\Sigma_c : \begin{cases} \dot{\zeta} = [J_c(\zeta) - \mathcal{R}_c(\zeta)]\frac{\partial H_c}{\partial \zeta}(\zeta) + g_c(\zeta)u_c \\ y_c = g_c^\top(\zeta)\frac{\partial H_c}{\partial \zeta}(\zeta) \end{cases}$$

for any skew–symmetric matrix $J_c(\zeta)$, any positive-semidefinite matrix $\mathcal{R}_c(\zeta)$, and any function $g_c(\zeta)$. The interconnection constraints are given by (16). The overall interconnected system is defined in the extended state space (x, ζ) and can be written as

$$
\begin{bmatrix} \dot{x} \\ \dot{\zeta} \end{bmatrix} = \begin{bmatrix} J(x) - \mathcal{R}(x) & -g(x)g_c^\top(\zeta) \\ g_c(\zeta)g^\top(x) & J_c(\zeta) - \mathcal{R}_c(\zeta) \end{bmatrix} \begin{bmatrix} \frac{\partial H}{\partial x}(x) \\ \frac{\partial H_c}{\partial \zeta}(\zeta) \end{bmatrix}. \tag{19}
$$

Notice that it still belongs to the class of port–controlled Hamiltonian models with total energy $H(x) + H_c(\zeta)$.

We introduce at this point the concept of Casimir functions [33,2], which are conserved quantities of the system *for any* choice of the Hamiltonian, and so are completely determined by the geometry (i.e., the interconnection structure) of the system. For ease of presentation, we keep the same notation we used in the previous subsection and look for Casimir functions of the form

$$
\mathcal{C}(x, \zeta) = F(x) - \zeta. \tag{20}
$$

Since the time derivative of these functions should be zero along the closed–loop dynamics for all Hamiltonians $H(x)$, this means that we are looking for solutions of the PDEs

$$
\left[\left(\frac{\partial F}{\partial x}(x) \right)^\top \vdots -I_m \right] \begin{bmatrix} J(x) - \mathcal{R}(x) & -g(x)g_c^\top(\zeta) \\ g_c(\zeta)g^\top(x) & J_c(\zeta) - \mathcal{R}_c(\zeta) \end{bmatrix} = 0. \tag{21}
$$

The following proposition was established in [17].

Proposition 3. The vector function (20) satisfies (21) (and thus is a Casimir function for the interconnected system (19)) *if and only if* $F(x)$ satisfies

$$
\left(\frac{\partial F}{\partial x}(x) \right)^\top J(x) \frac{\partial F}{\partial x}(x) = J_c(\zeta) \tag{22}
$$

$$
\mathcal{R}(x) \frac{\partial F}{\partial x}(x) = 0 \tag{23}
$$

$$
\mathcal{R}_c(\zeta) = 0 \tag{24}
$$

$$
\left(\frac{\partial F}{\partial x}(x) \right)^\top J(x) = g_c(\zeta)g^\top(x) \tag{25}
$$

In this case, the dynamics reduced to the set Ω (13) is a port–controlled Hamiltonian system of the form

$$
\dot{x} = [J(x) - \mathcal{R}(x)] \frac{\partial H_d}{\partial x}(x) \tag{26}
$$

with the shaped energy function $H_d(x) = H(x) + H_c[F(x) + \kappa]$.

5.4 Admissible Dissipation

Condition (23) of Proposition 3 characterizes the admissible dissipations for energy balancing PBC in terms of the coordinates where energy can be shaped. Indeed, if (23) holds, then

$$\mathcal{R}(x)\frac{\partial H_c(F)}{\partial x}(x) = 0$$

for any controller energy function H_c. Roughly speaking, this means that H_c should not depend on the coordinates where there is natural damping. The latter restriction can then be interpreted as: Dissipation in energy balancing PBC is admissible only on the coordinates that do not require "shaping of the energy."

Recall that in mechanical systems, where the state consists of position and velocities, damping is associated with the latter; hence it appears in the lower right corner of the matrix $\mathcal{R}(x)$. On the other hand, in position regulation, we are only concerned with potential energy shaping; thus the condition (23) will be satisfied. In the case of the series RLC circuit of the previous section, the resistance appears in a coordinate that did not need to be modified (i.e., the current x_2); whereas in the parallel RLC circuit, both coordinates have to be shaped.

6   Overcoming the Dissipation Obstacle

In Proposition 3 we have shown that under certain conditions, the interconnection of a port–controlled Hamiltonian plant with a port–controlled Hamiltonian controller leads to a reduced dynamics given by another port–controlled Hamiltonian system (26) with a shaped Hamiltonian. The reduction of the dynamics stems from the existence of Casimir functions that relate the states of the controller with those of the plant. In this section, we will show that, *explicitly* incorporating information on the systems state, we can shape the energy function without the need for Casimir functions. This will lead to the definition of a new class of PBCs that we call *interconnection and damping assignment* PBCs.

6.1   Control as a State–Modulated Source

To extend PBC to systems with *infinite dissipation*, we introduce two key modifications. First, since these systems cannot be stabilized by extracting a finite amount of energy from the controller, we consider the latter to be an (infinite energy) *source*; that is, a scalar system

$$\Sigma_c : \begin{cases} \dot{\zeta} = u_c \\ y_c = \frac{\partial H_c}{\partial \zeta}(\zeta) \end{cases} \tag{27}$$

with energy function

$$H_c(\zeta) = -\zeta. \tag{28}$$

Second, the classical unitary feedback interconnection (through the power port variables) imposes some very strict constraints on the plant and controller structures as reflected by the conditions (22)–(25). To provide more design flexibility, we propose to incorporate state information, which is done by coupling the source system with the plant via a *state-modulated* interconnection of the form

$$\begin{bmatrix} u(s) \\ u_c(s) \end{bmatrix} = \begin{bmatrix} 0 & -\beta(x) \\ \beta(x) & 0 \end{bmatrix} \begin{bmatrix} y(s) \\ y_c(s) \end{bmatrix}. \tag{29}$$

This interconnection is clearly power preserving. The overall interconnected system (18), (27), (28), (29) can be written as

$$\begin{bmatrix} \dot{x} \\ \dot{\zeta} \end{bmatrix} = \begin{bmatrix} J(x) - \mathcal{R}(x) & -g(x)\beta(x) \\ \beta^\top(x)g^\top(x) & 0 \end{bmatrix} \begin{bmatrix} \frac{\partial H}{\partial x}(x) \\ \frac{\partial H_c}{\partial \zeta}(\zeta) \end{bmatrix} \tag{30}$$

which is still a port–controlled Hamiltonian system with total energy $H(x) + H_c(\zeta)$. It is important to note that the x dynamics above describes the behavior of the system (18) with a static state feedback $u = \beta(x)$; hence our choice of the symbol β for the state-modulation function.

We have shown in [8] that the damping restriction (23) is a necessary condition for the existence of Casimir functions in this case as well. The key point here is that the energy of the x subsystem can be shaped and the port–controlled Hamiltonian structure preserved without generation of Casimir functions. Indeed, if (for the given $J(x), \mathcal{R}(x)$ and $g(x)$) we can solve the PDE

$$[J(x) - \mathcal{R}(x)]\frac{\partial H_a}{\partial x}(x) = g(x)\beta(x) \tag{31}$$

for some $\beta(x)$, then the plant dynamics will be given by (26) with energy function $H_d(x) = H(x) + H_a(x)$. If we can furthermore ensure that $H_d(x)$ has a minimum at the desired equilibrium, then the static state feedback control $u = \beta(x)$ will stabilize this point. Notice that there is no "finite dissipation" constraint for the solvability of (31); hence the new PBC design is, in principle, applicable to systems with infinite dissipation.

6.2   Parallel RLC circuit example

Before presenting the main result of this section, which is a systematic procedure for PBC of port–controlled Hamiltonian systems, let us illustrate the new energy shaping method with the parallel RLC circuit example. The dynamics of this circuit (11) can be written in port–controlled Hamiltonian form (18) with energy function (7) and the matrices

$$J = \begin{bmatrix} 0 & 1 \\ -1 & 0 \end{bmatrix}, \; \mathcal{R} = \begin{bmatrix} 1/R & 0 \\ 0 & 0 \end{bmatrix}, \; g = \begin{bmatrix} 0 \\ 1 \end{bmatrix}.$$

The PDE (31) becomes

$$-\frac{1}{R}\frac{\partial H_a}{\partial x_1}(x) + \frac{\partial H_a}{\partial x_2}(x) = 0$$

$$-\frac{\partial H_a}{\partial x_1}(x) = \beta(x).$$

The first equation can be trivially solved as

$$H_a(x) = \Phi(Rx_1 + x_2)$$

where $\Phi(\cdot) : \mathcal{R} \to \mathcal{R}$ is an arbitrary differentiable function, whereas the second equation defines the control law. We now need to choose the function Φ so that $H_d(x)$ has a minimum at the desired equilibrium point $x_* = (Cu_*, \frac{L}{R}u_*)$. For simplicity, we choose it to be a quadratic function

$$\Phi(Rx_1 + x_2) = \frac{K_p}{2}[(Rx_1 + x_2) - (Rx_{1*} + x_{2*})]^2 - Ru_*(Rx_1 + x_2)$$

which, as can be easily verified, ensures the desired energy shaping for all

$$K_p > \frac{-1}{(L + CR^2)}. \tag{32}$$

The assigned energy function, as expected, is quadratic in the increments

$$H_d(x) = (x - x_*)^\top \begin{bmatrix} \frac{1}{C} + R^2 K_p & RK_p \\ RK_p & \frac{1}{L} + K_p \end{bmatrix} (x - x_*) + \kappa.$$

Clearly, (32) is the necessary and sufficient condition for x_* to be a unique global minimum of this function. The resulting control law is a simple linear state feedback

$$u = -K_p[R(x_1 - x_{1*}) + x_2 - x_{2*}] + u_*.$$

6.3   Discussion

Remark 14. We should underscore that in the example above we did not need to "guess" candidate functions for $H_a(x)$ or $\beta(x)$. Instead, the solution of the PDE (31) provided a family of "candidates" parametrized in terms of the free function $\Phi(\cdot)$. The PDE, in turn, is uniquely determined by the systems interconnection, damping, and input matrices; we will show below that to provide more degrees of freedom to the design, we can also change the first two matrices. From this family of solutions, we then have to select one that achieves the energy shaping. Also, once a solution $H_a(x)$ is obtained, we know that the new energy function $H_d(x)$ will be nonincreasing, because $\dot{H}_d$ is nonpositive by construction. This situation should be contrasted with classical Lyapunov–based designs (or "standard" PBC, e.g., [9]), where we fix *a priori* the Lyapunov (energy) function—typically a quadratic in the increments—and then calculate the control law that makes its derivative negative definite. We claim that the proposed approach is more natural because, on the one hand, it is easier to incorporate prior knowledge in the choice of the desired interconnection and damping matrices; on the other hand, the resulting energy (Lyapunov) function will be specifically tailored to the problem.

Remark 15. Of course, stabilization of linear systems is possible using other, much simpler, methods. Our point is that, as we will show in the next subsection, the present procedure applies verbatim to the nonlinear case. Furthermore, even in the linear case, the technique allows us to design nonlinear controllers, which might be of interest to improve performance (e.g., assigning steeper Lyapunov functions for faster convergence or imposing certain shapes of the level sets to handle state or input constraints); see [22] for an example of the latter.

Remark 16. As discussed in [8] (see also Proposition 4 below), we do not even need to solve the PDE (31) for $H_a(x)$. Indeed, we can look for a solution of the problem directly in terms of $\beta(x)$, as follows. If $J(x) - \mathcal{R}(x)$ is invertible [see [8] and Proposition 4 below for the noninvertible case] it is well known that (31) has a solution *if and only if* the integrability conditions

$$\frac{\partial K}{\partial x}(x) = \left[\frac{\partial K}{\partial x}(x)\right]^{\top} \tag{33}$$

hold, where

$$K(x) \triangleq [J(x) - \mathcal{R}(x)]^{-1} g(x) \beta(x). \tag{34}$$

Given $J(x), \mathcal{R}(x)$, and $g(x)$, (33) defines a set of PDEs for $\beta(x)$. For instance, for the parallel RLC circuit example, we have that (33) is equivalent to

$$-\frac{1}{R}\frac{\partial \beta}{\partial x_1}(x) + \frac{\partial \beta}{\partial x_2}(x) = 0$$

whose solution yields directly the control law $\beta(x) = \Phi(Rx_1 + x_2)$. Although in this simple linear example both procedures lead to the same PDE, this will not be the case for the general nonlinear case. Furthermore, the importance of determining necessary and sufficient conditions for solvability can hardly be overestimated. We will elaborate further on these issues in the next subsection.

6.4 Assigning Interconnection and Damping Structures

In the previous subsections, we have shown that the success of our PBC design essentially hinges on our ability to solve the PDE (31). It is well known that solving PDEs is not easy. It is our contention that, for the particular PDE that we have to solve here, it is possible to incorporate prior knowledge about the system to simplify the task. More specifically, for port–controlled Hamiltonian models, besides the control law, we have the additional degrees of freedom of selecting the interconnection and damping structures of the closed–loop. Indeed, our energy shaping objective is not modified if, instead of (26), we aim at the closed–loop dynamics

$$\dot{x} = [J_d(x) - R_d(x)]\frac{\partial H_d}{\partial x}(x) \tag{35}$$

for some *new* interconnection $J_d(x) = -J_d^\top(x)$ and damping $R_d(x) = R_d^\top(x) \geq 0$ matrices. For this so–called interconnection and damping assignment PBC the PDE (31) becomes

$$[J(x) + J_a(x) - R(x) - R_a(x)]\frac{\partial H_a}{\partial x}(x) = \tag{36}$$

$$= -[J_a(x) - R_a(x)]\frac{\partial H}{\partial x}(x) + g(x)\beta(x)$$

where

$$J_a(x) \triangleq J_d(x) - J(x), \quad R_a(x) \triangleq R_d(x) - R(x)$$

are new design parameters that add more degrees of freedom to the solution of the PDE.

The proposition below (established in [8]) follows immediately from the derivations above. It is presented in a form that is particularly suitable for symbolic computations. We refer the interested reader to [8] for additional comments and discussions.

Proposition 4. Given $J(x), R(x), H(x), g(x)$, and the desired equilibrium to be stabilized x_*, assume we can find functions $\beta(x), R_a(x), J_a(x)$ such that

$$J(x) + J_a(x) = -[J(x) + J_a(x)]^\top$$
$$R(x) + R_a(x) = [R(x) + R_a(x)]^\top \geq 0$$

and a vector function $K(x)$ satisfying

$$[J(x) + J_a(x) - (R(x) + R_a(x))]K(x) = \tag{37}$$

$$= -[J_a(x) - R_a(x)]\frac{\partial H}{\partial x}(x) + g(x)\beta(x)$$

and such that the following conditions occur

(i) *(Integrability)* $K(x)$ is the gradient of a scalar function; that is, (33) holds.

(ii) *(Equilibrium assignment)* $K(x)$, at x_*, verifies

$$K(x_*) = -\frac{\partial H}{\partial x}(x_*).$$

(iii) *(Lyapunov stability)* The Jacobian of $K(x)$, at x_*, satisfies the bound

$$\frac{\partial K}{\partial x}(x_*) > -\frac{\partial^2 H}{\partial x^2}(x_*).$$

Under these conditions, the closed–loop system $u = \beta(x)$ will be a port–controlled Hamiltonian system with dissipation of the form (35), where $H_d(x) = H(x) + H_a(x)$ and

$$\frac{\partial H_a}{\partial x}(x) = K(x). \tag{38}$$

Furthermore, x_* will be a (locally) stable equilibrium of the closed loop. It will be *asymptotically* stable if, in addition, the largest invariant set under the closed–loop dynamics contained in

$$\left\{ x \in \mathcal{R}^n \cap \mathcal{B} \,\middle|\, \left[\frac{\partial H_d}{\partial x}(x)\right]^\top \mathcal{R}_d(x)\frac{\partial H_d}{\partial x}(x) = 0 \right\}$$

equals $\{x_*\}$.

Remark 17. From the following simple calculations

$$\dot{H}_d = u^\top y - \overbrace{\left[\frac{\partial H}{\partial x}(x)\right]^\top \mathcal{R}(x)\frac{\partial H}{\partial x}(x) + \dot{H}_a}^{\dot{H}} = \tag{39}$$

$$= -\left[\frac{\partial H_d}{\partial x}(x)\right]^\top \mathcal{R}_d(x)\frac{\partial H_d}{\partial x}(x)$$

and the fact that $\mathcal{R}_d(x) = \mathcal{R}_a(x) + \mathcal{R}(x)$, we have that

$$\dot{H}_a = -u^\top y - \left[2\frac{\partial H}{\partial x}(x) + \frac{\partial H_a}{\partial x}(x)\right]^\top \mathcal{R}(x)\frac{\partial H_a}{\partial x}(x) - \tag{40}$$

$$- \left[\frac{\partial H_d}{\partial x}(x)\right]^\top \mathcal{R}_a(x)\frac{\partial H_d}{\partial x}(x).$$

Consequently, if $\mathcal{R}_a(x) = 0$ and the natural damping $\mathcal{R}(x)$ satisfies the condition

$$\mathcal{R}(x)\frac{\partial H_a}{\partial x}(x) = 0,$$

then the new PBC is an energy balancing PBC. This is exactly the same condition that we obtained in the previous section.

Remark 18. In a series of papers, we have shown that, in many practical applications, the desired interconnection and damping matrices can be judiciously chosen by invoking physical considerations. The existing applications of interconnection and damping assignment PBC include mass–balance systems [18], electrical motors [19], power systems [20], magnetic levitation systems [21], underactuated mechanical systems [28], and power converters [22]. In the next section we present in detail a magnetic levitation system and refer the reader to the references cited above for additional examples that illustrate the generality of the new approach.

Remark 19. An interesting alternative to the Hamiltonian description of actuated mechanical systems is the *Lagrangian* description, with the Lagrangian being given by the *difference* of the kinetic and the potential energy. In this framework it is natural to pose the problem of when and how a state feedback for the actuation inputs can be designed such that the closed-loop system is again a Lagrangian system with a "desired" Lagrangian (as well as a desired damping). This line of research, called the technique of "controlled Lagrangians" was developed in a series of papers by Bloch et al. (e.g., [29,36], and followed up in [37] and [38]). The relation of these approaches to the approach of interconnection and damping assignment for port–controlled Hamiltonian systems taken in the present paper is rather straightforward. In particular, it is possible to show that modifying the *kinetic energy* of a mechanical system without affecting the potential energy nor the damping (as done in [29]) is tantamount—in our formulation—to selecting the closed–loop interconnection matrix as

$$J_d(q,p) = \begin{bmatrix} 0 & M_d^{-1}(q)M(q) \\ -M(q)M_d^{-1}(q) & Z(q,p) \end{bmatrix}$$

where $M_d(q)$, $M(q)$ are the closed–loop ("modified") and open–loop inertia matrices, respectively, and the elements of $Z(q,p)$ are computed as

$$Z(q,p)_{i,j} = -p^\top M^{-1}(q)M_d(q)\left[(M_d^{-1}M)._i,(M_d^{-1}M)._j\right](q) \tag{41}$$

with $(M_d^{-1}M)._i$ the i–th column of $M_d^{-1}M$ and $[\cdot,\cdot]$ the standard Lie bracket, see [2]. Furthermore, the addition of damping in the Lagrangian framework corresponds to damping assignment in the Hamiltonian case, while shaping the potential energy clearly fits within the shaping of the Hamiltonian. Hence we may conclude that the method of the "controlled Lagrangians" for actuated mechanical is a special case of our approach for the port–controlled Hamiltonian description of these systems. For example, in our approach the closed-loop interconnection matrix J_d can be chosen much more general than in (41). On the other hand, the freedom in choosing J_d may be so overwhelmingly rich that it is useful to have more specific subclasses of possible interconnection structure matrices like the one in (41) at hand. In general it seems of interest to investigate more deeply the embedding of the technique of controlled Lagrangians within our approach, also in relation to issues of "integrability", in particular the satisfaction of Jacobi-identity for J_d.

7 Magnetic Levitation System

7.1 Model

Consider the system of Fig. 3 consisting of an iron ball in a vertical magnetic field created by a single electromagnet. Here we adopt the standard assumption of unsaturated flux; that is, $\lambda = L(\theta)i$, where λ is the flux, θ is the difference between the position of the center of the ball and its nominal position, with the θ-axis oriented downward, i is the current, and $L(\theta)$ denotes the value of the inductance. The dynamics of the system is obtained by invoking Kirchoff's voltage law and Newton's second law as

$$\dot{\lambda} + Ri = u$$
$$m\ddot{\theta} = F - mg$$

where m is the mass of the ball, R is the coil resistance, and F is the force created by the electromagnet, which is given by

$$F = \frac{1}{2}\frac{\partial L}{\partial \theta}(\theta)i^2.$$

A suitable approximation for the inductance (in the domain $-\infty < \theta < 1$) is $L(\theta) = \frac{k}{1-\theta}$, where k is some positive constant that depends on the number of coil turns, and we have normalized the nominal gap to one.

To obtain a port-controlled Hamiltonian model, we define the state variables as $x = [\lambda, \theta, m\dot{\theta}]^\top$. The Hamiltonian function is given as

$$H(x) = \frac{1}{2k}(1 - x_2)x_1^2 + \frac{1}{2m}x_3{}^2 + mgx_2$$

and the port-controlled Hamiltonian model becomes

$$\dot{x} = \left(\underbrace{\begin{bmatrix} 0 & 0 & 0 \\ 0 & 0 & 1 \\ 0 & -1 & 0 \end{bmatrix}}_{J} - \underbrace{\begin{bmatrix} R & 0 & 0 \\ 0 & 0 & 0 \\ 0 & 0 & 0 \end{bmatrix}}_{\mathcal{R}} \right) \frac{\partial H}{\partial x}(x) + \underbrace{\begin{bmatrix} 1 \\ 0 \\ 0 \end{bmatrix}}_{g} u.$$

Given a constant desired position for the ball x_{2*}, the equilibrium we want to stabilize is $x_* = [\sqrt{2kmg}, x_{2*}, 0]^\top$.

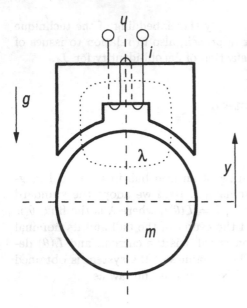

Fig. 3. Levitated ball $(y = \theta)$.

7.2   Changing the Interconnection

Next we show that, with the natural interconnection matrix of the system J, it is *not possible* to stabilize the desired equilibrium point with the proposed methodology; hence it is necessary to modify J. Toward this end, we observe that the key PDE to be solved (31) yields

$$(J - \mathcal{R})K(x) = g\beta(x) \Leftrightarrow \begin{cases} -RK_1(x) = \beta(x) \\ K_2(x) = 0 \\ K_3(x) = 0 \end{cases}$$

with $K(x)$ defined as (38). This means that the function $H_a(x)$ can only depend on x_1. Thus the resulting Lyapunov function would be of the form

$$H_d(x) = \frac{1}{2k}(1 - x_2)x_1^2 + \frac{1}{2m}x_3{}^2 + mgx_2 + H_a(x_1)$$

Even though, with a suitable selection of $H_a(x_1)$, we can satisfy the equilibrium assignment condition of Proposition 4, the Hessian will be defined as

$$\frac{\partial^2 H_d}{\partial x^2}(x) = \begin{bmatrix} \frac{(1-x_2)}{k} + \frac{\partial^2 H_a}{\partial x^2}(x_1) & -\frac{x_1}{k} & 0 \\ -\frac{x_1}{k} & 0 & 0 \\ 0 & 0 & \frac{1}{m} \end{bmatrix}$$

which is sign indefinite for all $H_a(x_1)$. It can actually be shown that the equilibrium is not stable.

The source of the problem is the lack of an *effective coupling* between the electrical and mechanical subsystems. Indeed, the interconnection matrix J only couples position with velocity. To overcome this problem, we propose to enforce a coupling between the flux x_1 and the velocity x_3; thus we propose the desired interconnection matrix

$$J_d = \begin{bmatrix} 0 & 0 & -\alpha \\ 0 & 0 & 1 \\ \alpha & -1 & 0 \end{bmatrix}$$

where α is a constant to be defined. Now, the key equation (37) becomes (with $\mathcal{R}_a = 0$)

$$-RK_1(x) = \frac{\alpha}{m}x_3 + \beta(x)$$

$$K_3(x) = 0$$

$$\alpha K_1(x) - K_2(x) = -\frac{\alpha}{k}(1 - x_2)x_1.$$

The first equation defines the control signal, whereas the last one can be readily solved (e.g., using symbolic programming languages) as

$$H_a(x) = \frac{1}{6k\alpha}x_1^3 + \frac{1}{2k}x_1^2(x_2 - 1) + \Phi(x_2 + \frac{1}{\alpha}x_1)$$

where $\Phi(\cdot)$ is an arbitrary continuous differentiable function. This function must be chosen to satisfy the equilibrium assignment and Lyapunov stability conditions of Proposition 4; that is, to assign a strict minimum at x_* to the new Lyapunov function

$$H_d(x) = \frac{1}{6k\alpha}x_1^3 + \frac{1}{2m}x_3^2 + mgx_2 + \Phi(x_2 + \frac{1}{\alpha}x_1).$$

It is easy to verify [21] that a suitable choice is given by

$$\Phi(x_2 + \frac{1}{\alpha}x_1) = mg[-(\tilde{x}_2 + \frac{1}{\alpha}\tilde{x}_1) + \frac{b}{2}(\tilde{x}_2 + \frac{1}{\alpha}\tilde{x}_1)^2]$$

where $\tilde{x}_i \overset{\Delta}{=} x_i - x_{i*}$, $i = 1, 2$, and $\alpha, b > 0$.

In conclusion, we have shown that the control law

$$u = \frac{R}{k}(1 - x_2)x_1 - K_p(\frac{1}{\alpha}\tilde{x}_1 + \tilde{x}_2) - \frac{\alpha}{m}x_3 - \frac{R}{\alpha}(\frac{1}{2k}x_1^2 - mg) \qquad (42)$$

stabilizes the equilibrium point x_* for all $K_p, \alpha > 0$, where we have defined a new constant K_p. It can be further established that stability is asymptotic, and an estimate of the domain of attraction can be readily determined.

7.3 Changing the Damping

A closer inspection of the control law (42) provides further insight that helps in its commissioning and leads to its simplification. The first right–hand term equals Ri; thus it cancels the voltage drop along the resistance. The second and third right–hand terms are linear proportional and derivative actions, respectively. Finally, the last term, which is proportional to acceleration, contains an undesirable nonlinearity that might saturate the control action. [We should note that the effect of the quadratic nonlinearity cannot be reduced without sacrificing the convergence rate, as can be seen from the dependence of the PD terms on α.] With the intent of removing this term, we propose to shuffle the damping, namely, to remove it from the electrical subsystem and add it up in the position coordinate; that is, we propose the added damping matrix

$$\mathcal{R}_a = \begin{bmatrix} -R & 0 & 0 \\ 0 & R_a & 0 \\ 0 & 0 & 0 \end{bmatrix}$$

where R_a is some positive number. Applying again the technique of Proposition 4, we can show that stabilization is possible with the simplified control law

$$u = \frac{R}{k}(1 - x_2)x_1 - K_p(\frac{1}{\alpha}\tilde{x}_1 + \tilde{x}_2) - \left(\frac{\alpha}{m} + K_p R_a\right) x_3.$$

where we have now defined $K_p \triangleq \frac{b\alpha}{R_a}$. Compare with (42).

8 Concluding Remarks

We have given a tutorial presentation of a control design approach for physical systems based on energy considerations that has been developed by the authors of the present article, as well as by some other researchers cited in the references, in the last few years. The main premise of this approach is that the fundamental concept of energy is lost in the signal processing perspective of most modern control techniques, hence we present an alternative viewpoint which focuses on interconnection. The choice of a suitable description of the system is essential for this research; thus we have adopted port–controlled Hamiltonian models which provide a classification of the variables and the equations into those associated with phenomenological properties and those defining the interconnection structure related to the exchanges of energy.

There are many possible extensions and refinements to the theory we have presented in this article. Some of these topics, and the lines of research we are pursuing to address them, may be found in [8]. Central among the various open issues that need to be clarified one finds, of course, the solvability of the PDE (37). Although we have shown that the added degrees of freedom

$(J_a(x), \mathcal{R}_a(x))$ can help us in its solution, it would be desirable to have a better understanding of their effect, that would lead to a more systematic procedure in their design. For general port–controlled Hamiltonian systems this is, we believe, a far–reaching problem. Hence, we might want to study it first for specific classes of physically–motivated systems.

Solving new problems is, of course, the final test for the usefulness of a new theory. Our list of references witnesses to the breadth of application of our approach, hence we tend to believe that this aspect has been amply covered by our work.

References

1. J. C. Willems, "Paradigms and puzzles in the theory of dynamical systems," *IEEE Trans. Automat. Contr.*, 36, pp. 259–294, 1991.
2. A. J., van der Schaft, L_2 -*Gain and Passivity Techniques in Nonlinear Control*, Springer–Verlag, Berlin, 1999.
3. M. Takegaki and S. Arimoto, "A new feedback method for dynamic control of manipulators," *ASME J. Dyn. Syst. Meas. Cont.*, 102, pp. 119-125, 1981.
4. E. Jonckheere, "Lagrangian theory of large scale systems," *Proc. European Conf. Circuit Th. and Design*, The Hague, The Netherlands, pp. 626–629, 1981.
5. J. J. Slotine, "Putting physics in control –The example of robotics," *IEEE Control Syst. Magazine*, 8, 6, pp. 12–17, 1988.
6. N. Hogan, "Impedance control: an approach to manipulation: part 1 - Theory," *ASME J. Dyn. Syst. Measure and Control*, 107, pp 1-7, March 1985.
7. R. Ortega and M. Spong, "Adaptive motion control of rigid robots: A tutorial," *Automatica*, 25, 6, pp. 877-888, 1989.
8. R. Ortega, A. van der Schaft, B. Maschke and G. Escobar, "Interconnection and damping assignment passivity based control of port–controlled Hamiltonian systems," *Automatica*, (to be published).
9. R. Ortega, A. Loria, P. J. Nicklasson and H. Sira–Ramirez, *Passivity-based Control of Euler-Lagrange Systems*, Springer-Verlag, Berlin, Communications and Control Engineering, Sept. 1998.
10. V. Petrovic, R. Ortega, A. Stankovic and G. Tadmor, "Design and implementation of an adaptive controller for torque ripple minimization in PM synchronous motors", *IEEE Trans. on Power Electronics*, Vol. 15, No. 5, Sept 2000, pp. 871–880.
11. H. Berghuis and H. Nijmeijer, "A passivity approach to controller-observer design for robots," *IEEE Trans. on Robotics and Automation*, 9, 6, pp. 740–754, 1993.
12. L. Lanari and J. Wen, "Asymptotically stable set point control laws for flexible robots," *Systems and Control Letters*, 19, 1992.
13. A. Sorensen and O. Egeland, "Design of ride control system for surface effect ships using dissipative control," *Automatica*, 31, 2, pp. 183–1999, 1995.
14. R. Prasanth and R. Mehra, "Nonlinear servoelastic control using Euler-Lagrange theory," *Proc. Int. Conf. American Inst. Aeronautics and Astronautics*, Detroit , pp. 837–847, Aug, 1999.

15. R. Akemialwati and I. Mareels, "Flight control systems using passivity–based control: Disturbance rejection and robustness analysis," *Proc. Int. Conf. American Inst. Aeronautics and Astronautics*, Detroit, Aug, 1999.

16. B. M. Maschke, R. Ortega and A. J. van der Schaft, "Energy–based Lyapunov functions for forced Hamiltonian systems with dissipation," *IEEE Conf. Dec. and Control*, Tampa, FL , Dec. 1998. Also *IEEE Trans. Automat. Contr.*, (to appear).

17. R. Ortega, A. van der Schaft, B. Maschke and G. Escobar, "Energy–shaping of port–controlled Hamiltonian systems by interconnection," *IEEE Conf. Dec. and Control*, Phoenix, AZ, Dec. 1999.

18. R. Ortega, A. Astolfi, G. Bastin and H. Rodriguez, "Output–feedback regulation of mass–balance systems," in *New Directions in Nonlinear Observer Design*, eds. H. Nijmeijer and T. Fossen, Springer–Verlag, Berlin, 1999.

19. V. Petrovic, R. Ortega and A. Stankovic, "A globally convergent energy-based controller for PM synchronous motors," *CDC'99*, Phoenix, AZ , Dec. 7–10, 1999. Also to appear in *IEEE Trans. on Control Syst. Technology*.

20. R. Ortega, M. Galaz, A. Bazanella and A. Stankovic, "Output feedback stabilization of the single–generator–infinite–bus system," (under preparation).

21. H. Rodriguez, R. Ortega and I. Mareels, "Nonlinear control of magnetic levitation sysstems via energy balancing," *ACC 2000*, Chicago, June 2000.

22. H. Rodriguez, R. Ortega, G. Escobar and N. Barabanov, "A robustly stable output feedback saturated controller for the boost DC–to–DC converter," *Systems and Control Letters*, 40, 1, pp. 1–6, 2000.

23. J. Scherpen, and R. Ortega, "Disturbance attenuation properties of nonlinear controllers for Euler–Lagrange systems," *Systems and Control Letters.*, 29, 6, pp. 300–308, March 1997.

24. A. Rodriguez, R. Ortega and G. Espinosa, "Adaptive control of nonfeedback linearizable systems," *11th World IFAC Congress*, Aug. 13-17, Tallinn, 1990.

25. P. Kokotovic and H. Sussmann, "A positive real condition for global stabilization of Nonlinear systems," *Systems and Control Letters*, 13, 4, pp. 125–133, 1989.

26. R. Sepulchre, M. Jankovic and P. Kokotovic, *Constructive Nonlinear Control*, Springer-Verlag Series on Communications and Control Engineering, Springer-Verlag, London, 1997.

27. M. Krstić. I. Kanellakopoulos and P. V. Kokotović, *Nonlinear and Adaptive Control Design*, Wiley, New York, 1995.

28. R. Ortega and M. Spong, "Stabilization of underactuated mechanical systems using interconnection and damping assignment," *IFAC Work. on Lagrangian and Hamiltonian Methods in Nonlinear Control*, Princeton, NJ, March 15–17, 2000.

29. A. Bloch, N. Leonhard and J. Marsden, "Controlled Lagrangians and the stabilization of mechanical systems," *Proc. IEEE Conf. Decision and Control*, Tampa, FL, Dec. 1998.

30. M. Spong and L. Praly, "Control of underactuated mechanical systems using switching and saturation," in *Control Using Logic–based Switching*, Ed. A. S. Morse, Springer, Lecture Notes No. 222, pp. 162–172, 1996.

31. D. Hill and P. Moylan, "The stability of nonlinear dissipative systems," *IEEE Trans. Automat. Contr.*, pp. 708–711, 1976.

32. M. Dalsmo and A.J. van der Schaft, "On representations and integrability of mathematical structures in energy-conserving physical systems," *SIAM J. on Optimization and Control*, Vol.37, No. 1, 1999.

33. J. Marsden and T. Ratiu, *Introduction to Mechanics and Symmetry*, Springer, New York, 1994.

34. R. Ortega, A. Loria, R. Kelly and L. Praly, "On output feedback global stabilization of Euler-Lagrange systems," *Int. J. of Robust and Nonlinear Cont.*, Special Issue on Mechanical Systems, Eds. H. Nijmeijer and A. van der Schaft, 5, 4, pp. 313-324, July 1995.

35. S. Stramigioli, B. M. Maschke and A. J. van der Schaft, "Passive output feedback and port interconnection," *Proc. 4th IFAC Symp. on Nonlinear Control Systems design*, NOLCOS'98, pp. 613–618, Enschede, July 1–3, 1998.

36. A.M. Bloch, N.E. Leonard, J.E. Marsden, "Potential shaping and the method of controlled Lagrangians," *38th Conf. Decision and Control*, Phoenix, Arizona, pp. 1652–1657, 1999.

37. D. Auckly, L. Kapitanski, W. White, "Control of nonlinear underactuated systems," *Communications on Pure and Applied Mathematics*, Vol. LIII, 2000.

38. J. Hamberg, "General matching conditions in the theory of controlled Lagrangians," *38th Conf. Decision and Control*, Phoenix, Arizona, pp. 2519–2523, 1999.

Geometric Modeling of mechanical Systems for Interactive Control

Stefano Stramigioli

Delft University of Technology, Delft, The Netherlands

1 Modeling of Mechanical Systems

Mechanical systems are in general much more complex than other physical systems because they bring with them the geometry of space. By modeling this geometry with proper tools like Lie groups [1] and screw theory[2], nice structures and properties can be specified for spatial mechanical systems. In this section we will review basic tools of Lie groups and screw theory for the modeling of rigid mechanical systems. In this work we explain mechanics using matrix Lie groups for didactical reasons. All the presented concept could be also introduced with more abstract Lie groups.

1.1 Introduction to Lie-groups

A manifold is intuitively a smooth space which is locally homeomorphic to $\mathbb{R}^n$ and brings with itself nice differentiability properties. Proper definitions of manifolds can be found on [3,4]. A group is an algebraical structure defined on a set. Definitions of groups can be found on any basic book of algebra.

A Lie group is a group, whose set on which the operation are defined is a manifold $\mathcal{G}$. This manifold $\mathcal{G}$ has therefore a special point 'e' which is the identity of the group.

Using the structure of the group, and by denoting the group operation as:

$$o : \mathcal{G} \times \mathcal{G} \to \mathcal{G} \; ;; \; (h, g) \mapsto h \circ g,$$

we can define two mappings within the group which are called respectively left and right mapping:

$$L_g : \mathcal{G} \to \mathcal{G} \; ; \; h \mapsto g \circ h \tag{1}$$

and

$$R_g : \mathcal{G} \to \mathcal{G} \; ; \; h \mapsto h \circ g \tag{2}$$

As we will see later the differential of these mappings at the identity, plays an important role in the study of mechanics.

The tangent space $T_e\mathcal{G}$ to $\mathcal{G}$ at e, which is indicated with $\mathfrak{g}$, has furthermore the structure of a Lie algebra which is nothing else than a vector space $\mathfrak{g}$ together with an internal, skew-symmetric operation called the commutator:

$$[,] : \mathfrak{g} \times \mathfrak{g} \to \mathfrak{g} ; \ (g_1, g_2) \mapsto [g_1, g_2] \tag{3}$$

For $\mathfrak{g}$ to be a Lie algebra, the commutator should furthermore satisfy what is called the Jacoby identity:

$$[g_1, [g_2, g_3]] + [g_2, [g_3, g_1]] + [g_3, [g_1, g_2]] = 0 \quad \forall g_1, g_2, g_3 \in \mathfrak{g} \tag{4}$$

Lie groups are important because we can use them as acting on a manifold $\mathcal{M}$, which in our case will be the Euclidean space. An action of $\mathcal{G}$ on $\mathcal{M}$, is a smooth application of the following form:

$$a : \mathcal{G} \times \mathcal{M} \to \mathcal{M}$$

such that

$$a(e, x) = x \quad \forall x \in \mathcal{M},$$

and

$$a(g_1, a(g_2, x)) = a(g_1 g_2, x) \quad \forall x \in \mathcal{M}, g_1, g_2 \in \mathcal{G}.$$

This means that an action is somehow compatible with the group on which it is defined.

1.2   Matrix Lie groups

For a lot of fundamental reasons like Ado's theorem [5], matrix algebras are excellent representatives for any finite dimensional group like the ones we need for rigid body mechanisms.

A matrix Lie group is a group whose elements are square matrices and in which the composition operation of the group corresponds to the matrix product. The most general real matrix group is $GL(n)$ which represents the group of non singular $n \times n$ real matrices. This is clearly a group since the identity matrix represents the identity element of the group, for each matrix, there is an inverse, and matrix multiplication is associative. We will now analyse more in detail features and operations of matrix Lie groups.

Left and Right maps If we consider a matrix Lie group $\mathcal{G}$, the operations of left and right translation clearly become:

$$L_G(H) = GH \quad \text{and} \quad R_G(H) = HG.$$

We can now consider how velocities are mapped using the previous maps. Suppose that we want to map a velocity vector $\dot{H} \in T_H\mathcal{G}$ to a velocity vector in $T_{GH}\mathcal{G}$ using the left translation and to a vector in $T_{HG}\mathcal{G}$ using right translation. We obtain:

$$(L_G)_*(H, \dot{H}) = (GH, G\dot{H}) \quad \text{and} \quad (R_G)_*(H, \dot{H}) = (HG, \dot{H}G)$$

In particular, if we take a reference velocity at the identity, we obtain:

$$(L_G)_*(I, T) = (G, GT) \quad \text{and} \quad (R_G)_*(I, T) = (G, TG)$$

where $T \in \mathfrak{g}$. With an abuse of notation, we will often indicate:

$$(L_G)_* T = GT \quad \text{and} \quad (R_G)_* T = TG$$

when it is clear that we consider mappings from the identity of the group. On a Lie group, we can define *left invariant* or *right invariant* vector fields. These vector fields are such that the differential of the left invariant and right invariant map leaves them invariant. If we indicate with

$$V : \mathcal{G} \to T\mathcal{G} \; ; \; x \mapsto (x, v)$$

a smooth vector field on the Lie group $\mathcal{G}$, we say that this vector field is left invariant if:

$$V(L_g(h)) = (L_g)_* V(h) \quad \forall g, h \in \mathcal{G},$$

and similarly it is right invariant if:

$$V(R_g(h)) = (R_g)_* V(h) \quad \forall g \in \mathcal{G}.$$

For a matrix group, if we take in the previous definitions $h = I$ we obtain respectively:

$$V(G) = GT_L \quad \text{and} \quad V(G) = T_R G$$

where we indicated the representative of the left and right invariant vector fields at the identity with T_L and T_R. We can conclude from this that any left or right invariant vector field is characterized completely by its value at the identity of the group. We could now ask ourself: what are the integrals of a left or invariant vector field? From what just said, the integral of a left invariant vector field, can be calculated as the integral of the following matrix differential equation:

$$\dot{G} = GT_L \Rightarrow G(t) = G(0)e^{T_L t} \tag{5}$$

where T_L is the value of the vector field at the identity. In a similar way, the integral of a right invariant vector field is:

$$\dot{G} = T_R G \Rightarrow G(t) = e^{T_R t}G(0). \tag{6}$$

From this it is possible to conclude that if we take an element $T \in \mathfrak{g}$, its left and right integral curves passing through the identity coincide and they represent the exponential map from the Lie algebra to the Lie group:

$$e : \mathfrak{g} \to \mathcal{G}; T \mapsto e^T.$$

It is easy to show, and important to notice, that integral curves passing through points $H = e^{T_1}$ of right and left invariant vector fields which have as representative in the identity T_2, are coincident iff $e^{T_1} e^{T_2} = e^{T_2} e^{T_1}$ which is true iff $[T_1, T_2] = 0$, where the last operation is the commutator of the Lie algebra. But how does the commutator look like for a matrix Lie algebra ? Being a Lie group a manifold, we can compute the Lie brackets of vector fields on the manifold. Furthermore, we know that elements of the Lie algebra $\mathfrak{g}$ have a left and right vector field associated to them. We can than calculate the Lie bracket of two left or right invariant vector fields, and if the solution is still left or right invariant, consider the value of the resulting vector field at the identity as the solution of the commutator. We will start with the left invariant case first. Consider we are in a point $G(t) \in \mathcal{G}$ at time t. If we have two left invariant vector fields characterized by $T_1, T_2 \in \mathfrak{g}$, the Lie bracket of these two vector fields, can be calculated by moving from $G(t)$ along the vector field correspondent to T_1 for $\sqrt{s}$ time, than along the one correspondent to T_2, than along $-T_1$ and eventually along $-T_2$. In mathematical terms we have:

$$G(t + \sqrt{s}) = G(t)e^{T_1\sqrt{s}} \to G(t + 2\sqrt{s}) = G(t + \sqrt{s})e^{T_2\sqrt{s}} \to$$
$$G(t + 3\sqrt{s}) = G(t + 2\sqrt{s})e^{-T_1\sqrt{s}} \to G(t + 4\sqrt{s}) = G(t + 3\sqrt{s})e^{-T_2\sqrt{s}} \to$$
$$G(t + 4\sqrt{s}) = G(t)e^{T_1\sqrt{s}}e^{T_2\sqrt{s}}e^{-T_1\sqrt{s}}e^{-T_2\sqrt{s}} \quad (7)$$

If we look at $\frac{d}{ds}G(t + 4\sqrt{s})\big|_{s=0}$, we can approximate the exponentials with the first low order terms and we obtain:

$$G(t + 4\sqrt{s}) \simeq G(t)\left(\left(I + T_1\sqrt{s} + \frac{T_1^2}{2}s\right)\left(I + T_2\sqrt{s} + \frac{T_2^2}{2}s\right)\right.$$
$$\left.\left(I - T_1\sqrt{s} + \frac{T_1^2}{2}s\right)\left(I - T_2\sqrt{s} + \frac{T_1^2}{2}s\right)\right)$$
$$\simeq G(t)(I + (T_1T_2 - T_2T_1)s + o(s)) \quad (8)$$

which implies

$$\frac{d}{ds}G(t + 4\sqrt{s})\bigg|_{s=0} = G(t)(T_1T_2 - T_2T_1).$$

From the previous equation, we can conclude that the resulting vector field is still left invariant and it is characterized by the Lie algebra element $T_1T_2 -$

$T_2 T_1$. We can therefore define the commutator based on left invariant vector fields as:

$$[T_1, T_2]_L = T_1 T_2 - T_2 T_1.$$

With similar reasoning, it is possible to show for right invariant vector fields that:

$$\frac{d}{ds} G(t + 4\sqrt{s}) \bigg|_{s=0} = (T_2 T_1 - T_1 T_2) G(t).$$

and therefore, in this case:

$$[T_1, T_2]_R = T_2 T_1 - T_1 T_2.$$

We have therefore that:

$$[T_1, T_2]_L = -[T_1, T_2]_R.$$

In the literature, $[,]_L$ is used as the standard commutator and we will adapt this convention.

Matrix Group Actions A group action we can consider for an n dimensional matrix Lie group is the linear operation on $\mathbb{R}^n$. We can therefore define as an action:

$$a(G, P) = GP \qquad G \in \mathcal{G}, P \in \mathbb{R}^n$$

It is easy to see that this group action trivially satisfies all the properties required.

Adjoint representation Using the left and right maps, we can define what is called the conjugation map as $K_g := R_{g^{-1}} L_g$ which for matrix groups results:

$$K_G : \mathcal{G} \to \mathcal{G} \; ; \; H \mapsto GHG^{-1}.$$

But what is the importance of this conjugation map ? To answer this question, we need the matrix group action. Suppose we have a certain element $H \in \mathcal{G}$ such that $Q = HP$ where $Q, P \in \mathbb{R}^n$. What happens if we move all the points of $\mathbb{R}^n$ and therefore also Q and P using an element of $\mathcal{G}$? What will the corresponding mapping of H look like ? If we have $Q' = GQ$ and $P' = GP$, it is straight forward to see that:

$$Q' = K_G(H)P'.$$

The conjugation map is therefore related to global motions or equivalently changes of coordinates. We clearly have that $K_G(I) = I$ and therefore the

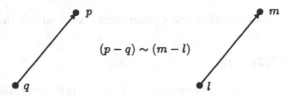

Fig. 1. The definition of a free vector

differential of $K_G()$ at the identity is a Lie algebra endomorphism. This linear map is called the Adjoint group representation:

$$Ad_G : \mathfrak{g} \to \mathfrak{g} ; T \mapsto GTG^{-1}.$$

The Adjoint representation of the group shows how an infinitesimal motion changes moving the references of a finite amount G. Eventually, it is possible to consider the derivative of the previous map at the identity

$$ad_T := \frac{d}{ds} Ad_{e^{Ts}}\bigg|_{s=0}.$$

This map is called the adjoint representation of the Lie algebra and it is a map of the form:

$$ad_T : \mathfrak{g} \to \mathfrak{g} \qquad T \in \mathfrak{g}$$

If we use the definitions we can see that:

$$\frac{d}{ds} Ad_{e^{T_1 s}} \cdot T_2 \bigg|_{s=0} = \frac{d}{ds} e^{T_1 s} T_2 e^{-T_1 s} \bigg|_{s=0} = T_1 T_2 - T_2 T_1 = [T_1, T_2]_L$$

which shows that:

$$ad_{T_1} T_2 = [T_1, T_2]_L \tag{9}$$

1.3 Euclidean space and motions

It is now possible to use the matrix Lie group concepts developed in the previous section for the study of rigid body motions. We will start by talking about Euclidean spaces.

An n dimensional Euclidean space $\mathcal{E}(n)$ is characterized by a *scalar product* which allows to define orthogonality of vectors and their lengths. We can consider the relative position of two points $p, q \in \mathcal{E}(n)$ as a vector $(p - q)$ directed from q to p in the usual way as reported in Fig. 1. In projective terms, such a vector can be interpreted as a vector belonging to the improper hyperplane [6]. The set of these *free vectors* will be indicated with $\mathcal{E}_*(n)$.

These vectors are called "free" because they are free to be moved around parallelly without being bound to a point (see Fig. 1).

If we constrain these vectors only to move along the line they span, we get the concept of *line vectors* which are characterized by a straight line in $\mathcal{E}$ plus a direction and a module. These kind of vectors are also called *rotors* for reasons which will become clear later on. A line vector is completely defined by a pair $(p, v) \in \mathcal{E} \times \mathcal{E}_*$ giving a point on the line and a vector specifying the direction and module. Clearly we have the following equivalence relation:

$$(p_1, v_1) \sim (p_2, v_2) \text{ iff } v_1 = v_2, \exists \lambda \text{ s.t. } (p_1 - p_2) = \lambda v_i.$$

The scalar product which characterizes the Euclidean space is a function of the following form:

$$\langle , \rangle : \mathcal{E}_*(n) \times \mathcal{E}_*(n) \to \mathbb{R} \; ; \; (v, w) \mapsto \langle v, w \rangle,$$

and satisfies the usual properties of an internal product. We can furthermore define the *distance* $d(p, q)$ of p from q in the usual way as as the length of $(p - q) \in \mathcal{E}_*$:

$$\| \cdot \| : \mathcal{E}_* \to \mathbb{R} \; ; \; v \mapsto \sqrt{\langle v, v \rangle},$$

and the orthogonality of vectors $v, w \neq 0$ in the following way:

$$v \perp w \Leftrightarrow \langle v, w \rangle = 0.$$

The cosine can be than defined as

$$\cos v \angle w := \frac{\langle v, w \rangle}{\| v \| \cdot \| w \|}.$$

Coordinate systems For the Euclidean space $\mathcal{E}(n)$, a coordinate system is an $(n + 1)$-tuple:

$$\Psi_o := (o, e_1, e_2, \ldots, e_n) \in \mathcal{E}(n) \times \underbrace{\mathcal{E}_*(n) \times \mathcal{E}_*(n)}_{n \text{ times}}$$

such that $e_1, \ldots, e_n$ are linear independent vectors and form therefore a base for $\mathcal{E}_*$. Furthermore, the coordinates systems is orthonormal if

$$\| e_i \| = 1 \quad \forall i \quad \text{(unit vectors)}$$

and

$$\langle e_i, e_j \rangle = 0 \quad \forall i \neq j \quad \text{(orthogonality)}.$$

The coordinates of a point $p \in \mathcal{E}$ are real numbers and calculated as:

$$x_i = \langle (p - o_i), e_i \rangle \in \mathbb{R} \quad \forall i.$$

In a similar way, the coordinates of a vector $v \in \mathcal{E}_*$ are

$$x_i = \langle v, e_i \rangle \in \mathbb{R} \quad \forall i.$$

If we consider the three dimensional Euclidean space $\mathcal{E}(3)$ we usually indicate for an orthonormal coordinate system $\hat{x} := e_1$, $\hat{y} := e_2$ and $\hat{z} := e_3$. In all what follows we will implicitly assume to be using orthonormal coordinate systems.

For any Cartesian coordinate system $\Psi_i = (o_i, \hat{x}_i, \hat{y}_i, \hat{z}_i)$, we can define a coordinate mapping as:

$$\psi_i : \mathcal{E}(3) \to \mathbb{R}^3 \; ; \; p \mapsto \begin{pmatrix} \langle (p - o_i), \hat{x}_i \rangle \\ \langle (p - o_i), \hat{y}_i \rangle \\ \langle (p - o_i), \hat{z}_i \rangle \end{pmatrix}$$

It is then possible to define a change of coordinates from a reference system Ψ_2 to a reference system Ψ_1:

$$p^2 = (\psi_2 \, o \, \psi_1^{-1})(p^1)$$

which is a mapping like

$$\mathbb{R}^3 \xrightarrow{\psi_1^{-1}} \mathcal{E}(3) \xrightarrow{\psi_2} \mathbb{R}^3 \; ; \; p^1 \mapsto p \mapsto p^2$$

Rotations We can now consider the changes of coordinates between two reference frames Ψ_1 and Ψ_2 which have the origin in common. We have that:

$$p^1 = \begin{pmatrix} x_1 \\ y_1 \\ z_1 \end{pmatrix} = \psi_1(p) = \begin{pmatrix} \langle (p - o_1), \hat{x}_1 \rangle \\ \langle (p - o_1), \hat{y}_1 \rangle \\ \langle (p - o_1), \hat{z}_1 \rangle \end{pmatrix}$$

or equivalently

$$(p - o_1) = x_1 \hat{x}_1 + y_1 \hat{y}_1 + z_1 \hat{z}_1$$

with $x_1, y_1, z_1 \in \mathbb{R}$ and $\hat{x}_1, \hat{y}_1, \hat{z}_1 \in \mathcal{E}_*$. Similarly, for $p^2 \in \mathbb{R}^3$, we have:

$$p^2 := \begin{pmatrix} x_2 \\ y_2 \\ z_2 \end{pmatrix} = \begin{pmatrix} \langle (p - o_2), \hat{x}_2 \rangle \\ \langle (p - o_2), \hat{y}_2 \rangle \\ \langle (p - o_2), \hat{z}_2 \rangle \end{pmatrix}$$

and since for hypothesis $o_1 = o_2$, using the expression of $(p - o_1)$ in Ψ_1, we have that

$$\begin{pmatrix} x_2 \\ y_2 \\ z_2 \end{pmatrix} = \begin{pmatrix} \langle x_1 \hat{x}_1 + y_1 \hat{y}_1 + z_1 \hat{z}_1, \hat{x}_2 \rangle \\ \langle x_1 \hat{x}_1 + y_1 \hat{y}_1 + z_1 \hat{z}_1, \hat{y}_2 \rangle \\ \langle x_1 \hat{x}_1 + y_1 \hat{y}_1 + z_1 \hat{z}_1, \hat{z}_2 \rangle \end{pmatrix} =$$

$$\begin{pmatrix} x_1 \langle \hat{x}_1, \hat{x}_2 \rangle + y_1 \langle \hat{y}_1, \hat{x}_2 \rangle + z_1 \langle \hat{z}_1, \hat{x}_2 \rangle \\ x_1 \langle \hat{x}_1, \hat{y}_2 \rangle + y_1 \langle \hat{y}_1, \hat{y}_2 \rangle + z_1 \langle \hat{z}_1, \hat{y}_2 \rangle \\ x_1 \langle \hat{x}_1, \hat{z}_2 \rangle + y_1 \langle \hat{y}_1, \hat{z}_2 \rangle + z_1 \langle \hat{z}_1, \hat{z}_2 \rangle \end{pmatrix}$$

which in matrix form gives:

$$\begin{pmatrix} x_2 \\ y_2 \\ z_2 \end{pmatrix} = \begin{pmatrix} \langle \hat{x}_1, \hat{x}_2 \rangle & \langle \hat{y}_1, \hat{x}_2 \rangle & \langle \hat{z}_1, \hat{x}_2 \rangle \\ \langle \hat{x}_1, \hat{y}_2 \rangle & \langle \hat{y}_1, \hat{y}_2 \rangle & \langle \hat{z}_1, \hat{y}_2 \rangle \\ \langle \hat{x}_1, \hat{z}_2 \rangle & \langle \hat{y}_1, \hat{z}_2 \rangle & \langle \hat{z}_1, \hat{z}_2 \rangle \end{pmatrix} \begin{pmatrix} x_1 \\ y_1 \\ z_1 \end{pmatrix}$$

We indicate the previous matrix with R_1^2 which represents a change of coordinates from Ψ_1 to Ψ_2. It is important to note that the columns of the matrix are the vector bases of Ψ_1 expressed in Ψ_2 and the rows are the vector bases of Ψ_2 expressed in Ψ_1. Due to the hypothesis on the ortho-normality of Ψ_1 and Ψ_2, we can conclude therefore that the matrix R_1^2 is an orthonormal matrix. For example, a change of coordinates due to a rotation of θ around $\hat{y}_1$ is such that:

$$R_1^2 = \begin{pmatrix} \cos(\theta) & 0 & \sin(\theta) \\ 0 & 1 & 0 \\ -\sin(\theta) & 0 & \cos(\theta) \end{pmatrix}$$

It is possible to see that the set of matrices satisfying $R^{-1} = R^T$ is a three dimensional matrix Lie group which is called *orthonormal* group and indicated with $O(3)$. Since $R^T R = I \ \forall R \in O(3)$, for the rule on the determinant of the product, it is clear that the determinant of any matrix in $O(3)$ can be ± 1. This shows that $O(3)$ is composed of two disjointed components, one whose matrices have determinant equal to -1 which is not a group by itself, and one called *special orthonormal group* indicated with $SO(3)$ of the matrices with determinant equal to 1 which is clearly a matrix Lie group:

$$SO(3) = \{ R \in \mathbb{R}^{3 \times 3} \ \text{s.t.} \ R^{-1} = R^T, \det R = 1 \}.$$

It is now possible to investigate how the elements of the Lie algebra of $SO(3)$, which is denoted with $\mathfrak{so}(3)$, look like. From the theory on left and right invariant vector fields, Eq. (5) and Eq. (6), we can see that elements of the algebra should look like $T_L = R^{-1} \dot{R}$ or like $T_R = \dot{R} R^{-1}$ for any curve $R(t) \in SO(3)$. But since $R^{-1} = R^T$, if we differentiate the equation $R^T R = I$ we obtain:

$$\dot{R}^T R + R^T \dot{R} = 0 \Rightarrow R^T \dot{R} = -(R^T \dot{R})^T$$

and therefore the matrix T_L is skew-symmetric. Similarly it can be shown that T_R is also skew-symmetric. This shows that $\mathfrak{so}(3)$ is the vector space of skew-symmetric matrices. Note that due to the structure of the matrix commutator of the algebra, the commutation of skew-symmetric matrices is still skew-symmetric.

There is a bijective relation between 3×3 skew-symmetric matrices and 3 vectors. We will use the following notation:

$$x = \begin{pmatrix} x_1 \\ x_2 \\ x_3 \end{pmatrix} \Rightarrow \tilde{x} = \begin{pmatrix} 0 & -x_3 & x_2 \\ x_3 & 0 & -x_1 \\ -x_2 & x_1 & 0 \end{pmatrix}.$$

This reflects the usual notation for the vector product in $\mathbb{R}^3$:

$$x \wedge y = \tilde{x}y \qquad \forall x, y \in \mathbb{R}^3.$$

It is straight forward to see that the algebra matrix commutator of $\mathfrak{so}(3)$ corresponds to the vector product of the corresponding three vectors:

$$[\tilde{\omega}_x, \tilde{\omega}_y] = (\omega_x \tilde{\wedge} \omega_y) \qquad \forall \omega_x, \omega_y \in \mathbb{R}^3.$$

The elements of $\mathfrak{so}(3)$ correspond to the angular velocity of the frames whose relative change of coordinates is represented by the matrix $R_1^2 \in SO(3)$. We will analyse in more detail the general case with also translations. Before proceeding with general motions, it is important to state that a lot of other representations of rotations exists beside $SO(3)$. Some examples are the group of unit quaternions which is isomorphic to another Lie group $SU(2)$ which is called *special unitary group*. They both double cover $SO(3)$ and they are topologically speaking simply connected. $SO(3)$ is instead not contractible and it is a typical example of a non simply connected manifold 'without holes'.

Vector product It is often a misconception that the vector product in three space is an extra structure which is defined on $\mathcal{E}_*$ beside the scalar product. This is NOT the case. The vector product is instead a consequence of the fact that for $\mathcal{E}$ the Lie group of rotations $SO(3)$ can be intrinsically defined. As we have just seen, the Lie algebra $\mathfrak{so}(3)$ has a commutator defined on it and therefore, if we find an intrinsic bijection between $\mathfrak{so}(3)$ and $\mathcal{E}_*$ we can transport the commutation operation defined in $\mathfrak{so}(3)$ to $\mathcal{E}_*$ and obtain the vector product.

It is possible to find two bijections in the following way. A vector of $\mathcal{E}_*$ is characterized by a direction d, an orientation v and a module m. An element of $\mathfrak{so}(3)$ represents an angular velocity. It is possible to see that any angular motion leaves a line d' invariant. This is the first step in this bijection: the vector of $\mathcal{E}_*$ which we associate to a vector in $\mathfrak{so}(3)$ will have as direction the line left invariant by this rotation ($d = d'$). We still need a module and an orientation. It is possible to show, that there exists a positive definite metric on $\mathfrak{so}(3)$ which is defined using what is called the *killing form* [7]. We can therefore choose as m the module of the angular velocity vector calculated using the metric on $\mathfrak{so}(3)$. The only choice we are left with is for the orientation. We clearly have two possible choices to orient the line. If we look at the rotation motion around its axis as a clockwise motion, we can orient the line as going away from us or as coming toward us. In the first case we say that we choose a right handed orientation and in the second case a left handed orientation.

Notice that if we look at this motion through a mirror, the orientation is changed because for the same rotational motion a line oriented toward the mirror will be seen through the mirror as a line oriented in the opposite direction. This is called a *reflection*.

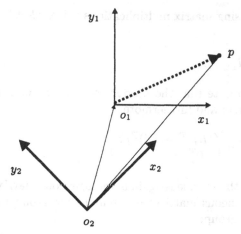

Fig. 2. A general change of coordinates

General motions Consider the coordinate systems reported in Fig. 2. What is an expression representing the change of coordinates $p^1 \to p^2$ where we denoted with $p^i \in \mathbb{R}^3$ the numerical expression of p in Ψ_i? First notice that $p^2 \in \mathbb{R}^3$ is the numerical vector of the coordinates of $(p - o_2) \in \mathcal{E}_*$ expressed in the frame Ψ_2. From Fig. 2 we can see that:

$$(p - o_2) = (p - o_1) + (o_1 - o_2).$$

Furthermore, it is possible to see that:

$$p^2 = R_1^2 p^1 + p_1^2$$

where p_1^2 is the vector $(o_1 - o_2)$ expressed in Ψ_2 and $R_1^2 \in SO(3)$ is the change of coordinate matrix if o_1 and o_2 would be coincident. We can also write the change of coordinates in matrix form:

$$\begin{pmatrix} p^2 \\ 1 \end{pmatrix} \begin{pmatrix} R_1^2 & p_1^2 \\ 0_3^T & 1 \end{pmatrix} \begin{pmatrix} p^1 \\ 1 \end{pmatrix}.$$

The previous matrix which will be denoted with $H_1^2 \in \mathbb{R}^{4 \times 4}$ is called a *homogeneous matrix*:

$$H_1^2 := \begin{pmatrix} R_1^2 & p_1^2 \\ 0_3^T & 1 \end{pmatrix},$$

and represents the change of coordinates from Ψ_1 to Ψ_2. Notice that we now have a four dimensional vector of coordinates for each point whose last component is equal to 1. This is interpretable in projective geometric terms as a projective point. Notice that sequential changes of coordinates can be now easily expressed using matrix multiplications using what is sometimes called the chain rule:

$$H_1^4 = H_3^4 H_2^3 H_1^2.$$

It is important to note that the product of homogeneous matrices is still a homogeneous matrix and furthermore:

$$H_2^1 = (H_1^2)^{-1} = \begin{pmatrix} (R_1^2)^T & -(R_1^2)^T p_1^2 \\ 0_3^T & 1 \end{pmatrix},$$

which shows that the inverse is again a homogeneous matrix. This shows that the set of homogeneous matrices is a matrix Lie group which is called the *special Euclidean* group:

$$SE(3) := \left\{ \begin{pmatrix} R & p \\ 0 & 1 \end{pmatrix} \text{ s.t. } R \in SO(3), p \in \mathbb{R}^3 \right\}.$$

Since $SE(3)$ is a matrix Lie group, we can map velocities $\dot{H}_i^j \in T_{H_i^j} SE(3)$ to $\mathfrak{se}(3)$ either with left or right translations. But how do the elements of $\mathfrak{se}(3)$ look like ? It is possible to see that:

$$\mathfrak{se}(3) = \left\{ \begin{pmatrix} \Omega & v \\ 0 & 0 \end{pmatrix} \text{ s.t. } \Omega \in \mathfrak{so}(3), v \in \mathbb{R}^3 \right\}.$$

Furthermore, the algebra commutator of $\mathfrak{se}(3)$ is such that since $\tilde{T}_i := \begin{pmatrix} \tilde{\omega}_i & v_i \\ 0 & 0 \end{pmatrix} \in \mathfrak{se}(3)$ then:

$$[\tilde{T}_1, \tilde{T}_2] = \begin{pmatrix} [\tilde{\omega}_1, \tilde{\omega}_2] & \tilde{\omega}_1 v_2 - \tilde{\omega}_2 v_1 \\ 0 & 0 \end{pmatrix}.$$

Twists We have seen that elements of $\mathfrak{se}(3)$ are 4×4 matrices of a specific form. We can uniquely associate to each of these matrices a six dimensional equivalent vector representation such that

$$T := \begin{pmatrix} \omega \\ v \end{pmatrix} \Rightarrow \tilde{T} = \begin{pmatrix} \tilde{\omega} & v \\ 0 & 0 \end{pmatrix} \in \mathfrak{se}(3).$$

We have therefore both a matrix and a vector representation for $\mathfrak{se}(3)$. Elements of $\mathfrak{se}(3)$ are called twists in mechanics and they represent the velocity of a rigid body motion geometrically. To understand this, we must look at the action of elements of $SE(3)$ on points of $\mathbb{R}^3$. Consider ANY point p not moving with respect to the reference frame Ψ_i. If we indicate with p^i its numerical representation in Ψ_i, this means that $\dot{p}^i = 0$. Take now a second reference Ψ_j

Transport	$\tilde{T}$	$\dot{H}_i^j$	$\dot{P}^j$	Notation
Left	$\tilde{T} := H_j^i \dot{H}_i^j$	$\dot{H}_i^j = H_i^j \tilde{T}$	$\dot{P}^j = H_i^j (\tilde{T} P^i)$	$\tilde{T}_i^{i,j}$
Right	$\tilde{T} := \dot{H}_i^j H_j^i$	$\dot{H}_i^j = \tilde{T} H_i^j$	$\dot{P}^j = \tilde{T} (H_i^j P^i)$	$\tilde{T}_i^{j,j}$

Table 1. The used notation for twists.

possibly moving with respect to Ψ_i. By looking at the change of coordinates and differentiating, we obtain:

$$\dot{P}^j = \dot{H}_i^j P^i \quad \text{where we denoted} \quad P^k := \begin{pmatrix} p^k \\ 1 \end{pmatrix} \quad k = i, j$$

and $H_i^j \in SE(3)$. We can now transport $\dot{H}_i^j$ to the identity either by left or right translation. If we do so, we obtain the two possibilities reported in Tab. 1. In the case we consider the left translation, working out the terms, we obtain:

$$\dot{P}^j = \underbrace{H_i^j (\tilde{T} P^i)}_{\dot{H}_i^j}$$

$$\begin{pmatrix} \dot{p}^j \\ 0 \end{pmatrix} = \begin{pmatrix} R_i^j & p_i^j \\ 0 & 1 \end{pmatrix} \begin{pmatrix} \tilde{\omega} & v \\ 0 & 0 \end{pmatrix} \begin{pmatrix} p^i \\ 1 \end{pmatrix} \Rightarrow \dot{p}^j = R_i^j (\omega \wedge p^i) + R_i^j v$$

and using the right translation instead we obtain:

$$\dot{P}^j = \underbrace{\tilde{T} (H_i^j P^i)}_{\dot{H}_i^j}$$

$$\begin{pmatrix} \dot{p}^j \\ 0 \end{pmatrix} = \begin{pmatrix} \tilde{\omega} & v \\ 0 & 0 \end{pmatrix} \begin{pmatrix} R_i^j & p_i^j \\ 0 & 1 \end{pmatrix} \begin{pmatrix} p^i \\ 1 \end{pmatrix} \Rightarrow \dot{p}^j = \omega \wedge (R_i^j p^i + p_i^j) + v$$

From the previous two expressions and Tab. 1, it is possible to see that $T_a^{b,c}$ represents the motion of Ψ_a with respect to Ψ_c expressed in frame Ψ^b. Note that v is NOT the relative velocity of the origins of the coordinate systems! This would not give rise to a geometrical entity! To understand this better, it is worth considering Chasles theorem. This theorem says that any rigid motion can be described as a pure rotation around an axis plus a pure translation along the same axis. Using the expression for twists in vector form, this can be mathematically expressed as:

$$\begin{pmatrix} \omega \\ v \end{pmatrix} = \|\omega\| \underbrace{\begin{pmatrix} \hat{\omega} \\ r \times \hat{\omega} \end{pmatrix}}_{\text{rotation}} + \alpha \underbrace{\begin{pmatrix} 0 \\ \hat{\omega} \end{pmatrix}}_{\text{translation}}$$

where $\omega = ||\omega||\hat{\omega}$. Analyzing the previous formula, it is possible to see that v is the velocity of an imaginary point passing through the origin of the coordinate system in which the twist is expressed and moving together with the object. The six vector representing the rotation is what we called previously a rotor and can be associated to a geometrical line, namely the line passing through r and spanned by ω which is left invariant by the rotation.

The theorem of Chasles is one of the two theorems on which screw theory is based because it gives to elements of $\mathfrak{se}(3)$ a real tensorial geometrical interpretation. This interpretation is the one of a *motor* or *screw* which are entities characterized by a geometrical line and a scalar called the pitch. This pitch relates the ratio of translation and rotation along and around the line.

Changes of coordinates for twists Using the left and right map, we have seen that:

$$\tilde{T}_i^{i,j} = H_j^i \dot{H}_i^j \text{ and } \tilde{T}_i^{j,j} = \dot{H}_i^j H_j^i.$$

It can be easily seen that

$$\tilde{T}_i^{j,j} = H_i^j \tilde{T}_i^{i,j} H_j^i,$$

and this gives an expression for changes of coordinates of twists. This clearly corresponds to the adjoint group representation introduced at pag.313. It is easier to work with the six dimensional vector form of twists and it is possible to see that we can find a matrix expression of the adjoint representation:

$$T_i^{j,j} = Ad_{H_i^j} T_i^{i,j}.$$

It is possible to proof that this matrix representation is:

$$Ad_{H_i^j} := \begin{pmatrix} R_i^j & 0 \\ \tilde{p}_i^j R_i^j & R_i^j \end{pmatrix}.$$

Notice that the change of coordinates can be seen as the change of coordinates of the geometrical line associated to the twist using Chasles' theorem. By differentiation of the previous matrix as a function of H_i^j we can also find an expression for its time derivative and the adjoint representation of the algebra ad. It can be shown that:

$$\left(\dot{Ad}_{H_i^j} \right) = Ad_{H_i^j} ad_{T_i^{i,j}} \quad \text{with} \quad T_i^{i,j} := H_j^i \dot{H}_i^j,$$

where

$$ad_{T_i^{k,j}} := \begin{pmatrix} \tilde{\omega}_i^{k,j} & 0 \\ \tilde{v}_i^{k,j} & \tilde{\omega}_i^{k,j} \end{pmatrix}$$

is the adjoint representation we where looking for. This can be easily checked by testing the relation proved in Eq. (9) for a general matrix Lie group.

Wrenches Twists are the generalization of velocities and are elements of $\mathfrak{se}(3)$. The dual vector space of $\mathfrak{se}(3)$ is called the dual Lie algebra and denoted with $\mathfrak{se}^*(3)$. It is the vector space of linear operators from $\mathfrak{se}(3)$ to $\mathbb{R}$. This space represents the space of 'forces' for rigid bodies which are called *wrenches*. The application of a wrench on a twists gives a scalar representing the power supplied by the wrench. A wrench in vector form will be a 6 dimensional row vector since it is a co-vector (linear operator on vectors) instead than a vector.

$$W = \begin{pmatrix} m & f \end{pmatrix}$$

where m represents a torque and f a linear force. For what just said we have:

$$\text{Power} = WT$$

where T is a twist of the object on which the wrench is applied. Clearly, to calculate the power, the wrench and the twist have to be numerical vectors expressed in the same coordinates and result

$$\text{Power} = WT = m\omega + fv$$

Another representation of a wrench in matrix form is:

$$\tilde{W} = \begin{pmatrix} \tilde{f} & m \\ 0 & 0 \end{pmatrix}.$$

How do wrenches transform changing coordinate systems ? We have seen that for twists:

$$T^{j,\bullet}_{\bullet} = Ad_{H^j_i} T^{i,\bullet}_{\bullet}$$

where

$$Ad_{H^j_i} := \begin{pmatrix} R^j_i & 0 \\ \tilde{p}^j_i R^j_i & R^j_i \end{pmatrix}.$$

Suppose to supply power to one body attached to Ψ_j by means of a wrench which represented in Ψ_j is W^j.

Changing coordinates from Ψ_j to Ψ_i the expression of the supplied power should stay constant and this implies that:

$$W^j T^{j,i}_j = W^j Ad_{H^j_i} T^{i,i}_j = (Ad^T_{H^j_i}(W^j)^T)^T T^{i,i}_j = W^i T^{i,i}_j$$

which implies that the transformation of wrenches expressed in vector form is:

$$(W^i)^T = Ad^T_{H^j_i}(W^j)^T.$$

Note that if the mapping $Ad_{H^j_i}$ was mapping twists from Ψ_i to Ψ_j, the transposed maps wrenches in the opposite direction: from Ψ_j to Ψ_i ! This is a direct consequence of the fact that wrenches are duals to twists.

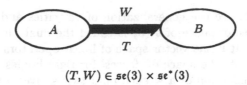

$$(T, W) \in \mathfrak{se}(3) \times \mathfrak{se}^*(3)$$

Fig. 3. The interaction between two mechanical systems

1.4   Power Ports

A basic concept which is needed to talk about interconnection of physical systems is the one of a power port [8],[9]. With reference to Fig. 3, a power port is the entity which describes the media by means of which subsystems can mutually exchange physical energy. Analytically, a power port can be defined by the Cartesian product of a vector space V and its dual space V^*:

$$P := V \times V^*.$$

Therefore, power ports are pairs $(e, f) \in P$. The values of both e and f (*effort* and *flow* variables) change in time and these values are shared by the two subsystems which are exchanging power through the considered port. The power exchanged at a certain time is equal to the intrinsic dual product:

$$\text{Power} = \langle e, f \rangle.$$

This dual product is intrinsic in the sense that elements of V^* are linear operators from V to $\mathbb{R}$, and therefore, to express the operation, we do not need any additional structure than the vector space structure of V.

To talk about the interconnection of mechanical systems, a proper choice is $V = \mathfrak{se}(3)$, the space of twists and $V^* = \mathfrak{se}^*(3)$ the space of wrenches.

1.5   Generalized Port Controlled Hamiltonian Systems

In the standard symplectic Hamiltonian theory, the starting point is the existence of a generalized configuration manifold Q. Based on Q, its co-tangent bundle T^*Q is introduced which represents the state space to which the configuration-momenta pair (q, p) belongs. It is possible to show that T^*Q can be naturally given a symplectic structure on the base of which the Hamiltonian dynamics can be expressed [10].

A limitation of this approach is that, by construction, the dimension of the state space T^*Q is always even. Moreover, it can be shown that in general the interconnection of Hamiltonian systems in this form does not originate a system of the same form.

These problems can be easily solved with the more general approach in the Poisson framework, [5] or more generally using Dirac structures [11]. In this chapter we will use the Poisson framework for the sake of simplicity. In general a GPCHS in the Poisson formulation is characterized by 4 elements: (a) a state manifold $\mathcal{X}$ which can be of any dimension, even or odd; (b) an interaction vector space V on which a power port is described as presented in Sec. 1.4; (c) a Poisson structure on $\mathcal{X}$; (4) a local vector bundle isomorphism [12] between $\mathcal{X} \times V$ and $T\mathcal{X}$. For the purposes of this work, it is sufficient to consider that a Poisson structure is characterized by a contravariant skew-symmetric tensor-field $J(x)$ defined on $\mathcal{X}$.

If we consider a chart ψ and the corresponding set of coordinates x for $\mathcal{X}$, and a base $B := \{b_1, \ldots, b_n\}$ for V, we can express a GPCHS with a set of equations of the following form:

$$\dot{x} = J(x)\frac{\partial H(x)}{\partial x} + g(x)u \tag{10}$$

$$y = g^T(x)\frac{\partial H(x)}{\partial x} \tag{11}$$

where u is a representation of an element of V in the base B, $J(x) = -J^T(x)$ is the skew-symmetric Poisson tensor describing the network structure and interconnection of the composing elements, $g(x)$ is the representation of the fiber bundle isomorphism describing how the system interacts with the external world and y is the representation of an element belonging to the dual vector space V^* in the dual base of B.

Any explicit physical conservative element can be given the previous representation. To account for dissipating elements, we can generalize the previous form considering a symmetric, semi-positive definite, two covariant tensor $R(x)$ which can be subtracted from $J(x)$:

$$\dot{x} = (J(x) - R(x))\frac{\partial H(x)}{\partial x} + g(x)u$$

$$y = g^T(x)\frac{\partial H(x)}{\partial x}. \tag{12}$$

With this new term, it can be seen that the change in internal energy is:

$$\dot{H} = \underbrace{y^T u}_{\text{supplied power}} - \underbrace{\left(\frac{\partial H}{\partial x}\right)^T R(x)\frac{\partial H}{\partial x}}_{\text{dissipated power}}.$$

Since $R(x)$ is positive semi-definite, this implies that the internal energy can only increase if power is supplied through the ports.

As an example, consider the interconnection shown in Fig. 4 of a mass representing a robot with the physical equivalent of a controller implementing damping injection as introduced in [13].

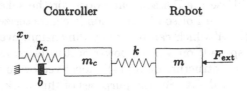

Fig. 4. A simple example of interconnection

The Generalized Hamiltonian model of the "robot" is:

$$\begin{pmatrix} \dot{x} \\ \dot{p} \end{pmatrix} = \begin{pmatrix} 0 & 1 \\ -1 & 0 \end{pmatrix} \begin{pmatrix} 0 \\ p/m \end{pmatrix} + \begin{pmatrix} 0 & 0 \\ 1 & 1 \end{pmatrix} \begin{pmatrix} F_{ext} \\ F_c \end{pmatrix}$$

$$\begin{pmatrix} \dot{x} \\ \dot{x} \end{pmatrix} = \begin{pmatrix} 0 & 1 \\ 0 & 1 \end{pmatrix} \begin{pmatrix} 0 \\ p/m \end{pmatrix}$$

where the port $(u_1, y_1) = (F_{ext}, \dot{x})$ represents the interaction port of the robot with the environment and $(u_2, y_2) = (F_c, \dot{x})$ the interaction port with the controller. The energy function is $H(p) = \frac{1}{2m} p^2$ where p is the momenta[1]. The physical system representing the "controller" of Fig. 4 can be instead represented using the following Generalized Hamiltonian equations with dissipation:

$$\begin{pmatrix} \dot{\Delta x}_c \\ \dot{p}_c \\ \dot{\Delta x} \end{pmatrix} = \begin{pmatrix} 0 & 1 & 0 \\ -1 & -b & 1 \\ 0 & -1 & 0 \end{pmatrix} \begin{pmatrix} k_c \Delta x_c \\ p_c/m_c \\ k \Delta x \end{pmatrix}$$

$$+ \begin{pmatrix} 0 & -1 \\ 0 & 0 \\ 1 & 0 \end{pmatrix} \begin{pmatrix} \dot{x} \\ \dot{x}_v \end{pmatrix}$$

$$\begin{pmatrix} -F_c \\ -F_v \end{pmatrix} = \begin{pmatrix} 0 & 0 & 1 \\ -1 & 0 & 0 \end{pmatrix} \begin{pmatrix} k_c \Delta x_c \\ p_c/m_c \\ k \Delta x \end{pmatrix}$$

where the port $(u_1, y_1) = (\dot{x}, -F_c)$ is used to express the interconnection with the "robot" and the port $(u_2, y_2) = (\dot{x}_v, -F_v)$ is used to express the interconnection with another system which turns out to be the supervisory module. the stored energy is:

$$H(\Delta x_c, p_c, \Delta x) = \frac{1}{2} k_c \Delta x_c^2 + \frac{1}{2} k \Delta x^2 + \frac{1}{2m} p_c^2.$$

---

[1] It is important to notice that $\frac{1}{2} mv^2$ is properly speaking called co-energy instead of energy because is a function of v which is not a physical state extensive variable [14]

It is shown in [13] that by choosing $m_c \ll m$ and $k \gg k_c$ for the controller, damping injection can be implemented with pure position measurements. Furthermore, actuator's saturation can be handled in a physical way by choosing a non linear spring k.

Any physical system can be modeled in the same way and this is the power of GPCHS and their importance to describe the proposed architecture.

1.6   Interconnection of GPCHSs

A very important feature of GPCHSs is that their interconnection is still a GPCHS. To show this, consider two GPCHSs:

$$\dot{x}_i = (J_i - R_i)\frac{\partial H_i}{\partial x_i} + (g_i^I \ \ g_i^O)\begin{pmatrix} u_i^I \\ u_i^O \end{pmatrix} \tag{13}$$

$$\begin{pmatrix} y_i^I \\ y_i^O \end{pmatrix} = \begin{pmatrix} (g_i^I)^T \\ (g_i^O)^T \end{pmatrix}\frac{\partial H_i}{\partial x_i} \qquad i = 1, 2. \tag{14}$$

The two systems can be interconnected through the interconnection ports (u_i^I, y_i^I) by setting:

$$u_1^I = y_2^I \qquad \text{and} \qquad u_2^I = -y_1^I \tag{15}$$

Note that the minus sign in the previous equation is necessary to be consistent with power: $P_i = \langle u_i^I, y_i^I \rangle$ is the input power of system i and, interconnecting the system throw the "I" ports, we clearly need $P_1 = -P_2$. It is possible to see that the interconnected system results:

$$\dot{x} = (J(x) - R(x))\frac{\partial H}{\partial x} + g(x)\begin{pmatrix} u_1^O \\ u_2^O \end{pmatrix} \tag{16}$$

$$\begin{pmatrix} y_1^O \\ y_2^O \end{pmatrix} = g^T(x)\frac{\partial H}{\partial x} \tag{17}$$

where $x = (x_1, x_2)^T$, $H(x) = H_1(x_1) + H_2(x_2)$ the sum of the two energies and

$$J(x) = \begin{pmatrix} J_1 & g_1^I(g_2^I)^T \\ -(g_2^I)(g_1^I)^T & J_2 \end{pmatrix},$$

$$R(x) = \begin{pmatrix} R_1 & 0 \\ 0 & R_2 \end{pmatrix},$$

and

$$g(x) = \begin{pmatrix} g_1^O \\ g_2^O \end{pmatrix}$$

where all dependencies of the matrices have been omitted for clarity.

It is possible to conclude that the interconnected system is therefore again a GPCHS with, as ports the remaining ports. Furthermore the total energy is the sum of the energies of the two systems.

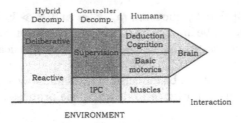

Fig. 5. A biological analogy of the control strategy

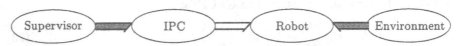

Fig. 6. The System Interconnection

2   The proposed control architecture

The main idea of the proposed scheme is to divide the control in two parts, one which controls the real time interaction passively and is called Intrinsically Passive Control (IPC) and one which takes care of the task decomposition and other planning issues. As shown in Fig. 5, the IPC has the role of the muscle spindles in biological systems, and the supervisor the neurological role of the brain.

2.1   The IPC

The IPC is interconnected with the supervisor and the robot through power ports as shown in Fig. 6 where bond-graph notation is used. Each bond is representing two elements belonging to a vector space and to its dual as explained in Sec. 1.4. With reference to Sec. 1.4, the power port corresponding to the interconnection with the robot is characterized by the vector space V being $T_q Q$ where Q is the robot configuration manifold and q is the current configuration and therefore the elements of the ports will be pairs of the form $(\dot{q}, \tau)$ where τ are the joint torques.

The other port of the IPC will be connected to the supervisor and in general will have a geometric structure such that:

$$V = \mathfrak{se}(3) \times \ldots \times \mathfrak{se}(3)$$

corresponding to a set of twists. This is the case in both [15] and [16].

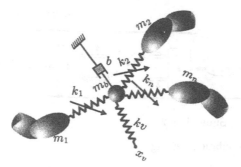

Fig. 7. Basic idea of an IPC for grasping tasks.

The IPC will therefore be characterized by the following differential equations:

$$\dot{x} = (J(x) - R(x))\frac{\partial H_c}{\partial x} + g(x) \begin{pmatrix} \dot{q} \\ T_1 \\ \vdots \\ T_n \end{pmatrix} \tag{18}$$

$$\begin{pmatrix} \tau \\ W_1 \\ \vdots \\ W_n \end{pmatrix} = g^T(x)\frac{\partial H_c}{\partial x} + B(x) \begin{pmatrix} \dot{q} \\ T_1 \\ \vdots \\ T_n \end{pmatrix} \tag{19}$$

where $J(x)$ is skew symmetric, $R(x)$ positive semi-definite, $T_1, \ldots, T_n$ are a set of interaction twists and $W_1, \ldots, W_n$ a set of the dual interaction wrenches. It has been shown in [15] that the feed-through term $B(x)$ is needed in tele-manipulation to adapt the impedance of the line.

An example of an IPC for the control of a robotic hand is reported in Fig. 7. The shown springs and the spherical object shown in the middle and called *the virtual object* are virtual and implemented by means of control in the IPC. The supervisor, by means of twists inputs $T_1, \ldots, T_n, T_v$ can change the rest length of the springs and the virtual position x_v. This schema has been also test experimentally and it has given very satisfactory results [16].

Intrinsic Passivity The most important feature of such a controller is that, in the case in which the supervisor would not supply power to the IPC by setting for example $T_1 = T_2 = \ldots = 0$, the physical robot together with the IPC have a certain amount of energy, namely $H_c + H$ where H is the mechanical energy of the physical robot and H_c the one of the IPC as it has been shown in Sec. 1.6. This energy cannot increase if no power is supplied by the environment. This is true in ANY situation like bouncing of the robot with an object or any kind of changes of contact situations! It is not necessary to discriminate between contact and no-contact situations!

Supervisor 1

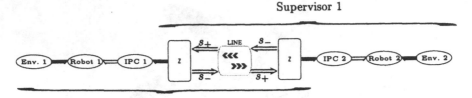

Supervisor 2

Fig. 8. A tele-manipulation Setting

2.2 The Supervisor

The Supervisor has the role of scheduling and planning and can be considered as having the role of the brain in the human analogy.

Following the grasping example of the previous section, to grasp and move an object, we have to

1. Open the hand
2. Move around the object
3. Close the hand (grasp)
4. Move the object
5. Open the hand

All this subtasks can be implemented by means of control signals to the IPC by the Supervisor which can supply a controlled amount of energy to the system in order to perform useful tasks.

A Tele-manipulation setting In a tele-manipulation setting, the presented architecture is still valid, but the role of the supervisor is taken over by a human on the other side of a transmission line.

A perfectly bilateral tele-manipulation system using the presented architecture is reported in Fig. 8. This system has been studied and implemented experimentally in [15].

The environment on one side is the environment to be manipulated and the one on the other side is the human which manipulates the system remotely.

The port variables are transformed to scattering variables (block Z) to preserve passivity even with time varying time delays due to the transmission line [15]. The supervisor on one side is in this case composed of the transmission line, the IPC, the Robot and the "Environment" of the other side.

Once again due to the consistent framework using power ports, passivity is preserved in any situation.

3   Summary

In the first part of the paper, we have introduced the theory of matrix Lie group to study motions of rigid body mechanisms. This allows to define a vector space $se(3)$ and its dual $se^*(3)$ which can be used together to talk geometrically about the port interconnection of mechanical parts.

In the second part a novel architecture has been presented which slightly resembles the human physiology. Real time behavior is controlled by the Intrinsically Passive Controller which corresponds to a virtual physical system. The IPC more or less resembles the role of muscle and spindles in biological systems. The IPC together with the robot can be seen as a pre-compensated robot. Due to the implementation of the IPC control as a power consistent interconnection with the robot to be controlled, passivity is ensured in ANY situations if no power is injected by the Supervisor. The structure of the IPC is corresponding to a Port Controlled Hamiltonian System with dissipation. An example of such an IPC has been given. An important point is that the IPC can be really designed using mechanical analogies like springs, dampers and masses and then implemented using the theory of interconnection of Port Controlled Hamiltonian Systems. Furthermore, it has been shown in [13] using the theory of Casimir functions that it is possible to implement the IPC with only measurements of positions q and not of velocities. With the presented strategy, it is not proper to talk about position or force control anymore, because what it is controlled is actually the behavior of the system and not the position or force at its interconnection port. This has the advantages of being very robust with respect to different materials and object with which the robot interact; just think about shaking hand to somebody, we never have a perfect model of the person we shake hand with, but the interaction is always well behaved.

References

1. Gilmore, R.: Lie Groups, Lie Algebras, and Some of Their Applications. John Wiley & Sons, New York (1974) ISBN 0-471-30179-5.
2. Ball, R.: A Treatise on the Theory of Screws. Cambridge at the University Press (1900)
3. Dubrovin, B., Fomenko, A., Novikov, S.: Modern Geometry - Methods and Applications, Part II:The geometry and topology of manifolds. Volume 104 of Graduate texts in mathematics. Springer Verlag, New-York (1985) ISBN 0-387-96162-3.
4. Boothby, W.M.: An Introduction To Differentiable Manifolds and Riemannian Geometry. 516'.36. Academic Press, Inc., New York (1975) ISBN 0-12-116050-5.
5. Olver, P.J.: Applications of Lie Groups to Differential Equations. Ii edn. Volume 107 of Graduate texts in mathematics. Springer Verlag, New-York (1993) ISBN 0-387-94007-3.
6. Stramigioli, S., Maschke, B., Bidard, C.: A Hamiltonian formulation of the dynamics of spatial mechanisms using lie groups and screw theory. In Lipkin,

H., ed.: Proceedings of the Symposium Commemorating the Legacy, Works, and Life of Sir Robert Stawell Ball, Cambridge (2000)

7. Selig, J.: Geometric Methods in Robotics. Monographs in Computer Sciences. Springer Verlag (1996) ISBN 0-387-94728-0.

8. Maschke, B.M., van der Schaft, A., Breedveld, P.C.: An intrinsic Hamiltonian formulation of network dynamics: Non-standard poisson structures and gyrators. Journal of the Franklin institute **329** (1992) 923–966 Printed in Great Britain.

9. Stramigioli, S.: From Differentiable Manifolds to Interactive Robot Control. PhD thesis, Delft University of Technology, Delft, The Netherlands (1998) ISBN 90-9011974-4, http://lcewww.et.tudelft.nl/~stramigi.

10. Arnold, V.: Mathematical Methods of Classical Mechanics. Ii edn. Springer Verlag (1989) ISBN 0-387-96890-3.

11. Dalsmo, M., van der Schaft, A.: On representations and integrability of mathematical structures in energy-conserving physical systems. SIAM Journal of Control and Optimization **37** (1999) 54–91

12. Abraham, R., Marsden, J.E.: Foundations of Mecahnics. Ii edn. Addison Wesley (1994) ISBN 0-8053-0102-X.

13. Stramigioli, S., Maschke, B.M., van der Schaft, A.: Passive output feedback and port interconnection. In: Proc. Nonlinear Control Systems Design Symposium1998, Enschede, The Netherlands (1998)

14. Breedveld, P.: Physical Systems Theory in Terms of Bond Graphs. PhD thesis, Technische Hogeschool Twente, Enschede, The Netherlands (1984) ISBN 90-90005999-4.

15. Stramigioli, S., van der Schaft, A., Maschke, B., Andreotti, S., Melchiorri, C.: Geometric scattering in tele-manipulation of port controlled hamiltonian systems. In: Proceedings of the 39th IEEE Conference on Decision and Control, Sydney (AUS), IEEE (2000)

16. Stramigioli, S., Melchiorri, C., Andreotti, S.: Passive grasping and manipulation. Submitted to the IEEE Transactions of Robotics and Automation (1999)

Lecture Notes in Control and Information Sciences

Edited by M. Thoma and M. Morari

Vol. 250: Corke, P. ; Trevelyan, J. (Eds)
Experimental Robotics VI
552pp: 2000 [1-85233-210-7]

Vol. 251: van der Schaft, A. ; Schumacher, J.
An Introduction to Hybrid Dynamical Systems
192pp: 2000 [1-85233-233-6]

Vol. 252: Salapaka, M.V.; Dahleh, M.
Multiple Objective Control Synthesis
192pp. 2000 [1-85233-256-5]

Vol. 253: Elzer, P.F.; Kluwe, R.H.;
Boussoffara, B.
Human Error and System Design and
Management
240pp. 2000 [1-85233-234-4]

Vol. 254: Hammer, B.
Learning with Recurrent Neural Networks
160pp. 2000 [1-85233-343-X]

Vol. 255: Leonessa, A.; Haddad, W.H.;
Chellaboina V.
Hierarchical Nonlinear Switching Control
Design with Applications to Propulsion
Systems
152pp. 2000 [1-85233-335-9]

Vol. 256: Zerz, E.
Topics in Multidimensional Linear Systems
Theory
176pp. 2000 [1-85233-336-7]

Vol. 257: Moallem, M.; Patel, R.V.;
Khorasani, K.
Flexible-link Robot Manipulators
176pp. 2001 [1-85233-333-2]

Vol. 258: Isidori, A.; Lamnabhi-Lagarrigue, F.;
Respondek, W. (Eds)
Nonlinear Control in the Year 2000
Volume 1
616pp. 2001 [1-85233-363-4]

Vol. 259: Isidori, A.; Lamnabhi-Lagarrigue, F.;
Respondek, W. (Eds)
Nonlinear Control in the Year 2000
Volume 2
640pp. 2001 [1-85233-364-2]

Vol. 260: Kugi, A.
Non-linear Control Based on Physical Models
192pp. 2001 [1-85233-329-4]

Vol. 261: Talebi, H.A.; Patel, R.V.;
Khorasani, K.
Control of Flexible-link Manipulators Using
Neural Networks
168pp. 2001 [1-85233-409-6]

Vol. 262: Dixon, W.; Dawson, D.A.;
Zergeroglu, E.; Behal,A.
Nonlinear Control of Wheeled Mobile Robots
216pp. 2001 [1-85233-414-2]

Vol. 263: Galkowski,K.
State-space Realization of Linear 2-D
Systems with Extensions to the General nD
(n>2) Case
248pp. 2001 [1-85233-410-X]